Practical Terahertz Electronics: Devices and Applications, Volume 1

Solid-state devices and vacuum tubes

Practical Terahertz Electronics: Devices and Applications, Volume 1

Solid-state devices and vacuum tubes

Vinod Kumar Khanna
Independent Researcher, Chandigarh, India
Retired Chief Scientist, CSIR-CEERI, Pilani, India

IOP Publishing, Bristol, UK

ISBN 978-0-7503-3171-5 (ebook)
ISBN 978-0-7503-3169-2 (print)
ISBN 978-0-7503-3172-2 (myPrint)
ISBN 978-0-7503-3170-8 (mobi)

DOI 10.1088/978-0-7503-3171-5

Version: 20211201

IOP ebooks

British Library Cataloguing-in-Publication Data: A catalogue record for this book is available from the British Library.

Published by IOP Publishing, wholly owned by The Institute of Physics, London

IOP Publishing, Temple Circus, Temple Way, Bristol, BS1 6HG, UK

US Office: IOP Publishing, Inc., 190 North Independence Mall West, Suite 601, Philadelphia, PA 19106, USA

To the memory of my respected parents

The late Shri Amarnath Khanna and Shrimati Pushpa Khanna

For their sacrifices and inspiration.

To my beloved grandson Hansh, daughter Aloka and wife Amita

For their affection and whole-hearted support.

Contents

Preface

We are very fortunate to be living in an era where we have witnessed several technological breakthroughs and experienced revolutionary changes in lifestyle and comforts. Notwithstanding our achievements, there still remains and very surprisingly too, a chunk of the electromagnetic spectrum which is full of promises but whose potential is still awaiting widespread technological utilization and commercialization for the benefit of civilization.

'Terahertz' is a very big unit of frequency equal to 10^{12} times the base unit called hertz. So when we talk about terahertz electronics we are concerned with very high frequencies in the parlance of electronics. Terahertz electronics is the electronics of extremely high frequencies. In electronics, we deal with two entities, charge carriers and electromagnetic waves. Oscillations in terahertz waves are so fast that electrons cannot follow them and therefore lag behind. So electronic devices become inefficient here.

The frequencies, which appear so high for electronics are too low for photonics. Frequency is linked to an important parameter 'energy'. A quantum of energy is called a photon. The photons have energies in accordance with their frequencies. Naturally so, because the faster a particle moves, the more is its kinetic energy. But here the situation is different compared to electronics. Energies of photons of terahertz frequencies are lower than desired values because photons of these energies produce effects of the same intensity or severity as thermal energy at room temperature. So the effects of terahertz photons become indistinguishable from those of thermal energy, creating chaos and confusion in existing photonic devices, which therefore become erratic and lose their meaningfulness in the terahertz range of frequencies. Hence, this range represents a bothersome zone both for electronics and photonics. Furthermore, these arguments lead us to think about frequency and energy of electrons/photons as the key parameters at the center of all discussions in terahertz electronics and optoelectronics.

The technologically awkward portion of electromagnetic spectrum known as terahertz gap, exhibits amazing properties. The electromagnetic waves of these frequencies are strongly attenuated by absorption by oxygen and water vapor in the atmosphere disallowing their use for medium and long distance communication but at the same time offering the capability to transmit vast amounts of data over short distances of a few meters to share the ever-increasing loads of expanding communication networks. High data rates are achievable with availability of wider bandwidths.

Terahertz radiation offers the opportunity to see through various non-metallic and nonpolar materials in dry state to reveal any prohibited object being transported past entry points at airports with malicious intentions or at crowded public places. Paper, cardboard, plastics, foam and clothing are transparent to terahertz waves. Unlike x-rays, this non-ionizing radiation is harmless to human tissue and DNA, encouraging its use in medical diagnostic imaging such as for epithelial cancer and tooth decay. It opens up immense opportunities in analytical chemistry allowing facile analysis of semiconductors, liquids and biological specimens. Prolific applications of terahertz radiation are expected in three premier technologies viz., communication, spectroscopy and imaging in the forthcoming years.

Realizing the immense potentialities of terahertz electronics, this field has of late captured the attention of scientists worldwide and some strides have been made during the last few decades. At the foundation of availing the fascinating prospects of terahertz waves, lie the various electronic devices that have been developed in the recent past to generate terahertz waves, control their properties and make them work in the designed plan to execute desired tasks. The devices of solid-state category cater mostly to low power applications. The high power zone is captured by vacuum tubes. The vacuum electronic devices benefit from the scattering-free medium of propagation of electrons, 'the vacuum', as opposed to collision-prone medium available to electrons in solid-state devices. This is the primary reason, which is at the root of all differences in properties of vacuum electronic and solid-state devices. The most remarkable characteristic of terahertz electronics is the vital contribution from photothermal and optoelectronic devices. Particularly interesting are the non-linear optical devices. It is this complete team of solid-state and vacuum electronic as well as photothermal and optoelectronic devices whose working principles, fabrication and operation constitute the theme of this book.

The 17 chapters of this book are subdivided into four parts. After introducing the subject in chapter 1, the ensuing six chapters (chapter 2–7) of Part I concentrate on solid-state devices. Then the next three chapters (chapters 8–10) of Part II treat vacuum electronic devices. For long the paucity of suitable optoelectronic devices has hindered progress of terahertz technology partly because of inadequate optical materials. Four chapters (chapters 11–14) of Part III deal at length with photothermal and optoelectronic devices. No field of science is complete without applications because applications determine the utility of the efforts and tell us how closely and deeply the scientific advances have touched our lives and improved our living styles. The closing chapters of the book (chapters 15–17) dwell at length on what terahertz electronics has done and is likely to do in the near future for betterment of human life.

Parts I and II constitute volume 1 of the book while parts III and IV are covered in volume II. Parts I and II of Volume 1 contain chapters 1–10. Parts III and IV of Volume 2 contain chapters 11–17, numbered as chapters 1–7 of this volume.

Two faces of any science are recognized, theoretical, concerning the knowledge gained about the laws of nature by an evidence-based systematic methodology of observation and experimentation, and practical, dealing with application of that knowledge for real-world uses. So our focus in this book is not only to know the truth about the behavior and properties of terahertz waves (theoretical aspects) but about how the knowledge of terahertz waves is beneficially used for their generation, modulation, transmission, detection, or performing any conceivable operation on them so that they can be controlled and manipulated as desired (practical aspects). Greater stress will be laid on the latter part. Numerical worked examples, conceptual and thought-provoking types, and those dealing with practical problems are therefore sprinkled all through the text.

The book is written from a practical engineering perspective. Therefore, dependency on detailed theoretical proofs or higher-level mathematics is avoided to prioritize the practical side of terahertz electronics. All through the book mathematical formulation has been restricted to closed form solutions. Device and process design involve extensive computer simulations involving complicated numerical modeling and optimization techniques. The analysis presented will serve as the first steps towards performing design of devices.

As this book primarily deals with operation of devices at high frequencies, it will be presumed that the reader has prior knowledge of fundamentals of device operation such as for Schottky diode, MOSFET, etc, and they will not be treated at length to make room for inclusion of relevant matter related to high frequencies. Assuming this familiarity with rudimentary knowledge regarding devices, we shall make only short digressions to succinctly touch upon the principles of these devices for sake of revision, and quickly move on to terahertz electronics.

Some flowers are late bloomers. Like these flowers, the terahertz part of the electromagnetic spectrum is a late blooming flower. But now that it has started to bloom, its fragrance will soon pervade the garden of electronics.

I hope the book will fulfill its goal of acquainting the reader with the up-to-date developments in this rapidly evolving, progressive discipline where invigorating innovations are expected every now and then. The author will deem his efforts amply rewarded if readers are able to get abreast with the latest information in the field and are able to tune up with the contemporary scientific advances.

Acknowledgements

I received generous help and support from all quarters during the course of book writing. I feel deeply overwhelmed with humility and gratefulness as I record this support.

Most importantly and above all, I am indebted to God, The Almighty for giving me the strength and wisdom to undertake, pursue and accomplish this task from chapter to chapter. Words become insufficient to convey gratitude for the benevolent blessings showered by God for a smooth journey towards execution of a task.

I acknowledge the valuable contributions of all authors and scientists whose work forms the subject matter of this book. These stalwarts are enlisted in the 'References sections' at the end of each chapter. The references are pointers to the original works and I urge the readers to read them thoroughly to appreciate the full potential of terahertz electronics. The book is intended to make the reading and understanding of original research papers in the field easier and enjoyable. Three cheers for the inventors!

The publishers, acquisitions and production editors and staff at IOP ebooks deserve my thanks for giving me the opportunity to write this book. Their keen enthusiasm and kind co-operation in every stage of this work is worthy of appreciation.

Last but not the least, 'a special thank-you' to my family. They have tolerated with great patience my pre-occupation with this work. Despite my busy schedule encroaching on the time to be spent with them, they have maintained their unfailing support all through the project.

Vinod Kumar Khanna
Chandigarh, India
Retired Chief Scientist, CSIR-Central Electronics Engineering Research Institute, Pilani and Professor, AcSIR, India

Author biography

Vinod Kumar Khanna

Introduction

Vinod Kumar Khanna is an independent researcher at Chandigarh, India. He is a retired Chief Scientist from Council of Scientific & Industrial Research (CSIR)-Central Electronics Engineering Research Institute (CEERI), Pilani-India, and retired Professor from Academy of Scientific & Innovative Research (AcSIR), Ghaziabad, India. He is a former Emeritus Scientist, CSIR and Professor Emeritus, AcSIR, India. His broad areas of research were the design, fabrication and characterization of power semiconductor devices, micro- and nanosensors.

Academic qualifications

He received the MSc Degree in Physics with specialization in Electronics from University of Lucknow in 1975 and PhD degree in Physics from Kurukshetra University in 1988 for the thesis entitled, 'Development, Characterization and Modeling of the Porous Alumina Humidity Sensor'.

Work experience and accomplishments

His research experience spans over a period of 40 years from 1977 to 2017. He started his career as a Research Assistant in the Department of Physics, University of Lucknow from 1977 to 1980. He joined CSIR-Central Electronics Engineering Research Institute, Pilani (Rajasthan) in April 1980. At CSIR-CEERI he worked on several CSIR-funded as well as sponsored research and development projects. His major fields of research included power semiconductor devices and microlectronics/MEMS and nanotechnology-based sensors and dosimeters.

In the power semiconductor devices area, he worked on the high-voltage and high-current rectifier (600 A, 4300 V) for railway traction, high voltage TV deflection transistor (5 A, 1600 V), power Darlington transistor for AC motor drives (100 A, 500 V), fast-switching thyristor (1300 A, 1700 V), power DMOSFET and IGBT. He contributed towards the development of sealed tube Ga/Al diffusion for deep junctions, surface electric field control techniques using edge beveling and contouring of large-area devices, and floating field limiting ring design, and characterization of minority-carrier lifetime as a function of process steps. He also contributed towards the development of P–I–N diode neutron dosimeter and PMOSFET-based gamma ray dosimeter.

In the area of sensor technology, he worked on the nanoporous aluminum oxide humidity sensor, ion-sensitive field-effect transistor-based microsensors for biomedical, food and environmental applications; microheater embedded gas sensor for automotive electronics, MEMS acoustic sensor for launch vehicles and capacitive MEMS ultrasonic transducer for medical applications.

Semiconductor facility creation and maintenance

He was responsible for setting up and looking after diffusion/oxidation facilities, edge beveling and contouring, reactive sputtering and carrier lifetime measurement facilities. As the Head of MEMS and microsensors group, he looked after the maintenance of six-inch MEMS fabrication facility for R&D projects as well as augmentation of processing equipment under this facility at CSIR-CEERI.

Scientific positions held

During his tenure of service at CSIR-CEERI from April 1980 till superannuation in November 2014, he was promoted to various positions including one merit promotion. He retired as Chief Scientist and Professor (AcSIR: Academy of Scientific & Innovative Research), and Head of MEMS & Microsensors Group. Subsequently, he worked for three years as Emeritus Scientist, CSIR and Emeritus Professor, AcSIR from November 2014 to November 2017. After completion of the emeritus scientist scheme, he now lives at Chandigarh. He is a passionate author, and enjoys reading and writing.

Membership of professional societies

He is a Fellow and life member of the Institution of Electronics & Telecommunication Engineers (IETE), India. He is a life member of Indian Physics Association (IPA), Semiconductor Society, India (SSI) and Indo-French Technical Association (IFTA).

Foreign travel

He is widely travelled. He participated and presented research papers in IEEE Industry Application Society (IEEE-IAS) Annual Meeting at Denver Colorado, USA, in September–October, 1986. His short-term research assignments include deputations to Technische Universität Darmstadt, Germany in 1999, at Kurt-Schwabe-Institut fur Mess-und Sensortechnik e.V., Meinsberg, Germany in 2008 and at Fondazione Bruno Kessler, Trento, Italy in 2011 under collaborative programmes. He was a member of the Indian Delegation to Institute of Chemical Physics, Novosibirsk, Russia in 2009.

Scholarships and awards

Awarded National Scholarship by the Ministry of Education & Social Welfare, Govt of India, on the basis of Higher Secondary result, 1970; CEERI Foundation

Day Merit Team Award for projects on fast-switching thyristor (1986); power Darlington transistor for transportation (1988), P–I–N diode neutron dosimeter (1992); and high-voltage TV deflection transistor (1994); Dr N G Patel Prize for best poster presentation in 12th National Seminar on Physics and Technology of Sensors 2007, BARC, Mumbai; CSIR-DAAD Fellowship in 2008 under Indo-German bilateral exchange programme of Senior scientists, 2008.

Research publications and books

He has published 192 research papers in leading peer-reviewed national/international journals and conference proceedings. Prior to the present book, he has authored 16 books, and has also contributed six chapters in edited books. He has five patents to his credit including two US patents.

About the Book

Practical Terahertz Electronics: Devices and Applications, Volume 1: Solid-state devices and vacuum tubes brings forth a refreshing change from the conventional style of presentation of subject matter. Rather than introducing the subject in the conventional descriptive style, it asks questions in the form of numerical problems, and tries to seek the solutions to these problems. These problem-solving examples integrated in the body of the text keep the readers' interest alive all throughout and make the learning process smoother and deeper in knowledge acquisition. The concepts are appreciated and assimilated with little effort and also in a long-lasting manner.

Perched on the crossroads of electronics and photonics, the enormous edifice of terahertz electronics has two distinct appearances depending upon the contributions from electronics and photonics. Further, two sprouts of the electronics component include solid-state and vacuum or gas-filled devices. In addition, the developing application areas of terahertz electronics impart utility to these activities for humanity. Accordingly, the book is subdivided into two parts devoted to electronic devices, a third part on photonic and optothermal devices, and a fourth part dealing with the applications resulting from these endeavors. The four parts of the book containing seventeen chapters present a seamless and cohesive treatment of this expansive technological field. Out of these four parts, the first two parts constitute the contents of Volume 1 while the last two parts will be presented in Volume 2 of the book subtitled: Optical devices and applications.

In line with the adage, 'a picture is worth a thousand words', the book is profusely illustrated with clear line diagrams for easy visualization of device structures and following their operational mechanisms.

The book will be immensely beneficial to both students and professional segments. The book will fulfill the requirements of graduate students in electronics and electrical engineering. Scientists engaged in research in this field as well as practicing engineers will find the book a useful resource providing up-to-date information in a concise format.

Abbreviations and acronyms

A	Ampere
A°	Angstrom unit
AC	Alternating current
ADP	Ammonium dihydrogen phosphate
AE	Antenna emitter
AFB1	Aflatoxin B1
AM	Amplitude modulation
AWGN	Additive white Gaussian noise
BCC	Basal cell carcinoma
BJT	Bipolar junction transistor
BWO	Backward wave oscillator
C	Coulomb
CB	Conduction band
CCD	Charge coupled device
CHIRP	Coherent heterointerfaces for reflection and penetration
cm	Centimeter
CMC	Ceramic matrix composite
CMOS	Complementary metal-oxide-semiconductor (FET)
CNL	Charge neutrality level
COM	Center of mass
C-RAN	Cloud or centralized radio access network
CRM	Cyclotron resonance maser
CT	Computed tomography
CW	Continuous wave (power)
2-D	Two-dimensional
3-D	Three-dimensional
DAST	4-dimethylamino-N-methyl-4-stilbazolium tosylate
dB	Decibel
dBi	Decibel isotropic
dBm	Decibel-milliwatt
DBQW	Double barrier quantum well
DBR	Distributed-Bragg-reflector (laser)
DC	Direct current
2DEG	Two-dimensional electron gas
DFB	Distributed-feedback (laser)
DFG	Difference frequency generation
DFT	Discrete Fourier transform
DHBT	Double heterojunction bipolar transistor
DIBL	Drain-induced barrier lowering
DNA	Deoxyribonucleic acid
DRIE	Deep reactive ion etching
EHF	Extremely high frequency
ELF	Extremely low frequency
eV	Electron volt
F	Farad
FEL	Free electron laser
FET	Field-effect transistor

fF	Femtofarad
FFT	Fast Fourier transform
FIR	Far infrared
FM	Frequency modulation
FMCW	Frequency modulated continuous wave (radar)
FOMs	Figures of merit
FPA	Focal plane array
fs	Femtosecond
FSPL	Free space path loss
FT	Fourier transform
FT-IR	Fourier-transform infrared (spectroscopy)
g	Gram
Gbps	Gigabits per second
GEO	Geosynchronous equatorial orbit
GeV	Giga electron volt
GHz	Gigahertz
GW	Gigawatt
HBT	Heterojunction bipolar transistor
HDPE	High-density polyethylene
HEMT	High-electron-mobility transistor
HF	High frequency
hPa	Hectopascal (I 1hPa = 100Pa)
HRFZ	High resistivity float zone (silicon)
Hz	Hertz
I	Intrinsic (region)
IC	Integrated circuit
IMPATT	IMPact ionization Avalanche Transit-Time (diode)
IR	Infrared
ISAR	Inverse synthetic-aperture radar
ITU	International Telecommunication Union
ITU-R	International Telecommunication Union Radiocommunication (Sector)
J	Joule
K	Kelvin
kA	Kiloampere
kg	Kilogram
kHz	Kilohertz
km	Kilometer
LAE	Large-area emitter
LED	Light-emitting diode
LINAC	Linear accelerator
LF	Low frequency
LO	Longitudinal optical (phonon)
LOS	Line of sight
LT-GaAs	Low temperature gallium arsenide
m	Meter
mA	Milli ampere
Mbps	Megabits per second
MEMS	Microelectromechanical systems
MESFET	Metal–semiconductor-field-effect transistor
MeV	Mega electronvolt

meV	Milli electronvolt
MF	Medium frequency
MHz	Megahertz
MIG	Magnetron injection gun
MIM	Metal–insulator–metal
MIR	Mid infrared
mol	Mole
MOSFET	Metal-oxide-semiconductor field-effect transistor
mS	Milli Siemens
ms	Milli seconds
mW	Milli Watt
mΩ	Milli ohm
N	Newton
NDR	Negative differential resistance
NDT	Non-destructive testing
NEP	Noise equivalent power
NETD	Noise equivalent temperature difference
NIR	Near infrared
nm	Nanometer
NMOS	N-channel metal-oxide–semiconductor (FET)
O/E	Optical-to-electrical (conversion)
OFHC	Oxygen-free high conductivity (copper)
OR	Optical rectification
Pa	Pascal
PCA	Photoconductive antenna
PD	Photodiode
PIN	P-type region/intrinsic region/N-type region (diode)
ps	Picosecond
PVDF	Polyvinylidene fluoride
pW	Picowatt
QCL	Quantum cascade laser
Q.O.	Quasi-optical (mode converter)
QoS	Quality of service
QW	Quantum well
QWITT	Quantum well injection transit time (diode)
RAN	Radio access network
RF	Radio frequency
RMS	Root-mean-square (current/voltage)
RMS SNR	Root-mean-square signal-to-noise ratio
RTD	Resonant tunneling diode
RX	Receiver
S	Siemen
s	Second
SASE	Self-amplified-spontaneous-emission (FEL)
SBD	Schottky barrier diode
SCEs	Short channel effects
SFG	Sum frequency generation
SHBT	Single heterojunction bipolar transistor
SHF	Super high frequency
SHG	Second harmonic generation
SHM	Simple harmonic motion

SI-GaAs	Semi-insulating gallium arsenide
SIN	Superconducting metal film/Insulating layer/ normal metal film
SIS	Superconducting metal film/insulating layer/ superconducting metal film
SLF	Super low frequency
SLM	Spatial light modulator
SNR	Signal-to-noise ratio
SOI	Silicon-on-insulator
sq	Square
SQUID	Superconducting quantum interference device
SS	Subthreshold swing
SSD	Self-switching diode
SWS	Slow wave structure
T	Tesla
Tbps	Terabits per second
TDS	Time-domain spectroscopy
TEA	Transversely excited atmosphere (laser)
TEM_{00} mode	Transverse electromagnetic mode of lowest order (fundamental mode having the same form as a Gaussian beam)
TE_{mn}	Transverse electric mode, m = number of half-wave patterns across the width of the rectangular waveguide and n = number of half-wave patterns across its height
TeraFET	Terahertz field-effect transistor
THz	Terahertz
THz-CW	Terahertz continuous wave (spectroscopy)
THz-TDS	Terahertz-time domain spectroscopy
TWT	Travelling wave tube
TX	Transmitter
UHF	Ultra high frequency
ULF	Ultra low frequency
UTC-PD	Unitravelling-carrier photodiode
UV	Ultraviolet
UV-LIGA	Ultraviolet-lithographie, galvanoformung, abformung (German): ultra-violet-lithography, electroplating, and molding
UV–vis	Ultraviolet–visible (spectroscopy)
V	Volt
VB	Valence band
VBM	Valence band maximum
VF	Voice frequency
VHF	Very high frequency
ViSAR	Video Synthetic aperture radar
VLF	Very low frequency
VUV	Vacuum ultraviolet (wavelength:100–300nm)
W	Watt
WiFi	Not an acronym, but trademarked term for a wireless local area networking (WLAN) technology, which is commonly thought as the shortened form of 'wireless fidelity'
XCOM	X ray communication
μK	Micro Kelvin
μm	Micron, micrometer
μs	Microsecond
μW	Microwatt

Chemical symbols

ADP:$NH_4H_2PO_4$	Ammonium dihydrogen phosphate
Al	Aluminum
AlAs	Aluminum arsenide
AlGaAs	Aluminum gallium arsenide
AlGaN	Aluminum gallium nitride
AlInAs	Aluminum indium arsenide
AlN	Aluminum nitride
Al_2O_3, AlO_x	Aluminum oxide
AO	Auramine O, a diarylmethane dye
AlSb	Aluminum antimonide
Au	Gold (Aurum)
Ba	Barium
$BaTiO_3$	Barium titanate
BCA	Barium calcium aluminate
BeO	Beryllium oxide
CCH	Chlortetracycline hydrochloride
CdTe	Cadmium telluride
CdZnTe	Cadmium zinc telluride
$C_{32}H_{16}CoN_8$	Cobalt phthalocyanine
CO_2	Carbon dioxide
$CsNO_3$	Caesium nitrate
CF_4	Carbon tetrafluoride
CH_3F	Fluoromethane
$^{13}CH_3F$	Fluoromethane-13C
CH_3OD	Methanol-OD or mono-deuteromethanol
CH_3OH	Methanol
Cl	Chlorine
C–O	Carbon–oxygen (bond)
CO_2	Carbon dioxide
Cr_2O_3	Chromium oxide
Cu	Copper
CuO	Copper (II) oxide or cupric oxide
DAST	4-dimethylamino-N-methyl-4-stilbazolium tosylate
GaAs	Gallium arsenide
GaInAs	Gallium indium arsenide
GaN	Gallium nitride
GaP	Gallium phosphide
GaSe	Gallium selenide
Ge	Germanium
H	Hydrogen
HCl	Hydrogen chloride gas, hydrochloric acid
HDPE	High-density polyethylene
HMX	High melting explosive: cyclotetramethylene-tetranitramine
InAlAs	Indium aluminum arsenide
InAs	Indium arsenide
InGaAs	Indium gallium arsenide
InGaAsP	Indium gallium arsenide phosphide

InGaN	Indium gallium nitride
InP	Indium phosphide
Ir	Iridium
KDP: KH_2PO_4	Potassium dihydrogen phosphate
$LiNbO_3$	Lithium niobate
$LiTaO_3$	Lithium tantalate
LT-GaAs	Low temperature gallium arsenide
MDMA	3,4-Methylenedioxymethamphetamine
Mo	Molybdenum
Nb	Niobium
NbN	Niobium nitride
NbTi	Niobium titanium
NbTiN	Niobium titanium nitride
Ni	Nickel
NiCr	Nickel–chromium (nichrome)
NiO	Nickel oxide
OFHC	Oxygen free high conductive (copper)
OH	Hydroxide
Os	Osmium
$PbTiO_3$	Lead titanate
Pd	Palladium
PETN	Pentaerythritol tetranitrate
PolySi	Polysilicon
Pt	Platinum
PVDF	Polyvinylidene fluoride
Re	Rhenium
RDX	Royal demolition explosive: cyclotrimethylenetrinitramine
Ru	Ruthenium
Si	Silicon
SI-GaAs	Semi-insulating GaAs
SiGe	Silicon Germanium
Si_3N_4	Silicon nitride
SiO_2	Silicon dioxide
SiON	Silicon-oxide-nitride, silicon oxynitride
TCsH	Tetracyclines hydrochloride
Ti	Titanium
$TiCl_4$	Titanium tetrachloride
TiO_2	Titanium oxide
TNT	Trinitrotoluene
Vox	Vanadium oxide
W	Tungsten (Wolfram)
ZnO	Zinc oxide
ZnSe	Zinc selenide
ZnTe	Zinc telluride

Mathematical notation

Roman alphabet

A	Junction area, Area, Voltage gain, A constant, Amplitude
a	Radius of the circular contact, Anode radius
A_E	Effective aperture area or aperture
$A_{effective}$	Effective Richardson constant for thermionic emission
A_i	Short-circuit current gain
A_l	The amplitudes
A_p	Power gain of a helix
$\|A_1(\omega)\|$, $\|A_2(\omega)\|$	Amplitudes of sine waves (angular frequency ω)
B	Magnetic field, Bandwidth, Rotational constant
$\mathbf{B}$	Magnetic field vector, Magnetic induction
b	Cathode radius
B_0	Peak magnetic field, Magnetic field at the cathode, Magnetic pole field
$B_{Effective}$	Effective bandwidth
B_{Peak}	Peak magnetic field at the resonant cavity
BW	Bandwidth
B_y	Magnetic field component in the Y-direction
C	Pierce gain parameter, Capacitance, Electronic heat capacity, Shannon information capacity
c	Velocity of light in vacuum
C_0	Capacitance in the resonant tunneling diode model, Geometrical depletion capacitance
c_1, c_2, c_3,….	First, second and third derivatives of polarization P with respect to electric field E
C_{BE}	Emitter-base capacitance
C, D	Constants
$C_{Depletion(BC)}$	Depletion capacitance across the collector-base junction
$C_{Depletion\ (BE)}$	Depletion capacitance across the emitter-base junction
$C_{Diff(BE)}$	Emitter-base diffusion capacitance
$C_{Gate\text{-}to\text{-}channel}$	Gate-to-channel capacitance
C_{GD}	Gate-drain capacitance
$C_{GD(Extrinsic)}$	Extrinsic gate-drain capacitance
$C_{GD(Intrinsic)}$	Intrinsic gate-drain capacitance
C_{GS}	Gate-source capacitance
C_{gSD}	Capacitance originating from the leakage of charging current for the gate capacitance $C_{GS}+C_{GD}$ through the output source-drain conductance g_{SD}
$C_{GS(Extrinsic)}$	Extrinsic gate-source capacitance
$C_{GS(Intrinsic)}$	Intrinsic gate-source capacitance
C_j	Junction capacitance
C_{j0}	Zero-bias junction capacitance
C_{jBE}	Base-emitter junction capacitance
C_{Miller}	Miller capacitance
$C_{overlap}$	Overlap capacitance
C_{ox}	Oxide capacitance per unit area

C_P	Parallel capacitance
$C_{\text{Parasitic}}$	Parasitic capacitance
C_Q	Quantum capacitance
C_S	Displacement capacitance
C_W	Quantum well capacitance
$\mathbf{D}$	Displacement vector
D	Diffusion coefficient, Length of the dipole, Distance between the transmitter and receiver, Diameter, Diameter of the aperture
d	Thickness of film, thickness of the dielectric, Diameter, Period of the planar periodic slow wave structure, Second-order nonlinear coefficient, Separation between the detectors, Distance between telescopes
D_1, D_2	Apertures of telescopes
d_1	Optical path length of the pump arm
d_2	Optical path length of the probe arm
D_e	Diffusion coefficient of electrons
d_{GC}, $d_{\text{Gate-Channel}}$	Separation between the gate and the channel
dI_{Wave}/dt	Wave intensity growth rate
D_n	Diffusion coefficient of electrons
dq	Infinitesimal charge
dW/dt	Rate of change of energy of an electron
E	Electric field, Energy
$\mathbf{E}$	Electric field vector
e	Electronic charge = 1.602×10^{-19}C
E_0	Peak electric field, Resonant energy level, Maximum value or amplitude of the wave, Ground state energy level
E_1, E_2, E_3	Energy levels in the quantum well
E_C	Energy of conduction band edge
E_{EB}	Electron beam energy
E_F	Fermi level, Fermi energy
E_G, E_g	Energy gap
E_j, $E_{J'}$	Rotational energy levels
E_{Kinetic}	Kinetic energy
E_{Peak}	Peak electric field
$E(t)$	Near IR pulse
E_{THz0}	Peak value of terahertz field
$E_{\text{THz}}(t)$	Electric field of the terahertz radiation pulse at time t, Continuous-wave terahertz field
$E_{\text{THz}}(t')$	Electric field of the terahertz radiation pulse at time t'
$e(t_n)$	A signal; amplitude of the input electric field at an instant of time t_n
E_x	Electric field component in the X-direction
$E(x, t)$	Electric field as a function of distance in X-direction and time t
$E(x, y, z)$	A periodic function of z with period L
$E(\omega)$	Fourier transform of a function of time $E(t)$
$\|E(\omega)\|$	Continuous amplitude spectrum of the wave
$E(\omega_k)$	Discrete Fourier transform of a signal $e(t_n)$
F	Force, RF noise figure
f	Frequency
F_{12}	Force acting on particle of mass m_2 due to particle of mass m_1
F_{21}	Force acting on particle of mass m_1 due to particle of mass m_2

f_C, $f_{Cut\text{-}off}$	Linear cut-off frequency
f_H	Upper cut-off or corner frequency
f_L	Lower cut-off or corner frequency
F_m	Magnetic field compression
f_{Max}, f_{max}	Maximum cut-off frequency, Maximum frequency of oscillation of a transistor
f_{Osc}	Linear oscillation frequency
f_p	Linear plasma frequency
f_r	Resonant frequency
$f_{Roll\text{-}off}$	Roll-off linear frequency
f_T	Tansition frequency or Cut-off frequency of a transistor
$f(x)$	Function of x
G	Electrical conductance, Differential conductance across the active region of RTD, Thermal conductance, Gap length, Gain of the antenna
GB	Gain-bandwidth product
G_D	Differential conductance
g_{DS}	Drain-source conductance
g_m	Transconductance
$g_{mintrinsic}$	Intrinsic transconductance
g_{oi}, $g_{oIntrinsic}$	Intrinsic output conductance
G_{RX}	Gain of receiver antenna
g_{SD}	Output source-drain conductance
g_{ss}	Small-signal gain of the planar undulator
G_{TX}	Gain of transmitter antenna
H	Irradiance, Magnetic field, Hamiltonian
h	Planck's constant = $6.626\ 07 \times 10^{-34}$ J.s = $4.135\ 6677 \times 10^{-15}$ eV.s
$\hbar$	Reduced Planck's constant = $1.054\ 5718 \times 10^{-34}$ J.s = $6.582\ 119\ 57 \times 10^{-16}$ eV. s
h_{FE}	DC common − emitter current gain of a transistor
h_{fe}	Small − signal common − emitter current gain of a transistor
I	Total current flowing through the diode, Moment of inertia, Intensity of terahertz radiation after traversing a distance x through the material
I_0	Reverse saturation current (dark current) of the diode, Beam current, Intensity of the incident terahertz radiation
I_A	Alfvén current
$I_{Average}$	Average intensity of electromagnetic wave
I_B	Base current of a transistor
I_{DS}	Drain−source current
I_E	Emitter current
I_{EB}	Average electron beam current
$I_{EB(Peak)}$	Peak electron beam current
I_C	Collector current of a transistor
i_i, i_{IN}	Input current
i_L	Output current
I_{Max}	Maximum current
$I_{Modulated}$	Intensity of modulated terahertz wave
I_n	Electron current
$I_{Opt}\ (t)$	Intensity of the optical beam at time t

I^0_{Opt}	Amplitude of the optical beam
i_{OUT}	Output current
$I_{\mathbf{P}}$, I_{Peak}	Peak current
i_p	Pyroelectric current
$I_{\mathrm{Probe}}(t)$	Intensity profile of the IR probe beam
I_{RMS}	Root-mean-square current
I_{S}	Saturation current
I_{THz}	Terahertz component of the photocurrent
$I_{\mathrm{Unmodulated}}$	Intensity of unmodulated terahertz wave
$I_{\mathbf{V}}$	Valley current
I_{Wave}	Intensity of the wave
$I_{\mathrm{Wave(0)}}$	Initial intensity of the wave at time t=0
$\mathbf{J}$	Current density vector
J	Rotational energy level
j	An operator which prefixed before a number identifies it as an imaginary number or imaginary part of a complex number
J_0, J_1	Bessel functions of the first kind of zeroth and first order
$JJ(\xi)$	Reduction factor of FEL interaction in the undulator
$J'_m(x)$	Bessel function of the first kind of order m
J_{n}	Current density of electrons
J_{W}	Instantaneous current density flowing through the quantum well
K	A dimensionless number called the undulator parameter
k	Momentum, Wavevector, Wavenumber, Force constant or spring constant
$k_{\mathbf{B}}$	Boltzmann constant=1.380 649×10^{-23}JK^{-1}=8.617 333 26×10^{-5}eVK^{-1}
k_{OH}	Force constant for strong (covalent) oxygen-to-hydrogen bonding
$k_{O.....H}$	Force constant for weak oxygen-to-hydrogen bonding
k_z, k_{zl}	Axial wave number/wave vector
k_{z0}	Wave number of the zeroth order spatial harmonic in the axial direction
$k_{\perp l}$	Transverse wave vector/wave number
$k(\Omega)$	Wave vector of the terahertz signal produced at angular frequency Ω
L	Length, Length of the quantum well, Drift length, Channel length, Gate length, Length of the microbridge, Angular momentum
l	Harmonic number, Length
$\hat{L}^2$	A quantum-mechanical operator (angular momentum operator)
$L_{\mathrm{Atmospheric}}$	Atmospheric loss
L_{B}	Width of the barrier layer
L–C	Inductance–capacitance
$L_{\mathrm{Collector}}$	Length of collector region
L_{D}	Width of the depletion region
L_{Dead}	Dead space
$L_{\mathrm{Diffusion}}$	Thermal diffusion length
L_{Emitter}	Emitter length
L_{G}	Gate length, Power gain length
L_{Q}	Quantum inductance
L_{RX}	Receiver losses
L_{S}	Inertial inductance
L_{Sat}	Power saturation length along the undulator
$L_{\mathbf{TX}}$	Transmitter losses

L_W	Width of the quantum well region
L_x	Length in the *X*-direction
L_y	Length in the *Y*-direction
L_x, L_y, L_z	Projections of angular momentum on the three axes
$\mathbf{M}$	Magnetization of the medium
M	Link margin
m	Mass of the electron, Modulation depth or modulation index, A quantum number, Number of times that the measurements are done
m_0	Rest mass of electron = 9.109×10^{-31} kg
m_1, m_2	Masses of particles
m^*, m_e^*	Effective mass of electron
M_A	Atomic mass
$m_{Effective}$	Effective mass of electron
m_H	Mass of hydrogen atom
m_h	Effective mass of holes
m_O	Mass of oxygen atom
$m_{Relativistic}$	Relativistic mass of electron
$m_{Rest\ mass}$	Rest mass of electron
N	Circuit length, Total number of free electrons, Number *N* of quantized levels used for data representation, Number of time samples
n	An ideality factor ~1–2 (also called quality factor or emission coefficient), Refractive index of a medium, Carrier concentration, An integer, Harmonic number, Number of detectors
N_A	Acceptor concentration in the semiconductor, Avogadro's number = 6.022×10^{23} mol^{-1}
N_C	Effective density of states in the conduction band
N_D	Donor concentration in the semiconductor
$n_{Extraordinary\ ray}$	Extraordinary refractive index
n_i	Intrinsic carrier concentration in the semiconductor
$n_{Ordinary\ ray}$	Ordinary refractive index
Noise_{RX}	Noise at receiver
$n_p(0)$	Electron concentration at the emitter-base boundary (x=0)
n_{p0}	Electron concentration in thermal equilibrium in the P-base
N_{Pump}	Carrier density at the semiconductor surface generated by optical pumping with laser
$n_p(x)$	Electron concentration in P-region as a function of distance x
$n_{Slow\ wave}$	Refractive index of slow wave structure
$n(t)$	Noise signal
N_u	Number of periods in the undulator
N_V	Effective density of states in the valence band, Number of electrons per atom
$N(\delta)$	Delta doping density
$n(\Omega)$	Refractive index of terahertz pulse at frequency Ω
$n_g(\omega)$	Group velocity refractive index of near IR pulse of frequency ω
$\mathbf{P}$	Polarization density of the medium
P	Power, Electromagnetic power, Total path length for the electromagnetic wave travelling along the helical path, Vapor pressure
p	Distance between the two adjoining turns of a helix (pitch of the helix), Momentum

$\mathbf{p}$	Momentum of the electron
P_0	Amplitude of the modulated terahertz radiation
Pa	Pascal
P_{AC}	AC power
P_{THz}^{AE}	Terahertz power of antenna emitter
P_{CW}	Continuous-wave output power
P_{dBm}	Power in decibel-milliwatt
P_{DC}	DC operating power, DC polarization
P_{Del}	Power delivered
P_{EB}	Electron beam power
$P_{Feedpoint}$	Input or feedpoint power
P_{FEL}	Power of free electron laser
P_{L}	Power delivered to the load
P_{Laser}	Power of the laser
$\mathbf{P}_{Linear}$	Linear polarization
P_{Max}, $P_{Maximum}$	Maximum power
P_{mW}	Power in milliwatt
P_{n}	Shot noise power of spontaneous radiation
$P_{Non\text{-}linear}$	Non-linear polarization
P_{R}	Received power
P_{S}	Power received from signal source
P_{s}	Spontaneous polarization
P_{Sat}	Power saturation level
$P_{Straight}$	Path length for the electromagnetic wave propagating along the straight line path
P_{T}	Transmitted power
P_{TX}	Power transmitted
P_{Watts}	Power in Watts
Q	Charge, Quality factor
q	Electronic charge = 1.602×10^{-19}C
Q_{B}	Base charge
R	Resistance, Radius, Bond length
r, r_0	Radius of a circle, Larmor radius
$\mathbf{r}_1$, $\mathbf{r}_2$	Position vectors of particles
$\mathbf{r}_{12}$	Vector sum of $\mathbf{r}_1$ and $\mathbf{r}_2$
r_1, r_2	Radii of circles
R_{bb}	Base resistance of a transistor
$R_{b,\ Contact}$	Contact resistance of base
$R_{b,\ gap}$	Emitter-base gap resistance
$R_{b,\ Sheet}$	Base sheet resistance
$R_{b,Spreading}$	Spreading resistance under the emitter
R–C	Resistance–capacitance
R_{C}, $R_{Collector}$	Collector resistance of a transistor
R_{D}	Series resistance of drain terminal
r_{d}	Dynamic resistance
R_{E}, $R_{Emitter}$	Emitter resistance of a transistor
$R_{Feedpoint}$	Feedpoint resistance
R_{L}	Load resistance
R_{Losses}	Resistance arising from losses
r_{max}	Channel capacity (data rate)

R_R	Radiation resistance of the antenna
$R_{Radiation\ (dipole)}$	Dipole radiation resistance
R_S	Series resistance, DC spreading resistance, series resistance of source terminal
R_s	Sheet resistance of the superconducting film
R_{Spr}	Spreading resistance
$r(t)$	Sum of transmitted signal and noise signal
R_{tt}	Transit-time negative resistance
R_W	Specific resistance of quantum well region
r, θ, ϕ	Radial distance, polar angle, and azimuthal angle
S	Instantaneous intensity of electromagnetic waves
s	Velocity of plasma waves, Velocity of electron, Harmonic number, Distance travelled by the wave front, Latency, length of the path
$S_{Average}$	Average intensity of electromagnetic waves
Signal_{RX}	Signal at receiver
S_T	Transmitted power density
$S(t)$	An instantaneous signal
$s(t)$	Transmitted signal
$S(t')$	Signal at a single time point t'
Sensitivity_{RX}	Receive sensitivity
T	Absolute temperature, Time period of the wave, Kinetic energy of a particle, Sampling interval or sampling period
$1/T$	Sampling rate
t	Time
t_1	Time during which the crest of a wave shifts from one point to another
t_2	Time during which the wave moves from one point to another
T_C	Transition (critical) temperature of the superconductor
t_C	Channel thickness
$T_{Collector}$	Thickness of collector region
$T_{Contact}$	Thickness of contact layer
$t_{Depletion\ (BC)tt}$	Time taken by electronic current to cross the collector-base depletion region
T_F	Fermi temperature
t_n	An instant of time
U	Unilateral power gain of a transistor
U_0	Positive DC gate-source voltage which can induce strong inversion
U_1	Amplitude or peak value of gate-source voltage
U_{GS}	Gate-source voltage
v, υ	Velocity of electrons
$\mathbf{v}$	Velocity vector
V_0	DC bias, Peak voltage, Beam voltage
V, V_a	Applied bias, Energy, Volume, Potential energy
v_1, v_2	Linear velocities of particles
V_{BE}	Base-emitter voltage
V_{bi}	Built-in potential of a diode/metal–semiconductor contact
$V_{bi}\vert_{BC}$	Built-in potential of collector-base junction
$V_{bi}\vert_{BE}$	Built-in potential of base-emitter junction
v_C	Rate of escape of electrons from the quantum well to the collector
V_{CE}	Collector–emitter voltage of a transistor

V_D	Voltage drop across the diode, Drain voltage
v_d	Electron drift velocity
V_{DC}	DC bias
V_{DS}	Drain-source voltage
$V_{D(Sat)}$	Saturation drain voltage
v_E	Rate of escape of electrons from the quantum well into the emitter
v_e	Velocity of electron beam
$v_e\|_{Approximate}$	Approximate velocity of electron
$v_e\|_{Relativistic}$	Relativistically obtained velocity of electron
V_G	Gate voltage
v_g	Group velocity
V_{GD}	Gate-drain voltage
V_{GS}	Gate-source voltage
$v_{High\ field\ domain}$	High field domain velocity
$v_{Incoming}$	Velocity of incoming waves
V_{Max}	Maximum voltage
V_{Noise}	Root mean square noise voltage
$v_{Outgoing}$	Velocity of outgoing waves
V_P, V_{Peak}	Peak voltage
v_p	Phase velocity
V_R	Voltage drop across the resistor *R*
V_{RMS}	Root-mean-square voltage
V_S	Source voltage
v_S	Saturated electron velocity in the drift region
v_{Sat}	Electron saturation velocity
V_{Signal}	Root-mean-square signal voltage
V_{Th}	Threshold voltage
V_V	Valley voltage
V_W	Instantaneous voltage across the quantum well
v_x	Component of electron velocity in the *X*-direction
v_z	Axial translational velocity of electron beam, component of electron velocity in the *Z*-direction
$v_\perp$	Perpendicular component of electron velocity **v** directed perpendicular to the magnetic field
$v_{\|\|}$	Component of electron velocity **v** directed parallel to the magnetic field
$v(\Omega)$	Velocity of terahertz waves of frequency Ω
W	Width of the depletion region, Thickness, Channel width, Work done
$W_{b,\ Contact}$	Base contact width
$W_{Collector}$	Width of collector region
$W_{Depletion\ (BC)}$, $W_{Depletion}\|_{BC}$	Width of the depletion region across the collector-base junction
$W_{Electrically\ active\ emitter}$	Width of electrically active emitter
$W_{Emitter} = W_{Emitter\ contact}$	Width of the emitter
$W_{Emitter\text{-}Base\ spacing}$	Emitter-base spacing
$W_{Emitter\ junction}$	Width of emitter junction
$W_{Optimum}$	Optimum thickness of depletion region
x	Aluminum fraction in $Al_xGa_{1-x}As$, extension/shortening of the spring

x_0	A particular x position
X_C	Capacitive reactance
x_d	Zero-bias depletion region width
$x_{Depletion}\|_{BE}$	Width of depletion region across the base-emitter junction
$x(\omega)$	Real part of the Fourier transform $E(\omega)$
$Y(\theta, \phi)$	Spherical harmonics
$y(\omega)$	Imaginary part of the Fourier transform $E(\omega)$
Z	Impedance
z	Distance in the Z-direction
Z_0	Impedance of the slow wave structure, Impedance or characteristic impedance of free space (wave impedance of a plane wave in free space)
$Z_{Bulk\text{-}Spreading}$	Bulk spreading impedance
Z_{in}	Input impedance
Z_S	Bulk spreading impedance
$Z_{Skin\text{-}effect}$	Skin-effect impedance
Z_{tt}	Transit-time specific impedance
Z_W	Specific impedance of the quantum well

Greek alphabet

α	Real part of the complex number $\alpha + j\beta$, Ratio of radial-to-axial velocity (pitch factor), Coupling coefficient
α_T	Base transport factor
β	Imaginary part of the complex number $\alpha + j\beta$
β_{AC}	Small signal common-emitter current gain of a transistor
β, β_{DC}	Large signal DC common-emitter current gain of a transistor
β_x	Twiss factor
β_z	Relative average velocity in the forward Z-direction
Γ	Pyroelectric coefficient
γ	A coefficient quantifying the influence of surface states, Lorentz factor
δ, δ_S	Skin depth
ΔE_C	Conduction band offset
ΔE_G	Bandgap difference
ΔE_V	Valence band offset
ΔI	(Peak current–valley current) of resonant tunneling diode
ΔI_B	Change in base current
ΔI_C	Change in collector current
$\Delta I_{EO}(t)$	Effective time-varying electro-optic signal
$<\Delta I_{EO}(t)>$	Time-averaged signal
ΔJ	Difference between rotational energy levels J and J'
Δk	Change in wave number
Δn	Refractive index change
$\Delta n\|_{Linear}$	Refractive index change in linear electro-optic effect
$\Delta n\|_{Quadratic}$	Refractive index change in quadratic electro-optic effect
Δt	Interval of time, Time taken by the electron to traverse a distance from a point A to an equivalent point B
$\delta\ (t)$	Delta function
$\Delta t'$	Time taken by the wave to traverse a distance from a point A to an equivalent point B (one undulator wavelength)
ΔV	(Peak voltage–valley voltage) of resonant tunneling diode
$\Delta \nu$	Measurement bandwidth, Frequency separation between any two consecutive absorption lines
$\Delta \nu_{THz}$	Frequency difference in terahertz range
$\Delta \omega$	Change in angular frequency
ε_0	Permittivity of free space = 8.854×10^{-12} Fm^{-1}
ε, ε', ε_r	Relative permittivity of a material
ε_B, $\varepsilon_{Barrier\ layer}$	Relative permittivity of barrier layer
ε_D	Relative permittivity of depletion region
$\varepsilon_{Effective}$	Effective permittivity
$\varepsilon_{n,x}$	Emittance
ε_{ox}	Dielectric constant of oxide film
ε_S	Dielectric constant of the semiconductor
$\varepsilon_{Semiconductor}$	Permittivity of the semiconductor
ε_W	Relative permittivity of quantum well region
η	Efficiency, Absorbance of the radiation-sensitive film, $\sqrt{\frac{\mu_0}{\varepsilon_0}}/n$, Mass attenuation or absorption coefficient

η_{FEL}	Optical conversion efficiency of a free electron laser
η_{OPT}	Output coupling efficiency of the optical cavity
θ	An angle, Angle of emission of the wavefront with respect to the axis of the electron beam, Polar angle, Angular resolution
$\theta_{Telescope}$	Angular resolution of the telescope
λ	Wavelength
λ_R	Material-specific correction factor
λ_u	Undulator wavelength in a laboratory frame of reference
λ_u^*	Undulator wavelength after length contraction, Wavelength in a moving frame of reference
μ	Mobility of charge carriers, Refractive index, Effective mass of carriers in the semiconductor, Reduced mass of particles of masses m_1 and m_2, Linear attenuation constant
μ_0	Permeability of free space or permeability constant = $4\pi \times 10^{-7}$ H/m = $1.256\ 637 \times 10^{-6}$ N A^{-2}
$\mu_{Effective}$	Effective permeability
μ_n	Electron mobility
$\mu_{O.....H}$	Reduced mass for hydrogen bonded oxygen and hydrogen atoms (O…H)
μ_r	Relative permeability of the material
ν	Frequency
$\bar{\nu}$	Wavenumber
ν_+	Sum frequency
ν_-	Difference frequency
ν_0	Central frequency
$\nu_1,\ \nu_2,\ \nu_3,\ \ldots$	Frequencies of absorption lines
ν_{Beats}	Beat frequency
$\nu_{Cut\text{-}off}$	Cut–off frequency
ν_{RC}	Cut–off frequency determined by resistance–capacitance time constant
ξ	A variable = $2\pi z/\lambda_u$
$\xi_{Roll\text{-}off}$	Roll–off factor
ρ	Resistivity, Pierce or FEL parameter, Density of a material, Free charge density, Mass density
$\rho_{b,Contact}$	Base contact resistivity
$\rho_{Collector}$	Resistivity of collector region
$\rho_{Contact(Specific)}$	Specific contact resistivity
σ	Conductivity, Conductance
σ_{DC}	DC electrical conductivity
$\sigma_{Interband}(\omega)$	Optical conductivity at frequency ω for transitions from one energy band to another band
$\sigma_{Intraband}(\omega)$	Optical conductivity at frequency ω for transitions in the same band
$\sigma_{Optimum}$	Optimum injection conductance
σ_x	Electron beam RMS transverse dimension
$\sigma(\omega)$	Optical conductivity due to absorption of waves of frequency ω
τ	Time, Time constant, Electron lifetime in the quantum well, Total time spent by plasma waves in the round trip of the channel, Momentum relaxation time for scattering of carriers
$\tau_{Backward}$	Time taken by plasma wave to travel from drain to source

τ_{Btt}	Base transit time
τ_C	Electron lifetime in the collector, Time spent in charging the depletion region of the collector
τ_E	Emitter charging time
τ_{ee}	Electron-electron interaction time
τ_{escape}	Escape time of the phonon
$\tau_{Extrinsic}$	Extrinsic delay
$\tau_{Forward}$	Time taken by plasma wave to travel from source to drain
$\tau_{Intrinsic}$	Intrinsic delay
$\tau_{Parasitic}$	Parasitic delay
τ_{pe}	Phonon-electron interaction time
τ_{RC}	Resistance-capacitance time constant
$\tau_{Relaxation}$	Momentum relaxation time
τ_t	Transit time of carriers
$\tau_{Thermal}$	Thermal response time
ϑ	Velocity of electromagnetic waves
ϕ	Phase angle, Azimuthal angle
ϕ_0	Energy of charge neutrality level reckoned from valence band maximum
ϕ_1	A phase term linked to the optical path length of the pump arm
ϕ_2	A phase term linked to the optical path length of the probe arm
ϕ_B	Zero-bias Schottky barrier height
ϕ_{bi}	Built-in potential at the metal-semiconductor interface
ϕ_M	Work function of the metal
ϕ_S	Work function of the semiconductor
$\phi(x, y, z)$	A periodic function of z with period L
$\phi(\omega)$	Continuous phase spectrum of the wave
$\phi_1(\omega)$, $\phi_2(\omega)$	Phases of the waves
χ, χ_S	Electron affinity of the semiconductor
$\chi^{(1)}$	Linear susceptibility
$\chi^{(1)}, \chi^{(2)}, \chi^{(3)}, \ldots$	First-order, second-order and third-order susceptibilities
χ_{mn}	nth zero of the derivative of the mth order Bessel function of the first kind
ψ	Opening angle for the radiation emitted by electron beam, Wave function
Ω	Ohm, Frequency of terahertz radiation pulse, Radian-frequency sampling interval, Beat frequency
ω	Angular frequency, Angular velocity
ω_0	Cut-off frequency of a waveguide
ω^*	Angular frequency after length contraction
Ω_C	Cyclotron frequency
$\omega_{Cut\text{-}off}$	Cut-off angular frequency
ω_D	Dielectric relaxation frequency
ω_{osc}	Angular oscillation frequency
ω_P	Plasma frequency
$\omega_{Roll\text{-}off}$	Roll-off angular frequency
ω_S	Carrier scattering frequency
ω_T	Angular transition or cut-off frequency

Part I

Solid-state electronic devices

IOP Publishing

Practical Terahertz Electronics: Devices and Applications, Volume 1
Solid-state devices and vacuum tubes

Vinod Kumar Khanna

Chapter 1

Terahertz electromagnetic waves

As terahertz waves are electromagnetic waves of terahertz frequencies, it is necessary to learn the basic concepts of electromagnetic waves and the subdivision of these waves into an electromagnetic spectrum in accordance with their frequencies. Terahertz electronics deals with the generation, control and detection of terahertz waves in a selected band of electromagnetic spectrum between 0.1 and 10 THz, which is given the name terahertz gap. This gap has for long been an ignored range of frequencies because of non-performance of traditional electronic and photonic devices in this frequency range. The book deals with the principles and practical realization of electronic devices designed to operate in the terahertz gap. The aims, scope and layout plan of the book are outlined and a brief description of contents of each chapter is presented.

In this chapter, we learn about the terahertz gap of frequencies, the much talked about last frontier of the electromagnetic spectrum, which holds tremendous potential for applications but has for long remained underexplored (Armstrong 2012). The executive plan of the book will be laid out.

1.1 What are terahertz waves?

'Terahertz' is a composite word made from 'Tera' and 'Hertz'. 'Tera' stands for 12th power of 10. Terahertz is a big unit of frequency which is 10^{12} times the base unit of frequency called hertz and representing cycles per second.

$$1 \text{ THz} = 10^{12} \text{ Hz} \tag{1.1}$$

or,

$$1 \text{ Hz} = 10^{-12} \text{ THz} \tag{1.2}$$

doi:10.1088/978-0-7503-3171-5ch1

When talking about frequency, we are addressing a periodically varying signal. What is the relevant signal here which vibrates so fast? Is it a mechanical vibration? No. It is the electromagnetic waves. So electromagnetic waves having frequencies in the terahertz range are at the focus of attention.

Are all terahertz frequencies of interest to us? Again, no. We are interested only in a segment or band of terahertz frequency spectrum from 0.1 to 10 THz. Why? Because this selected frequency band has for a long time evaded utilization due to various reasons, whereas all the frequency ranges above and below this band have been studied and applied to various uses. For this reason, this frequency band is popularly known as the 'terahertz gap' representing a gap in our knowledge about this regime. We shall elucidate later the causes for designating this band as terahertz gap.

Electromagnetic waves whose frequencies fall in the range 0.1–10 THz are called by various names such as submillimeter wave radiation, terahertz waves, tremendously high frequency waves, T-rays, T-light, T-lux, and so forth. In some places in the literature, the frequency band 0.3 to 3 THz is named as the terahertz gap as per the International Telecommunication Union (ITU) band of frequencies. But in this book we shall take this frequency interval as 0.1–10 THz. A major part of this interval overlaps with some portion of the far infrared (FIR) waves (0.3 to 20 THz).

The discussion so far has clarified two points: firstly, terahertz radiation consists of electromagnetic waves and secondly, these waves have very high frequencies as specified in the definition of terahertz gap. The moment we say that terahertz waves are high-frequency electromagnetic waves, the question is asked: What are electromagnetic waves? What makes frequency such an important parameter? So, we want to know more about (i) electromagnetic waves, and also about (ii) the part played by frequency in deciding the properties of the waves.

1.2 The electromagnetic waves

These are transverse waves comprising self-propagating, self-perpetuating, intercoupled, mutually perpendicular time-varying electric and magnetic fields. The electric and magnetic fields in these waves are in phase with each other, attaining maximum and minimum values at the same instant (Kimura 2017). Being transverse waves, the displacements of all points on the wave (the oscillations of electric and magnetic fields) take place along paths at right angles to the direction of wave motion (figure 1.1).

1.2.1 Production of electromagnetic waves

The next obvious question that boggles the mind is: How are such waves produced? Can a charge at rest called static charge produce electromagnetic waves? An example of charge at rest is the charge induced in a comb when rubbed with dry hair. No, static electricity cannot generate electromagnetic waves.

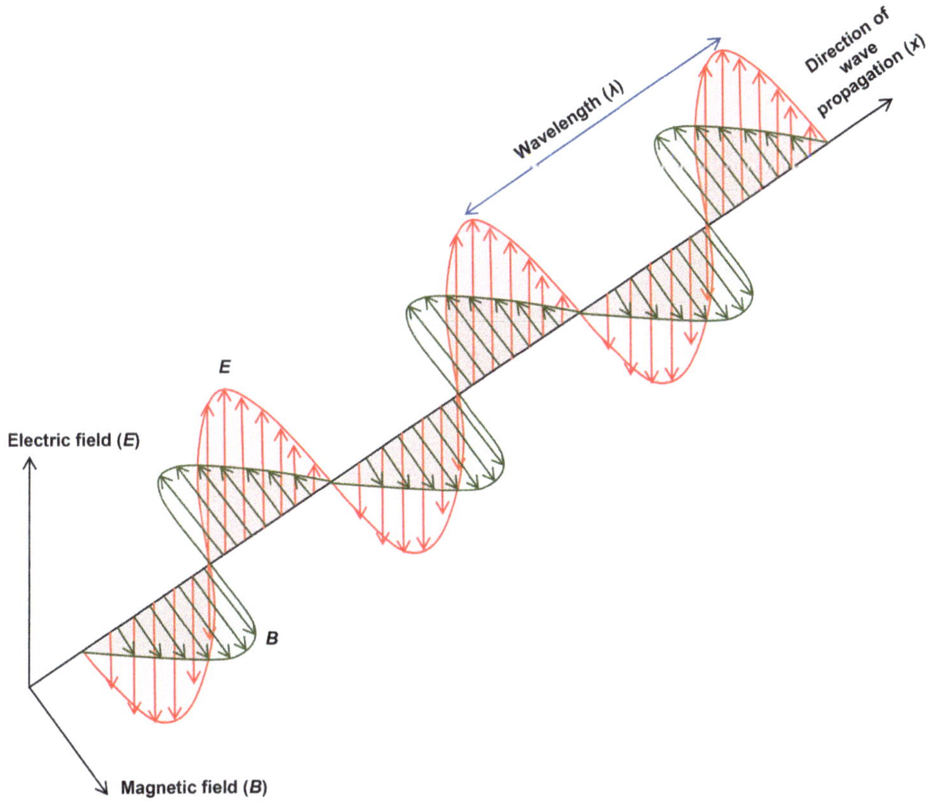

Figure 1.1. Illustration of an electromagnetic wave as made of two types of fields, the electric and magnetic fields. These fields are mutually coupled with each other. Besides being interlinked, they act at right angles to each other and to the direction of propagation of the wave.

Can a charge in motion at uniform velocity produce these waves? The direct electric current flowing in wires is an example. No, current electricity in this form is not a source of electromagnetic waves.

But no sooner than a charge undergoes acceleration or deceleration, electromagnetic waves are emitted. Thus whenever moving charges experience an increment or decrement in their velocities, they become sources of electromagnetic waves. The overall scenario is: a stationary charge produces an electric field. A charge in motion is associated with a magnetic field. An accelerated charge produces changing electric and magnetic fields. These changing electric and magnetic fields act co-operatively. The spatially varying electric field leads to a time-varying magnetic field. In reciprocation, the spatially-varying magnetic field results in a time-varying electric field. The sustained interplaying and regeneration of electric and magnetic fields constitutes an electromagnetic wave. The wave moves out into space carrying energy with it. The four equations propounded by the Scottish mathematical physicist James Clerk Maxwell and known as Maxwell's equations succinctly describe the creation of electric and magnetic fields by charges and their mutual interactions. These celebrated equations lay down the foundation of classical electrodynamics.

Example 1.1 What is peak value of magnetic field created by an electric field of amplitude 110 V m^{-1} in vacuum?

Solution: In a medium in which the velocity of electromagnetic waves is ϑ, the magnetic field B_y is connected to the electric field E_x as

$$B_y = \frac{E_x}{\vartheta} \tag{1.3}$$

In vacuum, the magnetic field B and electric field E are related through the velocity c of electromagnetic waves in vacuum by the formula

$$B = \frac{E}{c} \tag{1.4}$$

where the perpendicularity of directions of electric and magnetic fields is implied.
Hence,

$$B_{\text{Peak}} = \frac{E_{\text{Peak}}}{c} = \frac{110}{3 \times 10^8} = 3.67 \times 10^{-7}\ \text{T} \tag{1.5}$$

1.2.2 Oscillating charges as sources of electromagnetic waves

One common example of an accelerated charge is a charged particle oscillating about its equilibrium position. So electromagnetic waves can be generated by oscillating charges. What is the frequency of electromagnetic waves emitted by oscillating charges? It is the frequency of oscillation of the charges. By making charges oscillate at different frequencies, various types of electromagnetic waves can be generated. Terahertz waves are one form of electromagnetic waves that have frequency in the terahertz range.

1.2.3 Far-field and near-field interactions

Do electromagnetic waves need any help or do they in any way remain connected with the charges to whose acceleration they owe their origin? No, these waves are free to move without any influence from the charges from which they initially started. They may never interact with their parent charges. So, they are known as far field to discriminate them from phenomena such as electrostatic induction or electrodynamic induction, which are called near-field effects. Electromagnetic waves from a radio transmitter differ from the electric dipole fields in a transformer. Power density of far-field fades with distance according to an inverse square law while that of near-field decreases as inverse cube law.

1.2.4 Wave properties of electromagnetic waves

Over large distances and long time scales, electromagnetic waves display wave properties and suffer from reflection, refraction, interference, diffraction and polarization like any other waves. We know that the wavelength λ of a wave is inversely proportional to its frequency ν. In vacuum, the relationship between wavelength and frequency of electromagnetic waves is

$$\lambda = \frac{c}{\nu} \tag{1.6}$$

where c denotes the velocity of waves in vacuum.

The velocity ϑ of electromagnetic waves in a medium of refractive index n is related to their velocity c in free space or vacuum as

$$\vartheta = \frac{c}{n} \tag{1.7}$$

Needless to say that electromagnetic waves of different frequencies have different wavelengths.

Example 1.2 What is the apparent velocity of terahertz waves in a medium of refractive index 1.5?

Solution: Applying equation (1.7)

$$\vartheta = \frac{c}{n} = \frac{2.9979 \times 10^{8}}{1.5} = 1.9986 \times 10^{8}\ \mathrm{ms^{-1}} \tag{1.8}$$

Example 1.3 Find the peak electric and magnetic fields in a terahertz wave of average intensity 9 mW cm^{-2} propagating in: (a) a non-magnetic medium of dielectric constant 5, and (b) in air with $\varepsilon_r = 1$.

Solution: An electromagnetic wave propagating in a given direction carries electromagnetic energy with it (figure 1.2). Intensity of irradiance of a wave is the energy flow per unit area per unit time.

If an electric field E_x exists in a small region of space of relative permittivity ε_r, the region has an energy density or energy per unit volume:

$$\text{Energy density in the electric field} = \frac{1}{2}\varepsilon_0\varepsilon_r E_x^2 \tag{1.9}$$

where ε_0 is permittivity of free space. In the same way, a region of space in which there is a magnetic field B_y has an energy density:

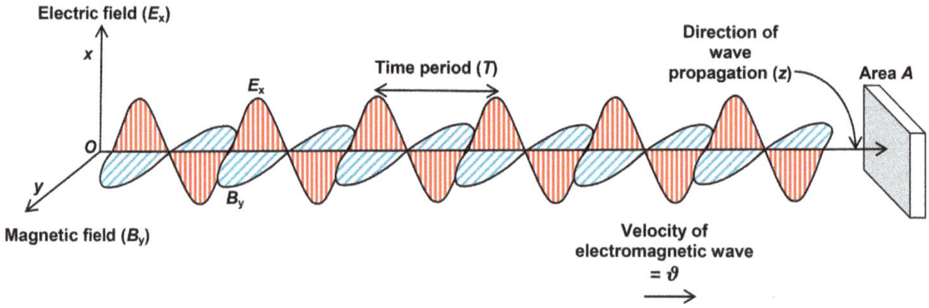

Figure 1.2. Finding the peak electric and magnetic fields in a terahertz wave of given average intensity.

$$\text{Energy density in the magnetic field} = \frac{1}{2\mu_0} B_y^2 \tag{1.10}$$

where μ_0 is permeability of free space. The velocity of terahertz waves is expressed as

$$\vartheta = \frac{c}{n} = \frac{1}{\sqrt{\mu_0 \varepsilon_0 \varepsilon_r}} \tag{1.11}$$

because

$$c = \frac{1}{\sqrt{\mu_0 \varepsilon_0}} \tag{1.12}$$

and

$$n = \sqrt{\varepsilon_r} \tag{1.13}$$

In a medium of dielectric constant ε_r, equation (1.3) between magnetic field B_y and electric field E_x is modified to

$$B_y = \frac{E_x}{\vartheta} = \frac{E_x}{\dfrac{1}{\sqrt{\mu_0 \varepsilon_0 \varepsilon_r}}} = \sqrt{\mu_0 \varepsilon_0 \varepsilon_r}\, E_x \tag{1.14}$$

where equation (1.11) has been applied. Substituting for B_y from equation (1.14) in equation (1.10) we get

$$\begin{aligned} \text{Energy density in the magnetic field} &= \frac{1}{2\mu_0} B_y^2 = \frac{1}{2\mu_0}\left(\sqrt{\mu_0 \varepsilon_0 \varepsilon_r}\, E_x\right)^2 \\ &= \frac{1}{2\mu_0} \times \mu_0 \varepsilon_0 \varepsilon_r E_x^2 = \frac{1}{2} \varepsilon_0 \varepsilon_r E_x^2 \end{aligned} \tag{1.15}$$

Comparing equations (1.15) and (1.9), we note that

Energy density in the magnetic field = Energy density in the electric field (1.16)

Hence,

$$\begin{aligned}&\text{Energy density in the electromagnetic field}\\&=\text{Energy density in the electric field}\\&+\text{Energy density in the magnetic field}\\&=2\times\text{Energy density in the electric field}\\&=2\times\frac{1}{2}\varepsilon_0\varepsilon_r E_x^2=\varepsilon_0\varepsilon_r E_x^2\end{aligned}\tag{1.17}$$

where equation (1.9) is used.

Let us calculate the energy incident on an area A placed perpendicular to the direction of propagation of the wave. In an interval of time Δt, the area A is crossed by a portion of the wave of spatial length $\vartheta\Delta t$. Then the volume crossed by the wave in this time interval is $A\ \vartheta\Delta t$, and

$$\begin{aligned}&\text{Energy in the electromagnetic field}\\&\quad=\text{Energy density in the electromagnetic field}\times\text{Volume}\\&\quad=\varepsilon_0\varepsilon_r E_x^2\times A\vartheta\Delta t\end{aligned}\tag{1.18}$$

The electromagnetic power flowing is

$$P=\frac{\text{Energy in the electromagnetic field}}{\text{Time}}=\frac{\varepsilon_0\varepsilon_r E_x^2\times A\vartheta\Delta t}{\Delta t}\tag{1.19}$$

and the electromagnetic power flowing per unit area is

$$\frac{P}{A}=\frac{\varepsilon_0\varepsilon_r E_x^2\times A\vartheta\Delta t}{A\Delta t}=\vartheta\varepsilon_0\varepsilon_r E_x^2\tag{1.20}$$

Applying equation (1.20), we can write

$$S=\vartheta\varepsilon_0\varepsilon_r E_x^2\tag{1.21}$$

where S represents the instantaneous intensity of electromagnetic waves. The average intensity is found by averaging the above intensity over one period of the wave by writing the electric field as a sinusoidally varying quantity with angular frequency ω

$$E_x=E_0\sin\omega t\tag{1.22}$$

where E_0 is the peak electric field.

The average intensity over the time period T of the wave is

$$\begin{aligned}
S_{\text{Average}} &= \frac{1}{T}\int_0^T Sdt = \frac{1}{T}\int_0^T \vartheta\varepsilon_0\varepsilon_r E_x^2 dt = \frac{1}{T}\int_0^T \vartheta\varepsilon_0\varepsilon_r E_0^2 \sin^2\omega t dt \\
&= \frac{1}{2T}\int_0^T \vartheta\varepsilon_0\varepsilon_r E_0^2 2\sin^2\omega t dt = \frac{\vartheta\varepsilon_0\varepsilon_r E_0^2}{2T}\int_0^T 2\sin^2\omega t dt \\
&= \frac{\vartheta\varepsilon_0\varepsilon_r E_0^2}{2T}\int_0^T (1-\cos 2\omega t)dt \\
&= \frac{\vartheta\varepsilon_0\varepsilon_r E_0^2}{2T}\int_0^T (dt-\cos 2\omega t dt) \\
&= \frac{\vartheta\varepsilon_0\varepsilon_r E_0^2}{2T}\int_0^T dt - \frac{\vartheta\varepsilon_0\varepsilon_r E_0^2}{2T}\int_0^T \cos 2\omega t dt = \frac{\vartheta\varepsilon_0\varepsilon_r E_0^2}{2T}|t|_0^T \\
&\quad - \frac{\vartheta\varepsilon_0\varepsilon_r E_0^2}{2T}\times\left|\frac{1}{2\omega}\sin 2\omega t\right|_0^T = \frac{\vartheta\varepsilon_0\varepsilon_r E_0^2}{2T}\{T-0\} \\
&\quad - \frac{\vartheta\varepsilon_0\varepsilon_r E_0^2}{2T\times 2\omega}|\sin 2\omega t|_0^T = \frac{\vartheta\varepsilon_0\varepsilon_r E_0^2}{2T}T - \frac{\vartheta\varepsilon_0\varepsilon_r E_0^2}{4T\omega}\{(\sin 2\omega T)-\sin 0\} \\
&= \frac{1}{2}\vartheta\varepsilon_0\varepsilon_r E_0^2 - \frac{\vartheta\varepsilon_0\varepsilon_r E_0^2}{4T\omega}(\sin 2\omega T - 0) \\
&= \frac{1}{2}\vartheta\varepsilon_0\varepsilon_r E_0^2 - \frac{\vartheta\varepsilon_0\varepsilon_r E_0^2}{4T\omega}\left(\sin 2\frac{2\pi}{T}T - 0\right)
\end{aligned} \tag{1.23}$$

where we have put

$$\omega = 2\pi \times \text{Frequency} = 2\pi \times \frac{1}{\text{Time period}} = \frac{2\pi}{T} \tag{1.24}$$

Equation (1.23) leads to

$$\begin{aligned}
S_{\text{Average}} &= \frac{1}{2}\vartheta\varepsilon_0\varepsilon_r E_0^2 - \frac{\vartheta\varepsilon_0\varepsilon_r E_0^2}{4T\omega}(\sin 4\pi) = \frac{1}{2}\vartheta\varepsilon_0\varepsilon_r E_0^2 - \frac{\vartheta\varepsilon_0\varepsilon_r E_0^2}{4T\omega}\times 0 \\
&= \frac{1}{2}\vartheta\varepsilon_0\varepsilon_r E_0^2
\end{aligned} \tag{1.25}$$

From equations (1.25) and (1.11),

$$S_{\text{Average}} = \frac{1}{2}\vartheta\varepsilon_0\varepsilon_r E_0^2 = \frac{1}{2}\frac{1}{\sqrt{\mu_0\varepsilon_0\varepsilon_r}}\varepsilon_0\varepsilon_r E_0^2 = \frac{\sqrt{\varepsilon_0\varepsilon_r}E_0^2}{2\sqrt{\mu_0}} \tag{1.26}$$

or,

$$E_0^2 = \frac{2\sqrt{\mu_0}\times S_{\text{Average}}}{\sqrt{\varepsilon_0\varepsilon_r}} = 2S_{\text{Average}}\sqrt{\frac{\mu_0}{\varepsilon_0\varepsilon_r}} \tag{1.27}$$

$$\therefore E_0 = \sqrt{2S_{\text{Average}}\sqrt{\frac{\mu_0}{\varepsilon_0\varepsilon_r}}} = \sqrt{2\frac{9\times10^{-3}}{(10^{-2})^2}\text{Wm}^{-2}\sqrt{\frac{1.2566\times10^{-6}\text{Hm}^{-1}}{8.854\times10^{-12}\varepsilon_r}}}$$
$$= \sqrt{18\times10^{-3+4}\sqrt{\frac{1.2566\times10^{-6+12}}{8.854\varepsilon_r}}} = \sqrt{\frac{180\times376.73}{\sqrt{\varepsilon_r}}} = \frac{260.41}{\sqrt{\sqrt{\varepsilon_r}}} = \frac{260.41}{\sqrt[4]{\varepsilon_r}} \tag{1.28}$$

$$B_0 = \frac{E_0}{\vartheta} = \frac{E_0}{\frac{1}{\sqrt{\mu_0\varepsilon_0\varepsilon_r}}} = \sqrt{\mu_0\varepsilon_0\varepsilon_r}E_0 = \frac{1}{c}\sqrt{\varepsilon_r}E_0 \tag{1.29}$$

from equation (1.11).

(a) In a medium of dielectric constant $\varepsilon_r = 5$

$$E_0 = \frac{260.41}{\sqrt[4]{5}} = \frac{260.41}{1.4953} = 174.15\ \text{Vm}^{-1} \tag{1.30}$$

by equation (1.28).

$$B_0 = \frac{1}{c}\sqrt{\varepsilon_r}E_0 = \frac{1}{2.9979\times10^8}\sqrt{5}\times260.4 = \frac{2.236\times260.4}{2.9979\times10^8}\text{T} = 194.22$$
$$\times10^{-8}\text{T} = 1.9422\times10^{-6}\text{T} \approx 1.94\times10^{-6}\text{T} = 1.94\ \mu\text{T} \tag{1.31}$$

by equation (1.29).

(b) In air, $\varepsilon_r = 1$ so that equations (1.30) and (1.31) give

$$E_0 = \frac{260.41}{\sqrt[4]{1}} = 260.4\ \text{Vm}^{-1} \tag{1.32}$$

$$B_0 = \frac{1}{c}\sqrt{\varepsilon_r}E_0 = \frac{1}{2.9979\times10^8}\sqrt{1}\times260.4$$
$$= 86.86\times10^{-8}\text{T} = 8.686\times10^{-7}\text{T} \tag{1.33}$$

1.2.5 Particle properties of electromagnetic waves

Over small distances and short time scales, electromagnetic waves exhibit particle properties as enshrined in quantum mechanics. They consist of discrete particles called photons, which are particles of zero rest mass. The energy E of a photon is related to its frequency ν through the Planck's constant h as (Klassen 2011)

$$E = h\nu \tag{1.34}$$

This equation implies that electromagnetic waves of different frequencies carry different amounts of energy.

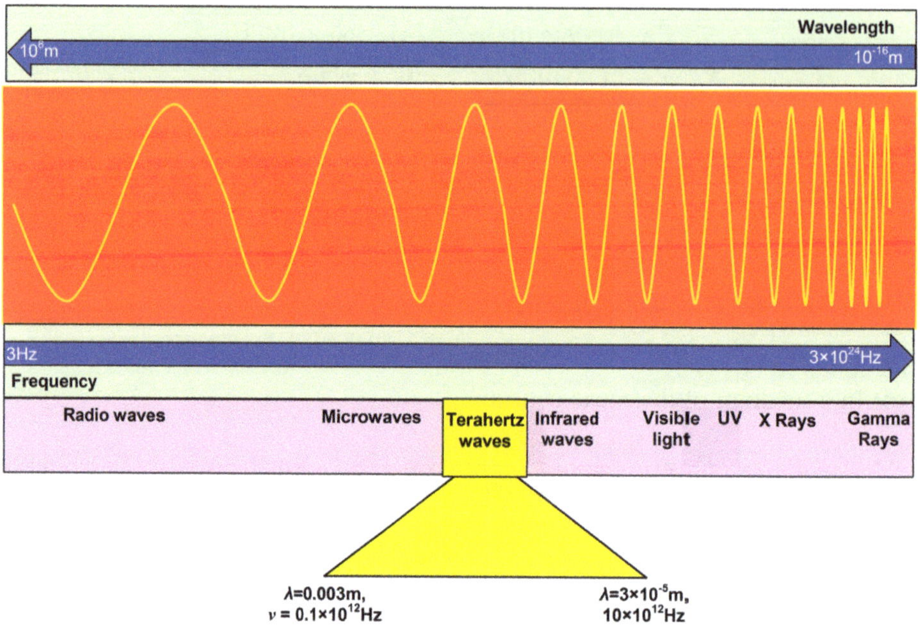

Figure 1.3. Terahertz waves in the electromagnetic spectrum.

1.3 Subdivisions of electromagnetic waves according to frequencies: the electromagnetic spectrum

From equations (1.6) and (1.34), it is clear that when we consider electromagnetic waves of various frequencies, two distinguishing characteristic parameters of these waves are wavelengths and energies. Another vital parameter concerns the nature of radiation, whether it can ionize matter or not because the ionization capability depends on the energy of the waves. Accordingly, there are ionizing and non-ionizing types of electromagnetic waves. On the basis of the range of frequencies from 0 to infinity (along with corresponding wavelengths and energies), the electromagnetic waves are subdivided into seven major segments called the radio waves, microwaves, infrared waves, visible light waves, ultraviolet rays, x-rays and gamma rays (figure 1.3). The entire range of frequencies constitutes the electromagnetic spectrum shown in table 1.1, of which these seven segments are constituent parts.

1.4 Location of terahertz gap in the international standard band designations

Table 1.2 shows the ITU radio bands spectrum, table 1.3 presents IEEE radar bands (microwaves) while table 1.4 displays the components of IR frequency spectrum. Tables 1.2–1.4 show the overlapping regions of terahertz gap with microwaves and infrared waves. These tables help us to visualize the space allocated to the terahertz gap in the IEEE and ITU-designated bands of electromagnetic spectrum. So we can

Table 1.1. Electromagnetic spectrum.

Sl. No.	Name of segment	Frequency range (Hertz = Hz)	Wavelength ($\lambda = c/\nu$) range (meter = m)	Photon energy ($E = h\nu$) range (electron volt = eV)	Nature of radiation
1.	Radio waves	3 Hz–3×10^{8}	1×10^{8}–1	1.24×10^{-14}–1.24×10^{-6}	Non-ionizing
2.	Microwaves	3×10^{8}–3×10^{11}	1–1×10^{-3}	1.24×10^{-6}–1.24×10^{-3}	
3.	Infrared waves	3×10^{11}–4.3×10^{14}	1×10^{-3}–7×10^{-7}	1.24×10^{-3}–1.778	
4.	Visible light waves	4.3×10^{14}–7.9×10^{14}	7×10^{-7}–3.8×10^{-7}	1.778–3.27	
5.	Ultraviolet rays	7.9×10^{14}–3×10^{16}	3.8×10^{-7}–1×10^{-8}	3.27–124	Ionizing
6.	X-rays	3×10^{16}–3×10^{19}	1×10^{-8}–1×10^{-11}	124–1.24×10^{5}	
7	Gamma rays	$>3\times10^{19}$	$<1\times10^{-11}$	$>1.24\times10^{5}$	

Table 1.2. ITU (International Telecommunication Union) radio bands spectrum (12 bands of which Sl. No. 1–8 come under RF spectrum of table 1.1, Sl. No. 9–11 fall under microwaves of table 1.1, and Sl. No. 12 is Terahertz band).

ITU band no.	Designation of band (abbreviation)	Frequency range (Hertz = Hz)	Wavelength ($\lambda = c/\nu$) range (meter = m)	Main applications
1.	Extremely low frequency (ELF)	3–30	1×10^8–1×10^7	Submarine communication
2.	Super low frequency (SLF)	30–300	1×10^7–1×10^6	Submarine communication
3.	Ultra low frequency (ULF)	300–3000	1×10^6–1×10^5	Submarine communication
4.	Very low frequency (VLF)	3000–3×10^4	1×10^5–1×10^4	Navigation
5.	Low frequency (LF)	3×10^4–3×10^5	1×10^4–1×10^3	Navigation
6.	Medium frequency (MF)	3×10^5–3×10^6	1×10^3–1×10^2	Amplitude modulation broadcasts
7.	High frequency (HF)	3×10^6–3×10^7	1×10^2–10	Shortwave broadcasts
8.	Very high frequency (VHF)	3×10^7–3×10^8	10–1	Frequency modulation, Television broadcasts
9.	Ultra high frequency (UHF)	3×10^8–3×10^9	1–0.1	Television broadcasts, RADAR, satellite communication
10.	Super high frequency (SHF)	3×10^9–3×10^{10}	0.1–0.01	RADAR, satellite communication
11.	Extremely high frequency (EHF)	3×10^{10}–3×10^{11}	0.01–0.001	RADAR, satellite communication
12.	Terahertz or tremendously high frequency	3×10^{11}–3×10^{12}	0.001–0.0001	Medical imaging to replace x-rays, spectroscopy, communications

Table 1.3. IEEE radar bands (microwaves). (Sl. No. 12 overlaps with terahertz gap).

Sl. No.	Designation of band	Frequency range (Hertz = Hz)	Wavelength ($\lambda = c/\nu$) range (meter = m)
1.	L band	1×10^9–2×10^9	0.30–0.15
2.	S band	2×10^9–4×10^9	0.15–0.075
3.	C band	4×10^9–8×10^9	0.075–0.0375
4.	X band	8×10^9–1.2×10^{10}	0.0375–0.025
5.	Ku band	1.2×10^{10}–1.8×10^{10}	0.025–0.0167
6.	K band	1.8×10^{10}–2.65×10^{10}	0.0167–0.0113
7.	Ka band	2.65×10^{10}–4×10^{10}	0.0113–0.0075
8.	V band	4×10^{10}–7.5×10^{10}	0.0075–0.004
9.	W band	7.5×10^{10}–1.10×10^{11}	0.004–0.0027
12.	mm	1.10×10^{11}–3.00×10^{11}	0.0027–0.000 99

Table 1.4. Infrared spectrum subdivision scheme. (Sl. No.1 overlaps with terahertz gap).

Sl. No.	Division name and abbreviation	Frequency range (Hz)	Wavelength ($\lambda = c/\nu$) range (m)	Photon energy ($E = h\nu$) range (eV)	Temperature ($T = E/k_B = h\nu/k_B$) range (K)
1.	Far infrared (FIR)	(1.199×10^{13}–7.495×10^{12}) to (1.499×10^{12}–8.566×10^{11})	(25×10^{-6}–40×10^{-6}) to (200×10^{-6}–350×10^{-6})	(0.0496–0.031) to (6.199×10^{-3}–3.5424×10^{-3})	(575.25–359.59) to (71.92–41.097)
2.	Mid infrared (MIR)	5.996×10^{13} to (1.199×10^{13}–7.495×10^{12})	5×10^{-6} to (25×10^{-6}–40×10^{-6})	0.248 to (0.0496–0.031)	2876.72 to (575.25–359.59)
3.	Near infrared (NIR)	(4.28×10^{14}–2.998×10^{14}) to 5.996×10^{13}	(0.7×10^{-6}–1×10^{-6}) to 5×10^{-6}	(1.7712–1.2398) to 0.248	(20 534.33–14 383.62) to 2876.72

realize where exactly this terahertz gap is located in the band assignments by standardization bodies.

1.5 Terahertz electronics

'Electronics' is a branch of physics and electrical engineering that deals with the generation and manipulation of electron streams for applications such as computation, communication and control systems. So our intent in this book is clear. We shall be presenting the ways of production and use of electromagnetic waves having frequencies lying between 0.1 and 10 THz in the context of electronics. Terahertz electronics is concerned with the controlling and applying of the fundamental negatively charged particles called electrons to various useful applications at terahertz frequencies. It tries to realize at terahertz frequencies what has hitherto been achieved up to several gigahertz, and to explore what new capabilities can be built in conventional electronics by extending its working range to such high frequencies.

1.6 The practical perspective of electronics

There are two aspects to all situations, theoretical and practical. The 'theoretical aspect' is convened with mind and imagination providing reasoning about how things work, and guides towards the direction of movement to achieve a goal. 'Practical' relates to translating ideas or thoughts into concrete action to build real-life objects. The 'practical aspect' happens when the theoretically thought-out plan is executed and physically realized in front of our eyes. 'Practicable' refers to any project or plan that is feasible or doable.

From a practical point of view, several questions crop up in our minds:

i. Why should we study terahertz electronics?
ii. What are the major hinderances which have obstructed the exploitation of terahertz part of electromagnetic spectrum for industrial applications?
iii. How are terahertz waves generated and put to various uses?
iv. What are the solid-state and vacuum electronic sources of terahertz waves, and what are their capabilities and limitations,
v. When radio waves and microwaves have established themselves as the vehicles for electronic telecommunication, what motivates us to look at terahertz waves for telecommunications engineering?
vi. In what respects does terahertz communication differ from infrared, visible light, x-ray or gamma ray communication on the higher frequency side or from microwave and radiowave communication on the lower frequency side?
vii. Already infrared and UV-visible spectroscopies have solved many analytical problems in chemistry. What inspires us to explore terahertz spectroscopy as a discipline?
viii. When ultrasound and x-ray imaging have been widely adopted, what is new about terahertz imaging?

We shall seek answers to such questions, and many more as we progress through the pages of this book. Throughout the book our eyes will be on two words 'terahertz frequency'. We shall be always striving to reach this frequency, work with waves at this frequency, manipulating, controlling and putting these waves to various uses. Another word that will receive repeated attention is 'power' because power is strongly tied to frequency. Whenever frequency is mentioned, the power delivered at that frequency is important because power capability decreases with rise of frequency.

1.7 Moving from conventional to terahertz electronics

We know and understand conventional electronics, and also some operations routinely performed in electronic circuits. Common device and circuit operations in electronics are extended to terahertz frequencies. Let us first relook at these operations in their primitive forms (figure 1.4). We shall be performing them in terahertz electronics in the book albeit in specially devised ways for the terahertz waves.

1.7.1 THz generation

This is the process of producing a continuous, repetitive alternating signal having terahertz frequency, such as a sine or square wave of the desired frequency using a unidirectional current from a DC source. The generation of alternating signal is carried out by an oscillator circuit (figure 1.4(a)).

1.7.2 THz amplification

This is the process of increasing the strength (current, voltage or power) of a terahertz frequency signal to generate a stronger version of the input signal by drawing energy from the power supply in the circuit (figure 1.4(b)).

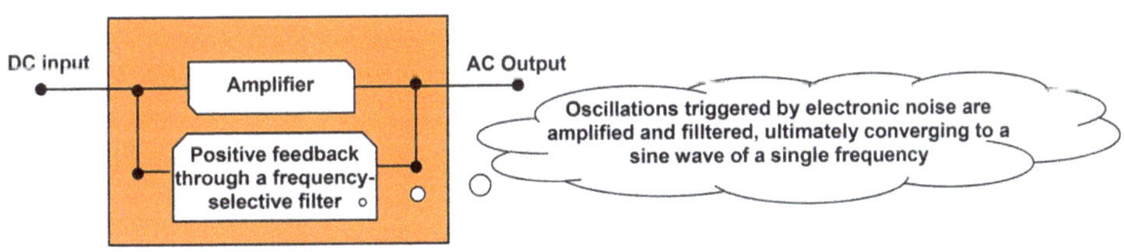

(a)

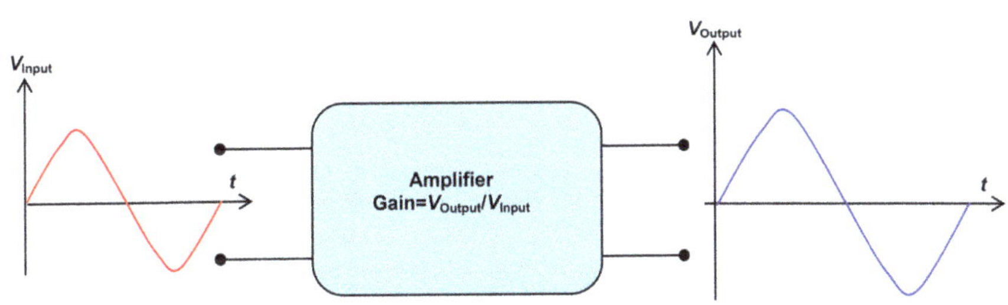

(b)

Figure 1.4. Preliminary operations in electronics: (a) amplification, (b) oscillation, (c) modulation, (d) demodulation, (e) frequency multiplication, (f) frequency mixing, (g) harmonic mixer and (h) subharmonic mixer.

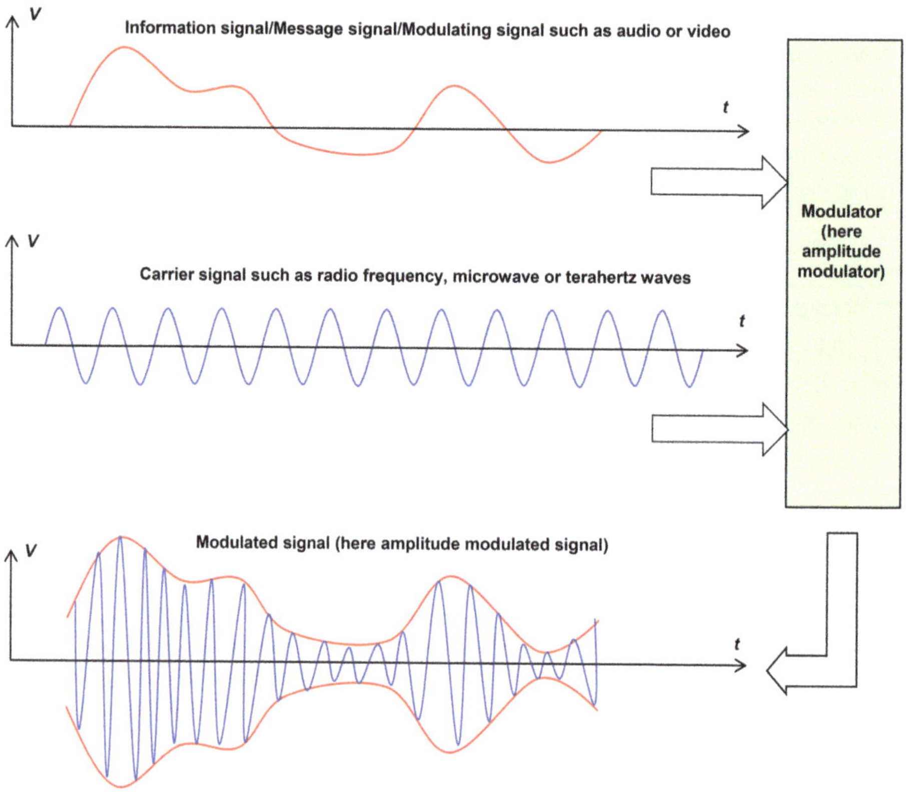

(c)

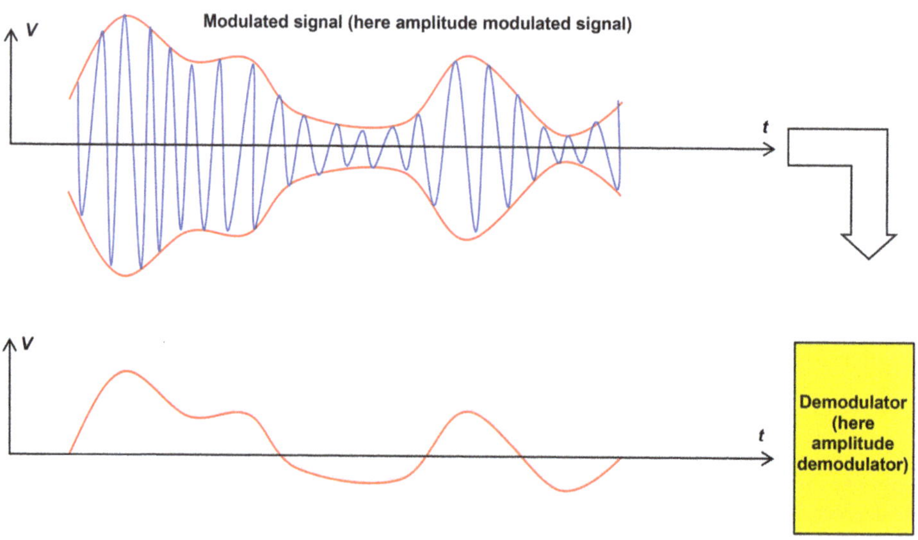

(d)

Figure 1.4. (Continued.)

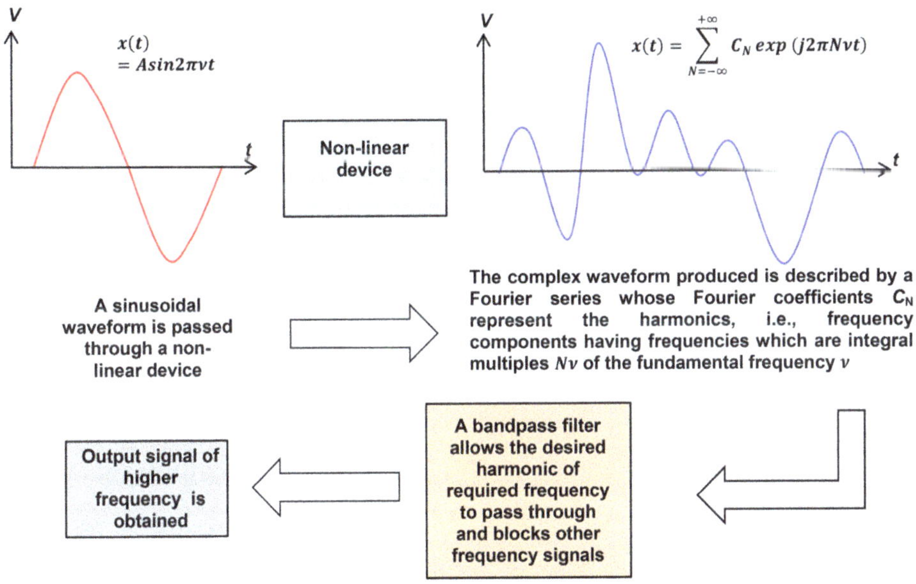

(e)

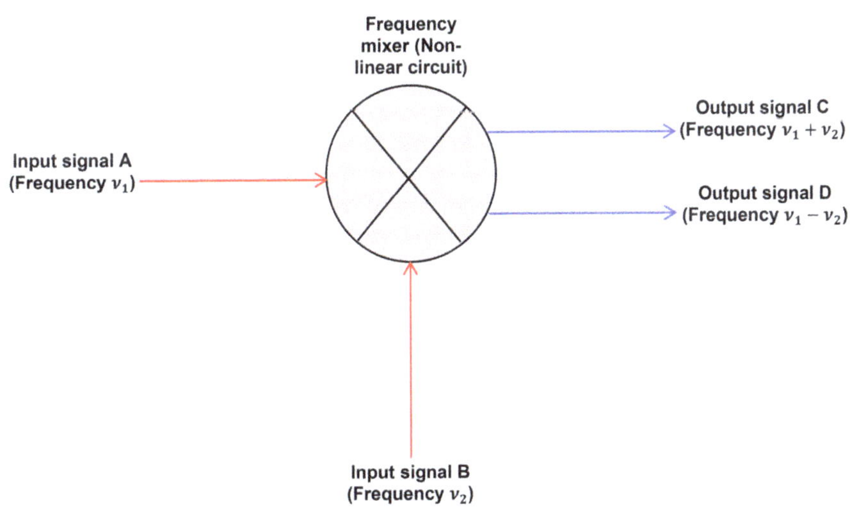

(f)

Figure 1.4. (Continued.)

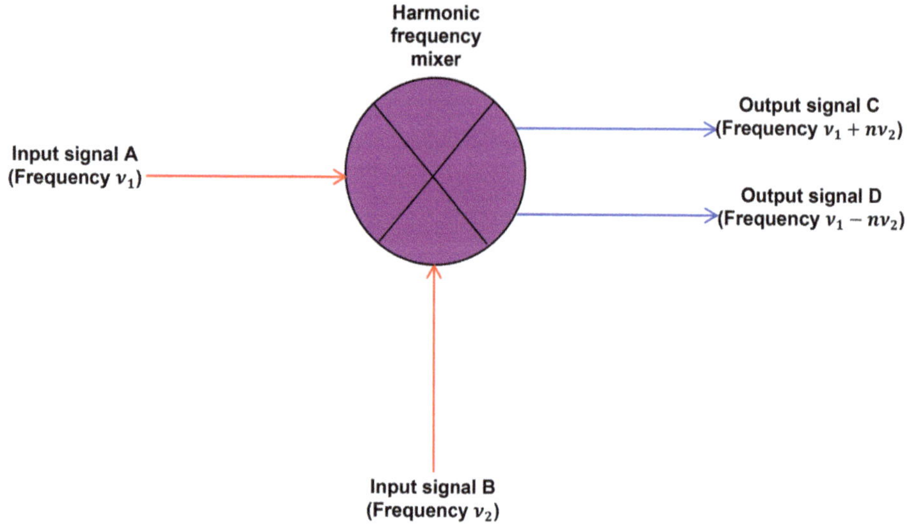

(g)

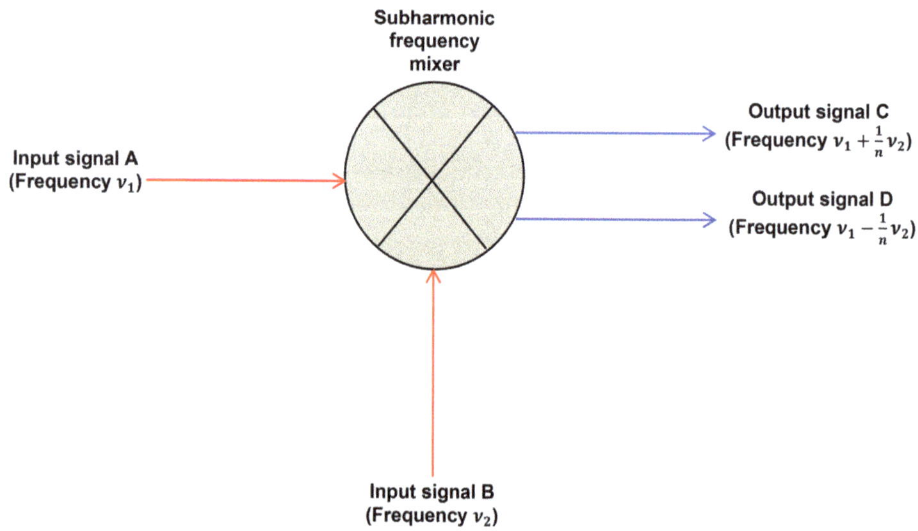

(h)

Figure 1.4. (Continued.)

1.7.3 THz modulation

This is the process by which some parameter of a periodic waveform having terahertz frequency (called the carrier signal), e.g., its amplitude, frequency or phase, is changed in accordance with the instantaneous values of the information signal for sending from a telecommunication transmitting station (figure 1.4(c)).

1.7.4 THz demodulation/THz detection

Opposite of demodulation, THz demodulation is the process of recovery or extraction of the information signal from the carrier signal having terahertz frequency at the receiving station (figure 1.4(d)).

It is the same operation as THz detection, the process of determining the presence of THz waves in a wireless or wired transmission medium by generating a DC signal proportional to the power of the input signal.

1.7.5 Frequency multiplication for terahertz generation

This is the process of generating an output signal whose frequency is a multiple of the frequency of input signal (figure 1.4(e)), e.g., by using a non-linear device for distorting the input signal whereby harmonics of input signal are produced at terahertz frequency. The fundamental and harmonics are passed through a bandpass filter which allows transit of only the signal of required frequency.

1.7.6 Frequency mixing for shifting signal frequency from one range to another

This is the production of signals of new frequencies by application of two input signals of different frequencies ν_1, ν_2 to a non-linear circuit (figure 1.4(f)). The newly generated signals have frequencies equal to the sum and difference of the frequencies of input signals, i.e., $\nu_1+\nu_2$ and $\nu_1-\nu_2$.

1.7.6.1 Harmonic mixer

This is a mixer in which the output sum and difference frequencies are produced at a harmonic multiple of frequency of one of the input signals, e.g. $\nu_1+n\nu_2$ and $\nu_1-n\nu_2$ where n is an integer (figure 1.4(g)).

1.7.6.2 Subharmonic mixer

Here the output frequencies are created at a harmonic submultiple of one of the inputs, e.g., (figure 1.4(h)).

$$C = \nu_1 + (1/n)\nu_2 \text{ and } D = \nu_1 - (1/n)\nu_2 \tag{1.35}$$

1.8 Peculiarities of the terahertz gap

The location of terahertz gap in the electromagnetic spectrum is now clear. It stands at the junction between microwave electronics and infrared ray photonics. It was mentioned earlier in sections 1.1, 1.6 that the terahertz gap has for several years been ignored. This avoidance has mainly been due to certain difficulties faced in putting electromagnetic waves of these frequencies to use. We are now in a position to investigate the underlying reason, which prohibited the utilization of these waves.

The range of frequencies popularly known as the terahertz gap has the microwave region of electromagnetic spectrum as its neighbor on the lower frequency side and infrared region of this spectrum as the neighbor on the high frequency side. Despite being nestled between two progressive disciplines, terahertz electronics has advanced slowly. The long-standing hurdle for terahertz electronics lies hidden in this frequency range itself. The terahertz frequency range is too high for electrons. The electron movements in devices are too slow to cope with these frequencies. This means that the electron motion in electronic devices failed to respond to the rapid changes in the terahertz waves. So the electronic devices have failed. The failure here is due to slower speed of movement.

Similarly, the photons are not sufficiently energetic to make their effects perceptible in the presence of room-temperature thermal energy. The implication is that the noise and the signal levels are same. Hence the signal becomes indistinguishable from the noise. In other words, the actual signal cannot be read clearly leading to erroneous performance. Hence the photonic devices did not work. The failure in this case arises from insufficient energy of photons.

Briefly, semiconductor devices like transistors have an upper limit of 0.3 THz and become markedly inefficient at 0.05 THz while the operational frequency of semiconductor lasers can be stretched downwards up to 30 THz (Sirtori 2002, Pawar *et al* 2013). A gap is therefore created in the non-overlapping region intervening between the operational regimes of the two technologies where available semiconductor technologies do not provide an efficient technique for generating electromagnetic waves from electricity. So workers from both microwaves and infrared sides considered terahertz gap as a problematic area over the years and tended to keep away from it.

Nevertheless, the terahertz gap can be closed or patched up by encroaching from both lower and higher frequency domains. This means that resources of both neighboring fields need to be pooled. Thinking a little more deeply, terahertz electronics is built by combining electronic and photonic or optoelectronic capabilities together steering the path at the confluence or merger of both fields. The intersection of inputs received from these disciplines and the cross-fertilization of ideas opens immense possibilities making electronics of the terahertz frequency band a fascinating field. It is the electronics of this frequency band, which is the center of attention all through this book.

1.9 Unique advantages of terahertz gap frequencies

As we rummage through the pages of this book, we shall find that the terahertz gap of frequencies is a band of frequencies which can be utilized to release the load on the telecommunication frequency bands which have been allotted for different types of applications and in which not much space is left to be assigned to any new application. Furthermore, the high frequencies of terahertz gap also provide wider bandwidths for communication. We shall also find that terahertz radiations are benign radiations which could be used for non-destructive testing and evaluation as well as for security checking at airports to detect hidden explosives and weapons. Equally interesting is the fact that many molecules have chemical fingerprints in the terahertz gap of frequencies enabling spectroscopic identification of many substances. To avoid repetition, we shall take up these topics at relevant places in the text.

Example 1.4 Convert the terahertz gap of frequencies (0.1 to 10 THz) into equivalent ranges of (i) times, (ii) wavelengths, (iii) wavenumbers, (iv) energies, and (v) temperatures. Tabulate the values of the variables calculated.

Solution: (i) Time t for a given frequency ν is

$$t = \frac{1}{\nu} \tag{1.36}$$

At frequency $\nu_1 = 0.1$ THz, the time is

$$t_1 = \frac{1}{\nu_1} = \frac{1}{0.1 \times 10^{12}} = 10 \times 10^{-12}\text{s} = 10 \text{ ps} \tag{1.37}$$

At frequency $\nu_2 = 10$ THz, the time is

$$t_2 = \frac{1}{\nu_2} = \frac{1}{10 \times 10^{12}} = 0.1 \times 10^{-12}\text{s} = 0.1 \text{ ps} \tag{1.38}$$

Hence the time range is

$$t_1,\ t_2 = 10 \text{ ps},\ 0.1 \text{ ps} \tag{1.39}$$

Wavelength λ for a given frequency ν is

$$\lambda = \frac{c}{\nu} \tag{1.40}$$

where c is the velocity of electromagnetic waves = 2.9979×10^8 ms^{-1}

At frequency $\nu_1 = 0.1$ THz, the wavelength is

$$\begin{aligned}\lambda_1 &= \frac{c}{\nu_1} = \frac{2.9979 \times 10^8}{0.1 \times 10^{12}} = 2.9979 \times 10 \times 10^{8-12} = 2.9979 \times 10^{-3}\text{m} \\ &= 2.9979 \times 10^{-3} \times 10^6\ \mu\text{m} = 2.9979 \times 10^3\mu\text{m} = 2997.9\ \mu\text{m} \approx 3000\ \mu\text{m}\end{aligned} \tag{1.41}$$

At frequency $\nu_2 = 10$ THz, the wavelength is

$$\begin{aligned}\lambda_2 &= \frac{c}{\nu_2} \frac{2.9979 \times 10^8}{10 \times 10^{12}} = 2.9979 \times 0.1 \times 10^{8-12} = 2.9979 \times 10^{-5}\text{m} \\ &= 2.9979 \times 10^{-5} \times 10^6\mu\text{m} = 29.979\ \mu\text{m} \approx 30\ \mu\text{m}\end{aligned} \tag{1.42}$$

Hence the wavelength range is

$$\lambda_1,\ \lambda_2\ = 3000\ \mu\text{m},\ \ 30\ \mu\text{m} = 3\ \text{mm},\ \ 30\ \mu\text{m} \tag{1.43}$$

Wavenumber $\bar{\nu}$ for a given frequency ν is

$$\bar{\nu}\ =\ \frac{\nu}{c} \tag{1.44}$$

At frequency $\nu_1 = 0.1$ THz, the wavenumber is

$$\begin{aligned}\overline{\nu_1} &= \frac{\nu_1}{c} = \frac{0.1\times 10^{12}}{2.9979\times 10^{8}} = 0.0333567\times 10^{12-8} = 0.0333567\times 10^{4}\text{m}^{-1}\\ &= 333.567\text{m}^{-1}\approx 333.57\text{m}^{-1}\end{aligned} \tag{1.45}$$

At frequency $\nu_2 = 10$ THz, the wavenumber is

$$\begin{aligned}\overline{\nu_2} &= \frac{\nu_2}{c} = \frac{10\times 10^{12}}{2.9979\times 10^{8}} = 0.333567\times 10^{12-8+1} = 0.333567\times 10^{5}\text{m}^{-1}\\ &= 33356.7\text{m}^{-1}\end{aligned} \tag{1.46}$$

Hence the wavenumber range is

$$\overline{\nu_1}\ = 333.57\text{m}^{-1},\ \overline{\nu_2}\ = 33356.7\text{m}^{-1} \tag{1.47}$$

Energy E for a given frequency ν is

$$E = h\nu \tag{1.48}$$

where h is Planck's constant = 6.626×10^{-34} J s.

At frequency $\nu_1 = 0.1$ THz, the energy is

$$\begin{aligned}E_1 &= h\nu_1 = 6.626\times 10^{-34}\times 0.1\times 10^{12} = 0.6626\times 10^{-34+12}\text{J} = 0.6626\times 10^{-22}\text{J}\\ &= \frac{0.6626\times 10^{-22}}{1.602\times 10^{-19}}\frac{\text{J}}{\text{C}} = 0.4136\times 10^{-22+19}\text{eV} = 0.4136\times 10^{-3}\text{eV} = 0.4136\ \text{meV}\end{aligned} \tag{1.49}$$

At frequency $\nu_2 = 10$ THz, the energy is

$$\begin{aligned}E_2 &= h\nu_2 = 6.626\times 10^{-34}\times 10\times 10^{12} = 6.626\times 10^{-34+12+1}\text{J} = 6.626\times 10^{-21}\text{J}\\ &= \frac{6.626\times 10^{-21}}{1.602\times 10^{-19}}\frac{\text{J}}{\text{C}} = 4.136\times 10^{-21+19}\text{eV} = 4.136\times 10^{-2}\text{eV}\\ &= 4.136\times 10^{-2}\times 10^{3}\text{eV} = 41.36\ \text{meV}\end{aligned} \tag{1.50}$$

Hence the energy range is

$$E_1\ = 0.4136\ \text{meV},\ E_2\ = 41.36\ \text{meV} \tag{1.51}$$

(v) Temperature of a body is related to chaotic agitation of particles in the body. It is a measure of the random motion of its constituent particles, the molecules and atoms. The Boltzmann constant has the physical significance that it gives a measure of the amount of heat energy corresponding to the haphazard or disorganized motion of particles and hence temperature of a substance.

Suppose temperature is measured in absolute scale and represented by the symbol T with unit Kelvin (K). Boltzmann constant k_B has the dimensions of [Energy] × [Temperature]$^{-1}$ with unit Joule K^{-1}. So the product k_BT has the dimensions of [Energy] × [Temperature]$^{-1}$× [Temperature]= [Energy] with unit Joule × K^{-1} × K = Joule. The value of this product at any temperature is usually referred to as thermal energy at that temperature, e.g., at 300 K

$$\begin{aligned}\text{Room-temperature thermal energy} &= k_BT = 1.38065 \times 10^{-23} \times 300\text{J}\\ &= \frac{1.38065 \times 10^{-23} \times 300}{1.602 \times 10^{-19}}\text{eV} = 258.55 \\ &\times 10^{-23+19}\text{eV} = 258.55 \times 10^{-4}\text{eV} \\ &= 0.025855\text{eV} \approx 0.0259\text{eV}\end{aligned} \tag{1.52}$$

The energy of electromagnetic radiation, expressed as the product of Planck's constant h and frequency ν of radiation, can be considered to be the same as thermal energy. The temperature obtained from the equality of the two energies can be said to be the temperature equivalent to radiation of that energy and frequency. In this way, the temperatures corresponding to electromagnetic waves of given frequencies can be determined. Hence waves of different frequencies/energies can be assigned various equivalent temperatures and can be specified in temperature units.

On these lines, temperature T for radiation of a given frequency ν is found by equating the thermal energy k_BT with energy E of the radiation given by equation (1.48) as (Fitch and Osiander 2004):

$$k_BT = h\nu \tag{1.53}$$

where k_B is Boltzmann constant = 1.380 65 × 10^{-23} JK^{-1}. From equation (1.53), it follows that

$$T = \frac{h\nu}{k_B} \tag{1.54}$$

At frequency ν_1 = 0.1 THz, the temperature is

$$\begin{aligned}T_1 &= \frac{h\nu_1}{k_B} = \frac{6.624 \times 10^{-34} \times 0.1 \times 10^{12}}{1.38065 \times 10^{-23}} = \frac{6.624 \times 10^{-34+12+23} \times 0.1}{1.38065} \\ &= 0.47977 \times 10^1 = 4.7977\text{ K} \approx 4.8\text{ K}\end{aligned} \tag{1.55}$$

At frequency ν_2 = 10 THz, the temperature is

$$\begin{aligned}T_2 &= \frac{h\nu_2}{k_B} = \frac{6.624 \times 10^{-34} \times 10 \times 10^{12}}{1.38065 \times 10^{-23}} = \frac{6.624 \times 10^{-34+12+23} \times 10}{1.38065} \\ &= 47.977 \times 10^1 = 479.77\text{ K} \approx 480\text{ K}\end{aligned} \tag{1.56}$$

Table 1.5. Chart of values of different variables at the two boundaries of the terahertz gap.

Sl. No.	Variable range	Frequency = 0.1 THz	Frequency = 10 THz
1.	Time	10 ps	0.1 ps
2.	Wavelength	3 mm	30 μm
3.	Wavenumber	333.57 m^{-1}	33 356.7 m^{-1}
4.	Photon energy	0.4136 meV	41.36 meV
5.	Temperature	−268 °C	207 °C

Hence the temperature range is

$$T_1 = 4.8 \text{ K}, \; T_2 = 480 \text{ K} \tag{1.57}$$

or,

$$T_1 = 4.8 - 273.15°\text{C}, \; T_2 = 480 - 273.15°\text{C} \tag{1.58}$$

or,

$$T_1 = -268.35\ °\text{C} \approx -268\ °\text{C}, \; T_2 = 480 - 273.15°\text{C} = 206.85\ °\text{C} \approx 207\ °\text{C} \tag{1.59}$$

The calculated values of the variables are given in table 1.5.

Example 1.5 The equivalent temperature of electromagnetic radiation emitted from a source is 500 K. What is the frequency of the source of radiation? Does this frequency fall in the regime of terahertz gap of frequencies?

Solution: Applying equation (1.53),

$$\nu = \frac{k_B T}{h} = \frac{1.38065 \times 10^{-23}\text{JK}^{-1} \times 500 \text{ K}}{6.624 \times 10^{-34}\text{JHz}^{-1}} = \frac{1.38065 \times 500}{6.624} \times 10^{-23+34}\text{Hz} \tag{1.60}$$
$$= 104.216 \times 10^{11}\text{Hz} = 10.422 \times 10^{12}\text{Hz} = 10.422 \text{ THz}$$

which exceeds 10 THz. Hence the electromagnetic radiation of given equivalent temperature falls above the terahertz range of frequencies (0.1–10 THz). It falls in the far infrared region (table 1.4) of the electromagnetic spectrum.

Example 1.6 The equivalent temperature of a source of electromagnetic radiation is 10^5 K? What is the frequency of radiation emitted by the source? In what portion of the electromagnetic spectrum can this source be placed looking at the classification of electromagnetic waves into different frequency ranges?

Solution: From equation (1.53),

$$\nu = \frac{k_B T}{h} = \frac{1.38065 \times 10^{-23} \times 10^5}{6.624 \times 10^{-34}} = 0.2084 \times 10^{-23+5+34} = 0.2084 \times 10^{16}$$
$$= 2.084 \times 10^{15}\text{Hz} \tag{1.61}$$

which falls in the ultraviolet region (7.9×10^{14} 3×10^{16} Hz: table 1.1) of the electromagnetic spectrum.

1.10 Organizational plan of the book

Our exploration of terahertz electronics is divided into four parts named as (figure 1.5): (i) solid-state electronic devices; (ii) vacuum electron devices; (iii) photothermal and optoelectronic devices; and (iv) applications. The ongoing chapter (chapter 1) provides background information on electromagnetic waves, electromagnetic spectrum, and the location of terahertz gap in this spectrum with respect to international standardizations, and introduces the subject matter of this book, namely the real-world aspects of terahertz electronics.

Part I on solid-state devices contains six chapters in addition to the present chapter. Three of these chapters are devoted to diodes and the remaining three to transistors. Chapter 2 describes Schottky, metal-insulator-metal, self-switching and geometric diodes. Resonant tunneling diodes are discussed in chapter 3. Impact ionization avalanche transit-time and intervalley transfer diodes are dealt with in chapter 4. In chapter 5, the hetrojunction bipolar transistors are treated. Chapter 6 touches upon the metal-oxide semiconductor field-effect transistors. A detailed presentation of high-electron mobility transistors is made in chapter 7.

Part II on vacuum electron devices consists of three chapters. Chapters 8, 9 and 10 are dedicated to systematic reviews of progress in backward wave oscillators, gyrotrons and free electron lasers.

Part III on photothermal and optoelectronic devices comprises four chapters. The first chapter of this part (chapter 11) addresses the devices working on thermal effects of terahertz radiation, viz., microbolometers, Golay cells and pyroelectric detectors. The last three chapters (chapters 12 to 14) are concerned with optical devices. Chapter 12 surveys the field of non-linear optics and explains the principles of optical rectification and difference frequency mixing, and their application in generation of terahertz pulses. Detection of terahertz waves is also elaborated. In chapter 13 the constructional features and operation of photoconductive antennas are described followed by their use in generation/detection of broadband terahertz pulses and narrowband continuous wave terahertz radiation. This description is followed by outlining photomixing methods of terahertz generation. Two types of lasers, which have made a far-reaching impact in the realm of terahertz electronics, viz., the quantum cascade and optically-pumped lasers constitute the topic of chapter 14.

Part IV is dedicated to applications of terahertz electronics for human welfare. Applications are the fruits reaped from intensive research investigations in any field. Technological breakthroughs have made the previously unused terahertz gap frequencies accessible for communications, spectroscopy and imaging. Terhertz electronics is likely to make a dent in these areas. The three chapters of this part

Figure 1.5. Structural organization of the book.

(chapters 15–17) provide an updated status of terahertz use in these three fields. Parts I and II (chapter 1–10) are covered in volume 1 of the book in your hands while parts III and IV (chapters 11–17) will be dealt with in Volume 2. The seven chapters of Volume 2 are renumbered as 1–7.

1.11 Discussion and conclusions

The preliminary information presented in this chapter was intended to familiarize the reader with the basic terminology of terahertz electronics. After finishing this chapter the location of terahertz gap in the electromagnetic spectrum is clarified with respect to standardized designations, its peculiar character is demystified, and unparalleled advantages are indicated. The knowledge acquired will be helpful in smoothening the journey ahead.

References

Armstrong C M 2012 The truth about terahertz *IEEE Spectr.* **49** 36–41

Fitch M J and Osiander R 2004 Terahertz waves for communications and sensing *Johns Hopkins APL Tech. Dig.* **25** 348–55

Kimura W D 2017 *Electromagnetic Waves and Lasers* (San Rafel, CA: Morgan & Claypool) pp 1-1–1-33

Klassen S 2011 The photoelectric effect: reconstructing the story for the physics classroom *Sci. Educ.* **20** 719–31

Pawar A Y, Sonawane D D, Erande K B and Derle D V 2013 Terahertz technology and its applications *Drug Invent. Today* **5** 157–63

Sirtori C 2002 Bridge for the terahertz gap *Nature* **417** 132–33

IOP Publishing

Practical Terahertz Electronics: Devices and Applications, Volume 1

Solid-state devices and vacuum tubes

Vinod Kumar Khanna

Chapter 2

Schottky barrier, metal–insulator–metal, self-switching and geometric diodes

A series of diodes covering the range from a few terahertz to optical frequencies is described. Current–voltage characteristics and the mathematical equation of a Schottky diode are explained. Applying its electrical equivalent circuit representation, series and dynamic resistances and junction capacitance of the diode are evaluated from structural parameters. The cut-off frequency of the diode is calculated for given geometrical dimensions. Impact of skin effect on series resistance and cut-off frequency of the diode is discussed. Limitations of the traditional model are highlighted and the necessary modifications to overcome its deficiencies are indicated. During fabrication, Schottky diode design using air bridge obviates parasitic resistance effects. Schottky diodes are used to construct frequency multipliers and mixers working at terahertz frequencies. Unconventional non-PN junction diodes based on quantum mechanical tunneling and self-switching mechanisms, as well as those on shape-dependent conduction are used for electromagnetic radiation detection in rectennas. A few research studies are alluded to.

In this chapter, we study in detail Schottky barrier diodes and enquire what design and process modifications must be implemented to make them work in the desired fashion at terahertz frequencies. The Schottky diode analysis is followed by non-PN junction diode concepts that stretch the frequency range up to far infrared.

2.1 Schottky diode principle and switching action

2.1.1 Working principle of the Schottky diode

In distinction to a P–N junction diode comprising a semiconductor–semiconductor junction between P-doped and N-doped materials, the Schottky diode is a metal–semiconductor junction made, for example, with metals like gold, platinum,

doi:10.1088/978-0-7503-3171-5ch2

palladium, etc contacting with moderately doped N-type silicon. These metals have a higher work function than silicon. The work function of a material is the minimum energy required to release an electron from the metal and transfer it to a point in vacuum. Since the work functions of the above metals are higher than that of silicon, therefore more energy is required to liberate an electron from the metal atom than from the silicon atom. Hence, thermal energy at room temperature can more easily release electrons from silicon than from metal.

What happens when a metal is joined to an N-type semiconductor? A metal–semiconductor contact is set up (figure 2.1). Now any mobile charges can cross the borderline of two materials and move to the opposite side. What are the mobile charges here? These are the free electrons on the semiconductor side. As soon as the contact has been established, the free electrons in the semiconductor begin crossing the borderline and moving to the metal side. While the electrons are transferred, positively charged donor ions are left behind in the semiconductor in vicinity of the joining boundary. These ions are produced from dopant-impurity atoms which have lost the electrons. As the region of semiconductor adjoining the metal is devoid of free electrons, a depletion region is formed on the semiconductor side.

What is the cumulative effect of this charge transfer? It leads to development of unbalanced positive and negative charges on the two sides with acquisition of a positive charge by the semiconductor side, and contrarily, a negative charge by the metal side. The charge imbalance caused by unneutralized charges engenders the creation of a potential difference at the interface between the metal and the semiconductor. The potential difference produced across the junction is called the built-in potential V_{bi} of the Schottky diode. It is expressed as the difference in work functions of the metal (ϕ_M) and semiconductor (ϕ_S):

$$eV_{bi} = \phi_M - \phi_S \tag{2.1}$$

Suppose an external voltage applied across the diode has such polarity that it opposes or neutralizes the effect of built-in potential. Then current flows through the diode. This is the state of forward-biased conduction. On reversing the direction of applied voltage, the polarity of this voltage aids the built-in potential. It therefore enlarges the depletion region width. Then the current flow decreases. This is reverse-bias condition. Thus a metal–semiconductor junction acts a rectifying junction.

Conventional current is permitted to flow from the metal side of a Schottky diode named the anode to its semiconductor side labelled as the cathode. It cannot flow in the opposite direction. Such current flow in one direction is called unilateral conduction. However, if such a junction is fabricated with a metal whose work function is lower than that of the semiconductor (figure 2.2), an ohmic or non-rectifying junction is realized. In an ohmic junction, the current flows easily in both directions, which is bilateral conduction.

An advantage of Schottky diode is its low forward drop due to a smaller knee voltage (0.15–0.45 V) than a P–N junction diode (0.6–1.7 V). But the even greater advantage is that it is much swifter in switching from the conducting to the

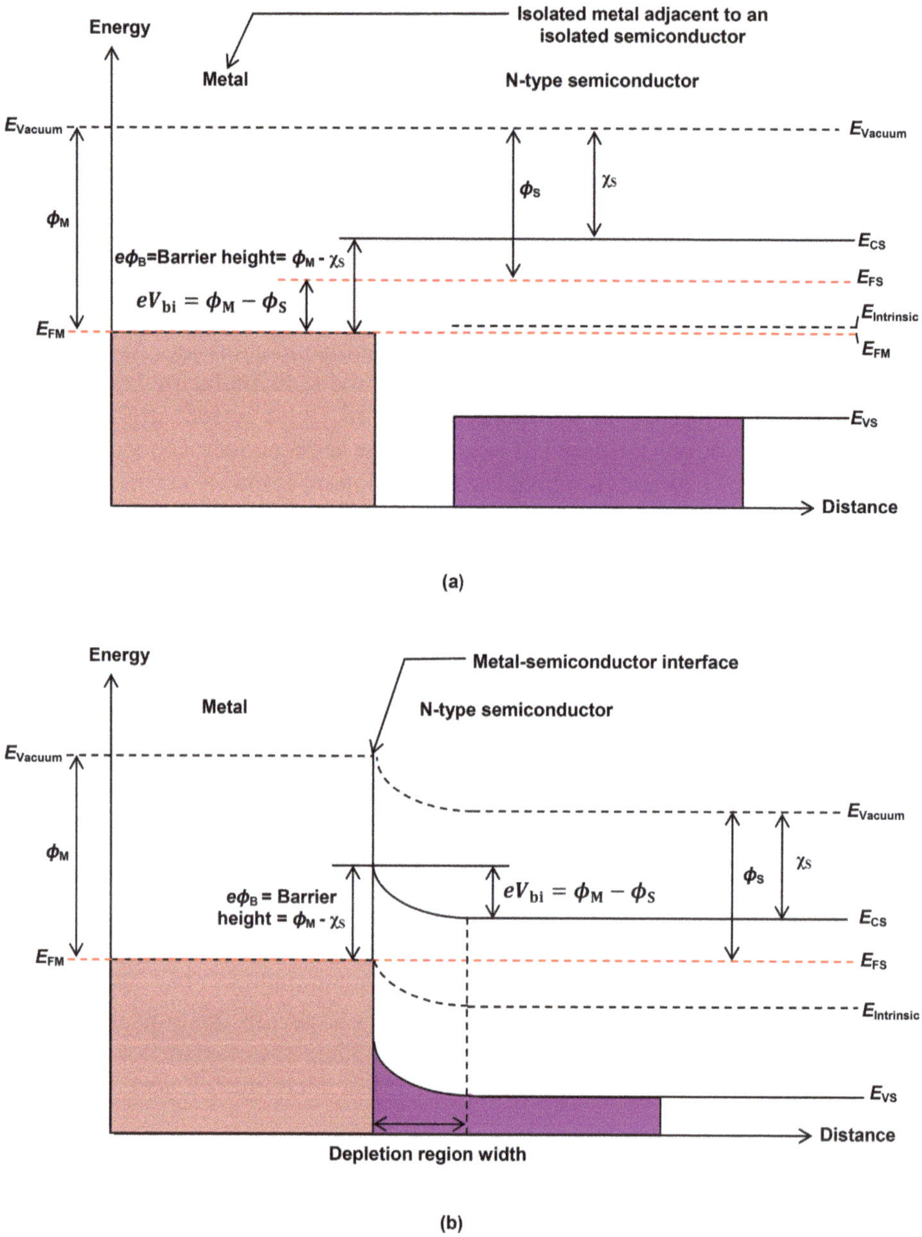

Figure 2.1. Energy-band diagram of a Schottky metal–semiconductor contact ($\phi_M > \phi_S$): (a) before contact and (b) after contact and at thermal equilibrium. ϕ_M and ϕ_S = the work functions of the metal and the semiconductor respectively; χ_S = the electron affinity of the semiconductor; ϕ_B = the barrier height at the metal–semiconductor interface; V_{bi} = the built-in potential of the metal–semiconductor junction; E_{Vacuum} = the energy of vacuum level; E_{CS}, E_{VS} = the energies of conduction and valence band edges; $E_{Intrinsic}$ = the intrinsic energy level; E_{FM}, E_{FS} = the energies of Fermi levels in the metal and the semiconductor. In (a) $E_{FM} \neq E_{FS}$ while in (b) $E_{FM} = E_{FS}$. In (b), the region of semiconductor contiguous to the metal surface loses electrons as indicated by the upward bending of the conduction and valence band edges of the semiconductor.

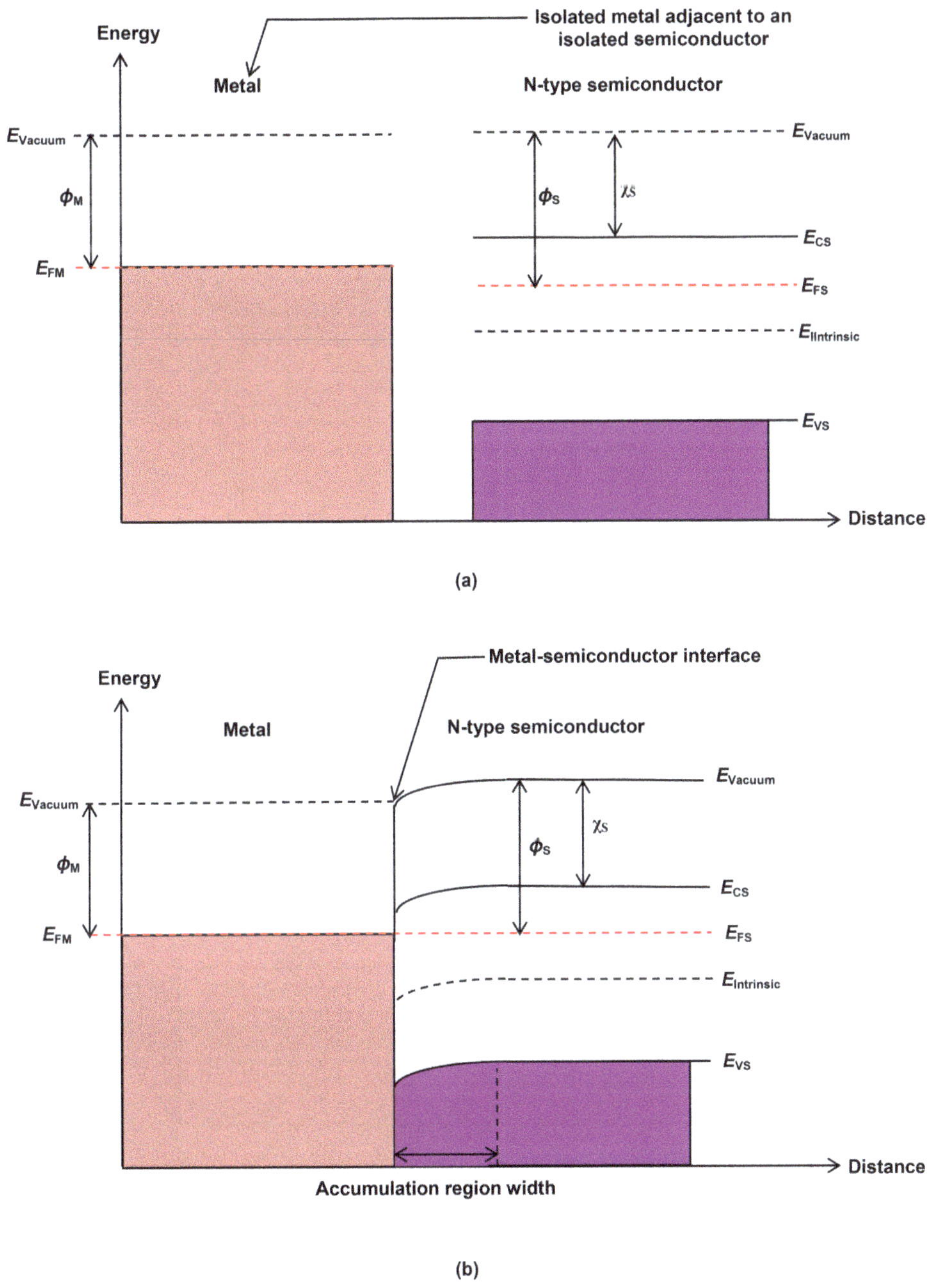

Figure 2.2. Energy-band diagram of an ohmic metal–semiconductor contact ($\phi_M < \phi_S$): (a) before contact and (b) after intimate contact and at thermal equilibrium. Notations are the same as used in figure 2.1. The transference of electrons from the metal to the semiconductor produces a downward band bending as opposed to the upward band bending in the case of Schottky contact.

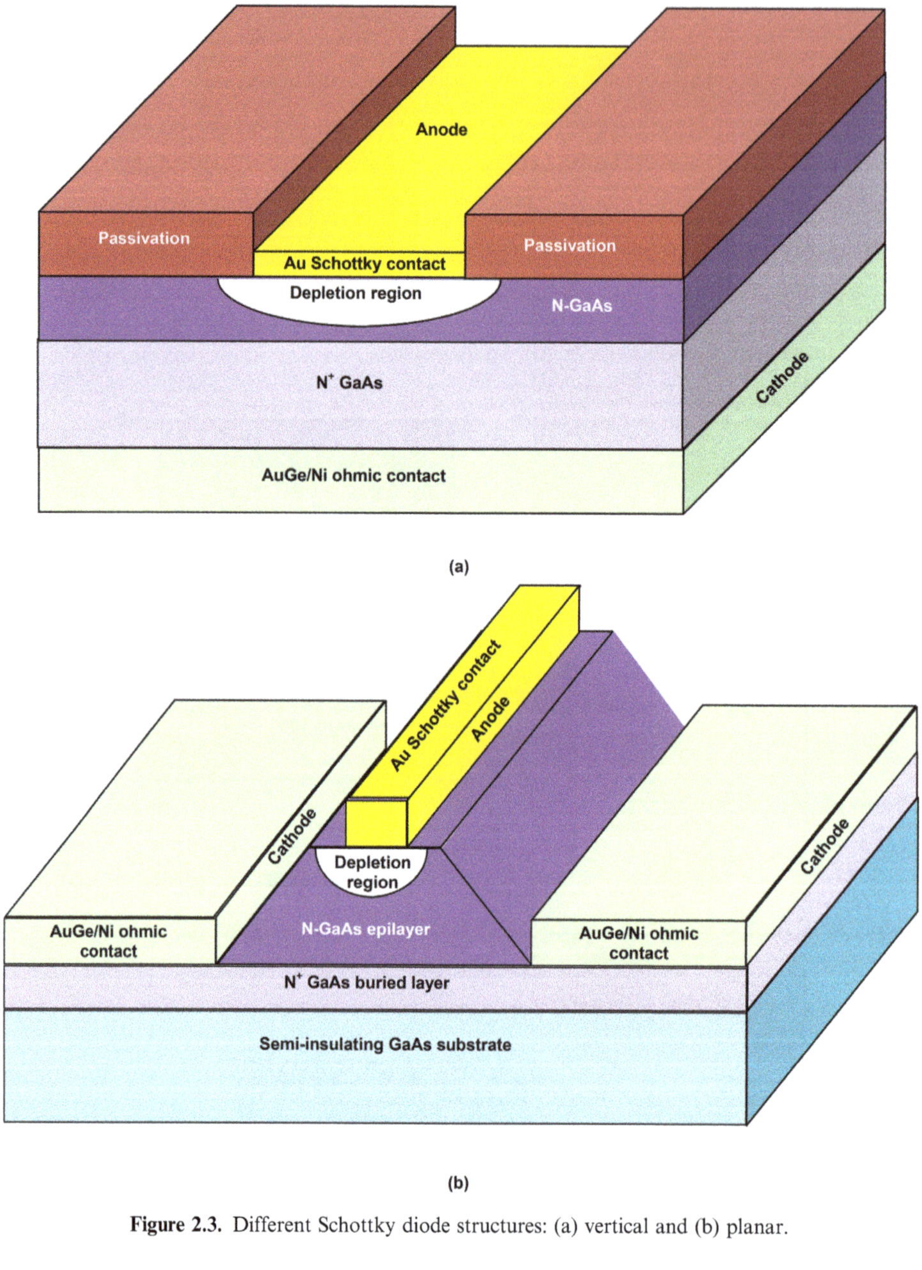

Figure 2.3. Different Schottky diode structures: (a) vertical and (b) planar.

non-conducting state than a P–N junction diode. This behavior is known as fast switching action.

The Schottky diode is fabricated in two familiar structures, namely, vertical and planar. GaAs diodes are illustrated in figure 2.3.

2.1.2 Competition between Schottky barrier diode (SBD) and P–N junction diode

The reason for selecting Schottky barrier diodes instead of P–N junction diodes is that they are majority-carrier devices. So unlike the P–N junction diodes, minority-carrier storage effects are absent in Schottky diodes. What is the benefit accrued by getting rid of these charge-storage effects? When a P–N junction diode is switched off, the minority-carrier holes injected in the N-region decay by recombination. The time consumed by the recombination process depends on the lifetime of excess carriers. Only after the excess carriers are fully annihilated is the current flow established in the opposite direction. Hence, minority-carrier storage effect is responsible for lengthening the reverse recovery time of the P–N junction diode. In a Schottky diode, there are no minority carriers, and so reverse recovery is almost instantaneous. Therefore, Schottky diodes are inherently fast-switching devices by virtue of their working mechanism offering switching times <100 ps as opposed to switching times of several microseconds to <100 ns for P–N junction diodes. This explains why Schottky diodes are deemed as natural choices when thinking about the terahertz frequency range.

The above explanation can also be made with respect to diffusion capacitance. A P–N junction diode is associated with two capacitances, namely diffusion capacitance related to minority-carrier charge storage and depletion capacitance linked to depletion-layer charge. Under forward bias, the capacitance of a P–N junction diode is dominated by diffusion capacitance. This diffusion capacitance inhibits its temporal response during switching to reverse bias. In a Schottky diode, diffusion capacitance is absent. So, a Schottky diode shows a faster response.

2.2 Current–voltage equation of a non-ideal Schottky-barrier diode (SBD)

The current–voltage characteristic of a Schottky diode (figure 2.4) is described by the equation (Tyagi 1984, Sze and Ng 2007)

$$I = I_0\left\{\exp\left(\frac{qV}{nk_BT}\right) - 1\right\} \tag{2.2}$$

where I stands for the total current flowing through the diode, I_0 is the reverse saturation current (dark), q is the electronic charge, V is the applied bias, k_B is Boltzmann constant, T is absolute temperature and n is an ideality factor ~1–2 (also called quality factor or emission coefficient).

For $n = 1$, the equation (2.2) reduces to the ideal Schottky diode equation:

$$I = I_0\left\{\exp\left(\frac{qV}{k_BT}\right) - 1\right\} \tag{2.3}$$

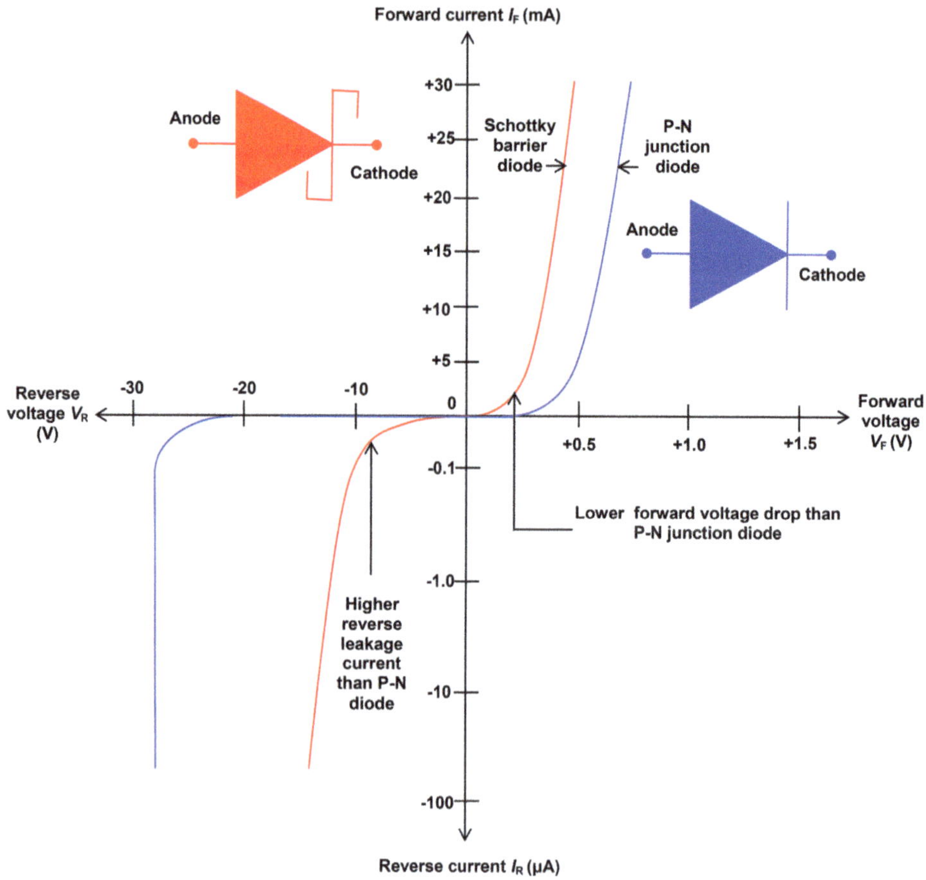

Figure 2.4. Current–voltage characteristics and circuit diagram symbols of Schottky barrier and P–N junction diodes.

Further,

$$I_0 = AA_{\text{Effective}}T^2 \exp\left(-\frac{q\phi_B}{k_B T}\right) \tag{2.4}$$

where A is the junction area, $A_{\text{effective}}$ is the effective Richardson constant for thermionic emission and ϕ_B is the zero-bias Schottky barrier height. Note that

$$A_{\text{Effective}} = \lambda_R A_0 \tag{2.5}$$

where

$$\lambda_R = \text{Material} - \text{specific correction factor} \tag{2.6}$$

and

$$
\begin{aligned}
A_0 &= \frac{4\pi m^* q k_B^2}{h^3} \\
&= \frac{4 \times 3.14159 \times 9.109 \times 10^{-31}\,\text{kg} \times 1.602 \times 10^{-19}\,\text{C} \times (1.3806 \times 10^{-23})^2 (\text{JK}^{-1})^2}{(6.62\,607 \times 10^{-34})^3 (\text{J. s})^3} \\
&= \frac{349.525 \times 10^{-31-19-46}}{290.9163 \times 10^{-102}} \frac{\text{kg} \times \text{C} \times \text{J}^2\text{K}^{-2}}{\text{J}^3 \times \text{s}^3} = 1.201\,462\,4 \times \frac{10^{-96}}{10^{-102}} \frac{\text{kg} \times \text{C} \times \text{K}^{-2}}{\text{J} \times \text{s}^3} \\
&= 1.201\,462\,4 \times 10^6 \frac{\text{kg} \times \text{C} \times \text{K}^{-2}}{\text{kg} \times \text{m}^2 \times \text{s}^{-2} \times \text{s}^3} = 1.201\,462\,4 \times 10^6 \frac{\text{C}}{\text{s}} \times \frac{1}{\text{m}^2 \times \text{K}^2} \\
&= 1.201\,462\,4 \times 10^6\,\text{A} \times \frac{1}{\text{m}^2 \times \text{K}^2} = 1.2 \times 10^6\,\text{A. m}^{-2}\text{. K}^{-2}
\end{aligned} \tag{2.7}
$$

where electrons are considered to be 'free electrons' with effective mass m^* of the electron assumed as its rest mass:

$$m^* = m_0 \tag{2.8}$$

The symbol m_0 represents the rest mass of electrons and h is Planck's constant. The values of effective Richardson constant for P- type and N-type silicon are different. They are found to be

$$|A_{\text{Effective}}|_{\text{P-type}} = 32\,\text{A. cm}^{-2}\text{. K}^{-2}, \quad |A_{\text{Effective}}|_{\text{N-type}} = 112\,\text{A. cm}^{-2}\text{. K}^{-2} \tag{2.9}$$

2.3 Components of the traditional equivalent circuit of a Schottky-barrier diode

The circuit (figure 2.5) consists of a DC spreading resistance in series with a parallel connection of dynamic resistance r_d and junction capacitance C_j; the spreading resistance constitutes the series resistance R_S of the diode.

At a high angular frequency ω, the capacitive reactance $X_C = 1/(\omega C_j)$ due to junction capacitance C_j dominates over the dynamic resistance and equals the series resistance R_S, i.e.,

$$\frac{1}{\omega C_j} = R_S \tag{2.10}$$

or,

$$\omega = \frac{1}{R_S C_j} \tag{2.11}$$

The accompanying linear frequency is called the cut-off frequency and is given by

$$2\pi f_C = \frac{1}{R_S C_j} \tag{2.12}$$

or,

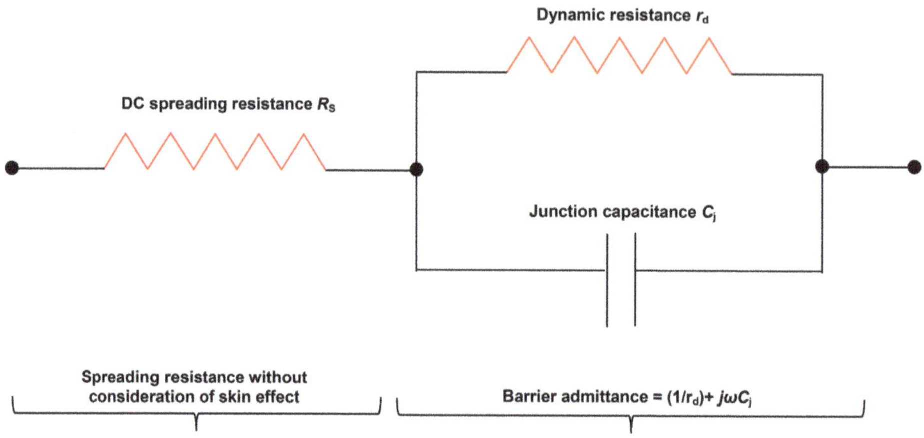

Figure 2.5. Conventional electrical equivalent circuit model of a Schottky barrier diode containing two resistances and one capacitance.

$$f_C = \frac{1}{2\pi R_S C_j} \tag{2.13}$$

Since the capacitance C_j is bias-dependent, the bias for C_j calculation is taken as zero. Then cut-off frequency of the SBD is the frequency obtained from the series resistance R_S and the zero-bias junction capacitance C_{j0}.

$$f_C = \frac{1}{2\pi R_S C_{j0}} \tag{2.14}$$

2.3.1 Series resistance (R_S) of the SBD

Consider a circular contact diode (figure 2.6). If a is the radius of the circular contact, the spreading resistance due to the constriction of lines of current flowing adjoining the contact is

$$R_S = \frac{1}{4\sigma a} \tag{2.15}$$

where σ is the electrical conductivity of the semiconductor.

2.3.2 Dynamic resistance (r_d) of the SBD

Rewriting the ideal diode equation (2.3) as

$$I = I_0 \exp\left(\frac{qV}{k_B T}\right) - I_0 \tag{2.16}$$

and differentiating equation (2.16) with respect to voltage V we get

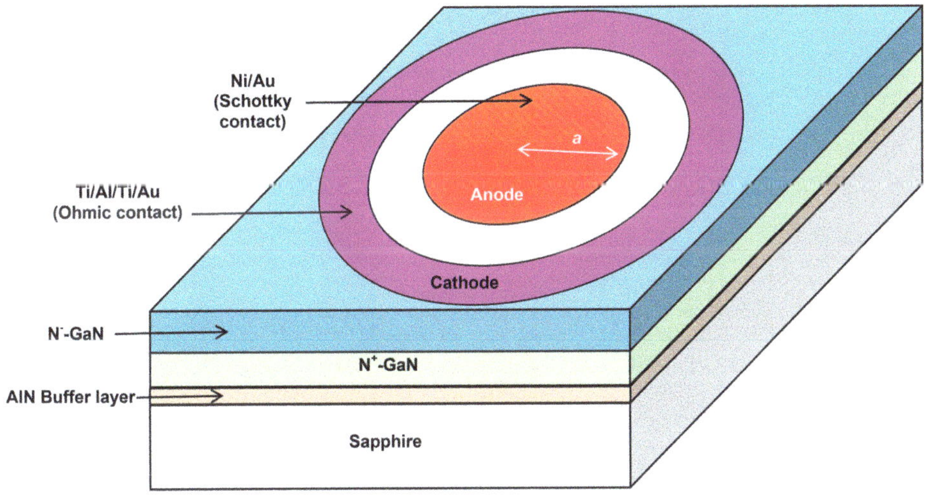

Figure 2.6. A circular-contact GaN Schottky-barrier diode.

$$\begin{aligned}\frac{\partial I}{\partial V} &= I_0 \exp\left(\frac{qV}{k_B T}\right)\frac{d}{dV}\left\{\left(\frac{qV}{k_B T}\right)\right\} - \frac{dI_0}{dV} = I_0\left(\frac{q}{k_B T}\right)\exp\left(\frac{qV}{k_B T}\right) - 0 \\ &= \left(\frac{q}{k_B T}\right) \times I_0 \exp\left(\frac{qV}{k_B T}\right) = \left(\frac{q}{k_B T}\right) I\end{aligned} \tag{2.17}$$

where we have used the approximation

$$I_0 \exp\left(\frac{qV}{k_B T}\right) \approx I \tag{2.18}$$

Equation (2.17) may be expressed as

$$\frac{\partial V}{\partial I} = \frac{\frac{k_B T}{q}}{I} = \frac{\frac{1.3806 \times 10^{-23} \times 300}{1.602 \times 10^{-19}}}{I} = \frac{258.54 \times 10^{-4}}{I} \tag{2.19}$$

Hence the diode looks like a resistor of dynamic resistance

$$r_d = \frac{258.54 \times 10^{-4}}{I} \tag{2.20}$$

2.3.3 Junction capacitance (C_j) of the SBD

When the diode is reverse biased, a depletion region is formed on the semiconductor side. The width W of this depletion region is given by (figure 2.7)

$$W = \sqrt{\frac{2\varepsilon_0 \varepsilon_S(\phi_{bi} - V_a)}{qN_D}} \tag{2.21}$$

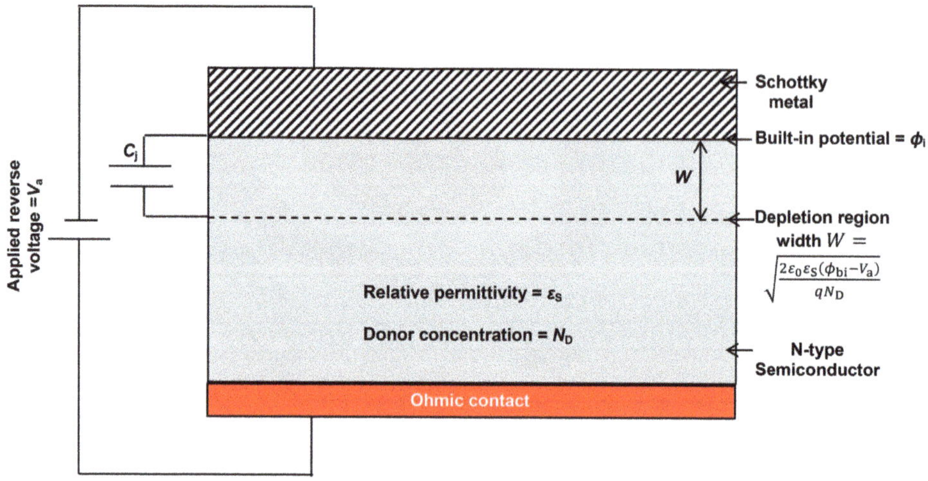

Figure 2.7. Relation of depletion region width W in a reverse-biased Schottky diode to the structural parameters. The junction capacitance C_j associated with the depletion region is also shown.

where ε_0 is the permittivity of free space, ε_S is the dielectric constant of the semiconductor, V_a is the applied voltage, ϕ_{bi} is the built-in potential at the metal–semiconductor interface, q is the elementary charge and N_D is the donor concentration of the semiconductor.

The built-in potential ϕ_{bi} is

$$\phi_{bi} = \phi_B - \frac{k_B T}{q}\ln\left(\frac{N_C}{N_D}\right),\ q\phi_B = (\phi_M - \chi)\text{eV for N-type semiconductor} \quad (2.22)$$

and

$$\phi_{bi} = \phi_B - \frac{k_B T}{q}\ln\left(\frac{N_V}{N_A}\right),\ \phi q_B = \left(\chi + \frac{E_G}{q} - \phi_M\right)\text{eV for P-type semiconductor} \quad (2.23)$$

where ϕ_B is the barrier height in Volts; N_C, N_V are the effective densities of states in the conduction and valence bands, respectively, at 300 K: $N_C = 2.82 \times 10^{19}$ cm^{-3}, $N_V = 1.83 \times 10^{19}$ cm^{-3} for silicon; ϕ_M is the work function of the metal = 4.25 eV for aluminum, 5.3 eV for gold; χ is the electron affinity of the semiconductor = 4.05 eV for Si, 4.07 eV for GaAs; and E_G is the energy bandgap of the semiconductor.

If A is the surface area of the diode, the total charge Q in the depletion region is

$$\begin{aligned} Q &= -\text{Area} \times \text{Electronic charge} \times \text{Doping concentration} \\ &\quad \times \text{Depletion region width} = -AqN_D W \\ &= -AqN_D \times \sqrt{\frac{2\varepsilon_0\varepsilon_S(\phi_{bi} - V_a)}{qN_D}} = -A\sqrt{2qN_D\varepsilon_0\varepsilon_S(\phi_{bi} - V_a)} \\ &= -A\sqrt{2qN_D\varepsilon_0\varepsilon_S} \times \sqrt{\phi_{bi} - V_a} \end{aligned} \quad (2.24)$$

The junction capacitance is

$$C_j = \frac{dQ}{dV_a} = \frac{d}{dV_a}\left\{-A\sqrt{2qN_D\varepsilon_0\varepsilon_S} \times \sqrt{\phi_{bi} - V_a}\right\} = -A\sqrt{2qN_D\varepsilon_0\varepsilon_S}\frac{d}{dV_a}\left(\sqrt{\phi_{bi} - V_a}\right)$$
$$= -A\sqrt{2qN_D\varepsilon_0\varepsilon_S} \times \frac{1}{2} \times (\phi_{bi} - V_a)^{1/2-1} \times -1 = A\sqrt{\frac{qN_D\varepsilon_0\varepsilon_S}{2(\phi_{bi} - V_a)}} \tag{2.25}$$

and the capacitance at $V_a = 0$ is

$$C_{j0} = A\sqrt{\frac{qN_D\varepsilon_0\varepsilon_S}{2(\phi_{bi} - 0)}} = A\sqrt{\frac{qN_D\varepsilon_0\varepsilon_S}{2\phi_{bi}}} \tag{2.26}$$

Dividing equation (2.25) by equation (2.26) we get

$$\frac{C_j}{C_{j0}} = A\sqrt{\frac{qN_D\varepsilon_0\varepsilon_S}{2(\phi_{bi} - V_a)}} \times \frac{1}{A}\sqrt{\frac{2\phi_{bi}}{qN_D\varepsilon_0\varepsilon_S}} = \sqrt{\frac{1}{\phi_{bi} - V_a} \times \phi_{bi}} = \sqrt{\frac{\phi_{bi}}{\phi_{bi} - V_a}}$$
$$= \sqrt{\frac{1}{1 - \frac{V_a}{\phi_{bi}}}} = \frac{1}{\sqrt{1 - \frac{V_a}{\phi_{bi}}}} \tag{2.27}$$

or,

$$C_j = \frac{C_{j0}}{\sqrt{1 - \frac{V_a}{\phi_{bi}}}} \tag{2.28}$$

2.4 Cut-off frequency of the circular-contact SBD

This is given by

$$f_C = \frac{1}{2\pi R_S C_{j0}} = \frac{1}{2\pi\left(\frac{1}{4\sigma a}\right)A\sqrt{\frac{qN_D\varepsilon_0\varepsilon_S}{2\phi_{bi}}}} = \frac{2\sigma a}{\pi \times \pi a^2}\sqrt{\frac{2\phi_{bi}}{\varepsilon_0\varepsilon_S qN_D}}$$
$$= \frac{2qN_D\ \mu_n}{\pi^2 a}\sqrt{\frac{2\phi_{bi}}{\varepsilon_0\varepsilon_S qN_D}} = \frac{2\mu_n}{\pi^2 a}\sqrt{\frac{2\phi_{bi}qN_D}{\varepsilon_0\varepsilon_S}} \tag{2.29}$$

where a is the anode radius; equations (2.15) and (2.26) are applied together with the conductivity equation

$$\sigma = qN_D\,\mu_n \tag{2.30}$$

where μ_n is the electron mobility. Equation (2.29) provides useful pointers towards attaining higher cut-off frequencies of Schottky diodes:

(i) The cut-off frequency is inversely proportional to the contact radius a. Hence, the smaller the contact area, the higher is the frequency.

(ii) The cut-off frequency is directly proportional to the product (μ_n, $N_D^{1/2}$), i.e., it can be increased by using a semiconductor material with higher mobility and with heavier doping.

Example 2.1. An Au/N-GaAs Schottky diode is fabricated on an N-GaAs epilayer of doping concentration 2×10^{18} cm^{-3}. The anode diameter is 0.4 μm and cathode diameter is 1 μm. Find the cut-off frequency of the diode by taking the electron mobility as 2700 cm^2 (V–s)$^{-1}$, dielectric constant of GaAs as 10.89, work function of gold $\phi_{Au} = 5.1$ eV, electron affinity of gallium arsenide $\chi_{GaAs} = 4.07$ eV, and effective conduction band density of states = 4.7×10^{17} cm^{-3}. What is the series resistance?

Solution: Here, $a = (0.4/2)$μm = 0.2 μm = 0.2×10^{-6} m, $b = (1/2)$ μm = 0.5×10^{-6} m, $N_D = 2 \times 10^{18}$ cm^{-3} = $2 \times 10^{18} \times 10^6$ m^{-3}, $N_C = 4.7 \times 10^{17}$ cm^{-3} = $4.7 \times 10^{17} \times 10^6$ m^{-3}, μ_n = 2700 cm^2 (V–s)$^{-1}$ = 2700$\times 10^{-4}$ m^2 (V–s)$^{-1}$.

We shall apply equation (2.22). The Schottky barrier height ϕ_B at the Au/GaAs interface is given by

$$\begin{aligned} q\phi_B &= \text{Work function of gold} - \text{Electron affinity of GaAs} = \phi_{Au} - \chi_{GaAs} \\ &= 5.1 - 4.07 = 1.03 \text{ eV} \end{aligned} \tag{2.31}$$

or,

$$\phi_B = 1.03 \text{ V} \tag{2.32}$$

By equation (2.22)

$$\begin{aligned} \phi_{bi} &= \phi_B - \frac{k_B T}{q}\ln\left(\frac{N_C}{N_D}\right) = 1.03 - \frac{1.380\,65 \times 10^{-23} \times 300}{1.602 \times 10^{-19}} \\ &\ln\left(\frac{4.7 \times 10^{17} \times 10^6}{2 \times 10^{18} \times 10^6}\right) = 1.03 - 258.55 \times 10^{-4} \times \ln(0.235) = 1.03 \\ &- 258.55 \times 10^{-4} \times (-1.448) = 1.03 + 0.037 = 1.067 \text{ V} \end{aligned} \tag{2.33}$$

$$\begin{aligned} \therefore f_C &= \frac{2\mu_n}{\pi^2 a}\sqrt{\frac{2\phi_{bi} q N_D}{\varepsilon_0 \varepsilon_S}} = \frac{2 \times 2700 \times 10^{-4}}{(3.141\,59)^2 \times 0.2 \times 10^{-6}} \\ &\times \sqrt{\frac{2 \times 1.067 \times 1.602 \times 10^{-19} \times 2 \times 10^{18} \times 10^6}{8.854 \times 10^{-12} \times 10.89}} \\ &= \frac{2 \times 2700 \times 10^{-4+6}}{9.8696 \times 0.2} \times \sqrt{\frac{2 \times 1.067 \times 1.602 \times 2 \times 10^{-19+18+6+12}}{8.854 \times 10.89}} \\ &= 2.7357 \times 10^5 \times 0.842 \times 10^8 = 2.3 \times 10^{13} \text{ Hz} = 23 \times 10^{12} \text{ Hz} \\ &= 23 \text{ THz} \end{aligned} \tag{2.34}$$

where equation (2.29) has been applied.

From equation (2.15), the series resistance is

$$R_S = \frac{1}{4\sigma a} = \frac{1}{4 \times 2 \times 10^{18} \times 10^{6} \times 1.602 \times 10^{-19} \times 2700 \times 10^{-4} \times 0.2 \times 10^{-6}} \quad (2.35)$$
$$= 1.445 \times 10^{-4} \times 10^{-18-6+19+4+6} = 14.45\ \Omega$$

Example 2.2. Ti/Pt/Au/N–GaAs Schottky contacts were prepared by sputtering and electron beam evaporation (Sehgal *et al* 2003). Low barrier heights obtained were explained in terms of a barrier height inhomogeneity model. The barrier heights of sputtered samples were 0.67 eV and those of electron beam evaporated samples were 0.85 eV. Taking all remaining parameters as in example 2.1, find the cut-off frequencies for Schottky contacts made by: (a) sputtering and (b) electron beam evaporation.

Solution: Barrier height affects the built-in potential ϕ_{bi}:

From equation (2.33)

$$|\phi_{bi}|_{\text{Sputtering}} = 0.67 + 0.037 = 0.707\ \text{V} \quad (2.36)$$

From equation (2.34)

$$|f_C|_{\text{Sputtering}} = 2.7357 \times 10^5 \times \sqrt{\frac{2 \times 0.707 \times 1.602 \times 2 \times 10^{-19+18+6+12}}{8.854 \times 10.89}}$$
$$= 2.7357 \times 10^5 \times \sqrt{0.0469\,866 \times 10^{17}} = 2.7357 \times 10^5$$
$$\times 0.6855 \times 10^8 = 1.875 \times 10^{13} = 18.75 \times 10^{12} = 18.75\ \text{THz} \quad (2.37)$$

Applying equation (2.33)

$$|\phi_{bi}|_{\text{Electron beam}} = 0.85 + 0.037 = 0.887\ \text{V} \quad (2.38)$$

Applying equation (2.34)

$$|f_C|_{\text{Electron beam}} = 2.7357 \times 10^5 \times \sqrt{\frac{2 \times 0.887 \times 1.602 \times 2 \times 10^{-19+18+6+12}}{8.854 \times 10.89}}$$
$$= 2.7357 \times 10^5 \times \sqrt{0.058\,949\,3 \times 10^{17}} = 2.7357 \times 10^5$$
$$\times 0.767\,78 \times 10^8 = 2.1 \times 10^{13} = 21 \times 10^{12} = 21\ \text{THz} \quad (2.39)$$

Example 2.3. In Au/N-GaAs Schottky diodes fabricated by thermal evaporation of gold film, the Schottky barrier height was experimentally measured to be 0.735 eV at 301 K (Hudait *et al* 2001). If the diodes have an area of 200 μm × 200 μm, and the reverse saturation current is 134.4 pA, find the effective Richardson constant for N–GaAs.

Solution: Here $A = 200 \times 10^{-6}$ m $\times$ 200×10^{-6} m, $T = 301$ K, $\phi_B = 0.735$ eV

Putting the given values in equation (2.4) we get

$$\begin{aligned} 1.344 \times 10^{-10}\,\text{A} &= 200 \times 10^{-6}\,\text{m} \times 200 \times 10^{-6}\,\text{m} \\ &\quad \times A_{\text{Effective}}(301)^2\,\text{K}^2 \exp\left(-\frac{1.602 \times 10^{-19} \times 0.735}{1.3806 \times 10^{-23} \times 300}\right) \\ &= 4 \times 10^{-8} A_{\text{Effective}} \times 9.0601 \times 10^4 \exp(-0.00284 \times 10^4)\,\text{m}^2\,\text{K}^2 \\ &= 36.2404 \times 10^{-4} A_{\text{Effective}} \times 4.635 \times 10^{-13}\text{m}^2\text{K}^2 \\ &= A_{\text{Effective}} 167.974 \times 10^{-17}\text{m}^2\,\text{K}^2 \end{aligned} \tag{2.40}$$

$$\begin{aligned} \therefore A_{\text{Effective}} &= \frac{1.344 \times 10^{-10}\,\text{A}}{167.974 \times 10^{-17}\text{m}^2\text{K}^2} = \frac{1.344 \times 10^{-10+17}}{167.974}\text{Am}^{-2}\text{K}^{-2} \\ &= 0.008 \times 10^7\,\text{Am}^{-2}\text{K}^{-2} = 8 \times 10^4\quad \text{Am}^{-2}\text{K}^{-2} \end{aligned} \tag{2.41}$$

2.5 Consideration of skin effect for series resistance calculation

At high frequencies, the skin effect significantly influences the current flow by restricting the current to the outermost layer of current-carrying regions. Due to the small skin depth, the effective cross-section of the conductor is reduced. This reduction of cross-section is observed as an increase in series resistance R_S. The increased series resistance adversely impacts the cut-off frequency of SBD.

The skin effect is not included in the traditional model. Therefore, the actual cut-off frequency is lower than that predicted by this model. The influence of frequency on series resistance of SBD is analyzed by Dickens (1967). The series resistance is given by the real part of the impedance Z

$$Z = \frac{(1+j)}{2\pi\sigma\delta_S}\ln\left(\frac{b}{a}\right) + \frac{1}{2\pi\sigma a}\tan^{-1}\left(\frac{b}{a}\right) \tag{2.42}$$

as

$$\begin{aligned} R_S = Re(Z) &= \text{Re}\left\{\frac{(1+j)}{2\pi\sigma\delta_S}\ln\left(\frac{b}{a}\right) + \frac{1}{2\pi\sigma a}\tan^{-1}\left(\frac{b}{a}\right)\right\} \\ &= \text{Re}\left\{\frac{1}{2\pi\sigma\delta_S}\ln\left(\frac{b}{a}\right) + \frac{1}{2\pi\sigma a}\tan^{-1}\left(\frac{b}{a}\right) + j\frac{1}{2\pi\sigma\delta_S}\ln\left(\frac{b}{a}\right)\right\} \\ &= \frac{1}{2\pi\sigma\delta_S}\ln\left(\frac{b}{a}\right) + \frac{1}{2\pi\sigma a}\tan^{-1}\left(\frac{b}{a}\right) \end{aligned} \tag{2.43}$$

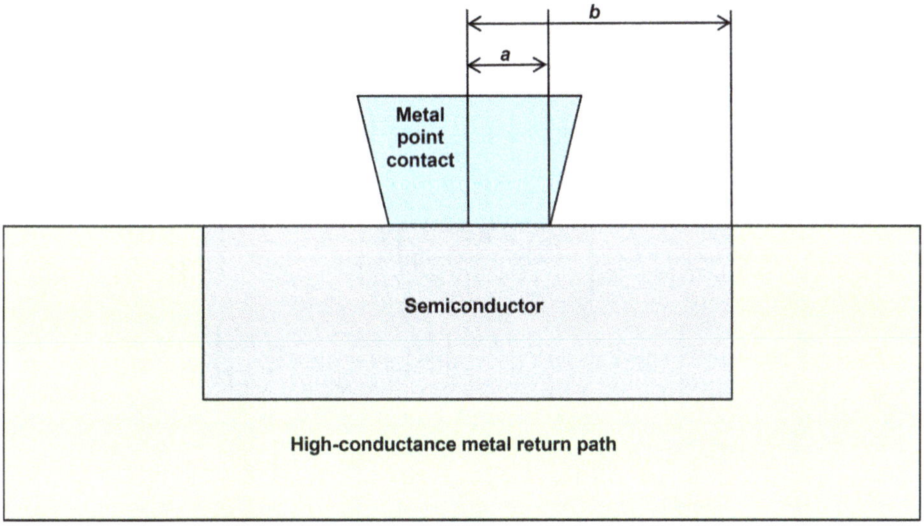

Figure 2.8. Point contact metal-semiconductor junction.

where a, b are the anode and cathode radii, respectively, of a point-contact diode configuration (figure 2.8); and the skin depth δ_S is a frequency dependent variable

$$\begin{aligned}\delta_S &= \sqrt{\frac{2}{\omega_C \mu_0 \sigma}} = \sqrt{\frac{2}{2\pi f_C (4\pi \times 10^{-7})\sigma}} \\ &= \sqrt{\frac{2}{2 \times 3.141\,59 f_C (4 \times 3.141\,59 \times 10^{-7})\sigma}} = \frac{5.033 \times 10^2}{\sqrt{f_C \sigma}}\end{aligned} \tag{2.44}$$

where we have put the magnetic permeability of free space $\mu_0 = 4\pi \times 10^{-7}$ N A^{-2}.

In equation (2.43), when

$$\frac{b}{a} \rightarrow \infty, \ \tan^{-1}\left(\frac{b}{a}\right) \rightarrow \frac{\pi}{2} \tag{2.45}$$

Also, as

$$\omega \rightarrow 0, \ \delta_S \rightarrow \infty \tag{2.46}$$

Then equation (2.42) reduces to

$$Z = \frac{(1+j)}{2\pi\sigma \times \infty} \ln\left(\frac{b}{a}\right) + \frac{1}{2\pi\sigma a}\left(\frac{\pi}{2}\right) = 0 + \frac{1}{2\pi\sigma a} \times \frac{\pi}{2} = \frac{1}{4\sigma a} \tag{2.47}$$

which is the same as equation (2.15) for the DC case.

Substituting for δ_S from equation (2.44) in equation (2.43) we get

$$\begin{aligned} R_S &= \frac{1}{2\pi}\left\{\frac{1}{\sigma\frac{5.033\times10^2}{\sqrt{f_C\sigma}}}\ln\left(\frac{b}{a}\right)+\frac{1}{\sigma a}\tan^{-1}\left(\frac{b}{a}\right)\right\} \\ &= \frac{1}{2\times3.141\,59}\left\{\frac{\sqrt{f_C\sigma}}{5.033\times10^2\sigma}\ln\left(\frac{b}{a}\right)+\frac{1}{\sigma a}\tan^{-1}\left(\frac{b}{a}\right)\right\} \\ &= 0.159\left\{1.987\times10^{-3}\sqrt{\frac{f_C}{\sigma}}\ln\left(\frac{b}{a}\right)+\frac{1}{\sigma a}\tan^{-1}\left(\frac{b}{a}\right)\right\} \\ &= 3.159\times10^{-4}\sqrt{\frac{f_C}{\sigma}}\ln\left(\frac{b}{a}\right)+\frac{0.159}{\sigma a}\tan^{-1}\left(\frac{b}{a}\right) \end{aligned} \tag{2.48}$$

Example 2.4. Find the cut-off frequency of the Schottky diode of example 2.1 using the modified equation (2.48) including the skin effect. Find also the skin depth and series resistance.

Solution: For this diode, equation (2.30) gives

$$\begin{aligned} \sigma &= qN_D\mu_n = 1.602\times10^{-19}\times2\times10^{18}\times10^6\times2700\times10^{-4} \\ &= 8650.8\times10^{-19+18+6-4} = 8650.8\times10\,\Omega^{-1}\text{m}^{-1} \\ &= 865\,08\Omega^{-1}\text{m}^{-1} \end{aligned} \tag{2.49}$$

Applying equation (2.48) we get

$$\begin{aligned} R_S &= 3.159\times10^{-4}\sqrt{\frac{f_C}{\sigma}}\ln\left(\frac{b}{a}\right)+\frac{0.159}{\sigma a}\tan^{-1}\left(\frac{b}{a}\right) \\ &= 3.159\times10^{-4}\sqrt{\frac{f_C}{865\,08}}\ln\left(\frac{0.5\times10^{-6}}{0.2\times10^{-6}}\right) \\ &\quad+\frac{0.159}{865\,08\times0.2\times10^{-6}}\tan^{-1}\left(\frac{0.5\times10^{-6}}{0.2\times10^{-6}}\right) \\ &= \frac{3.159\times10^{-4}\times0.9163}{294.122}\sqrt{f_C}+9.1899\times10^{-6}\times10^6\times68.199 \\ &= 9.84\times10^{-7}\sqrt{f_C}+6.27\times10^2 \end{aligned} \tag{2.50}$$

By equation (2.26)

$$C_{j0} = A\sqrt{\frac{qN_{\mathrm{D}}\varepsilon_0\varepsilon_{\mathrm{S}}}{2\phi_{\mathrm{bi}}}} = 3.141\,59 \times (0.2 \times 10^{-6})^2$$
$$\sqrt{\frac{1.602 \times 10^{-19} \times 2 \times 10^{18} \times 10^{6} \times 8.854 \times 10^{-12} \times 10.89}{2 \times 1.067}} \qquad (2.51)$$
$$= 0.1257 \times 10^{-12} \times \sqrt{\frac{1.602 \times 2 \times 8.854 \times 10.89 \times 10^{-19+18+6-12}}{2 \times 1.067}}$$
$$= 0.1257 \times 10^{-12} \times 3.805 \times 10^{-3} = 4.78 \times 10^{-16}\ \mathrm{F}$$

Equations (2.14), (2.50) and (2.51) give

$$f_{\mathrm{C}} = \frac{1}{2\pi R_{\mathrm{S}} C_{j0}} = \frac{1}{2 \times 3.141\,59\left(9.84 \times 10^{-7}\sqrt{f_{\mathrm{C}}} + 6.27 \times 10^{2}\right)4.78 \times 10^{-16}}$$
$$= \frac{1}{30.03 \times 10^{-16}\left(9.84 \times 10^{-7}\sqrt{f_{\mathrm{C}}} + 6.27 \times 10^{2}\right)} \qquad (2.52)$$
$$= \frac{1}{2.955 \times 10^{-21}\sqrt{f_{\mathrm{C}}} + 1.883 \times 10^{-12}}$$

or,

$$f_{\mathrm{C}}\left(2.955 \times 10^{-21}\sqrt{f_{\mathrm{C}}} + 1.883 \times 10^{-12}\right) = 1 \qquad (2.53)$$

or,

$$2.955 \times 10^{-21} f_{\mathrm{C}}\sqrt{f_{\mathrm{C}}} + 1.883 \times 10^{-12} f_{\mathrm{C}} = 1 \qquad (2.54)$$

Put

$$f_{\mathrm{C}} = 1 \times 10^{12}\ \mathrm{Hz} \qquad (2.55)$$

Then the left-hand side is

$$2.955 \times 10^{-21} \times 10^{12}\sqrt{10^{12}} + 1.883 \times 10^{-12} \times 10^{12} = 2.955 \times 10^{-3} + 1.883$$
$$= 1.886 >> 1 \qquad (2.56)$$

Put

$$f_{\mathrm{C}} = 5.5 \times 10^{11}\ \mathrm{Hz} \qquad (2.57)$$

Then the left-hand side is

$$2.955 \times 10^{-21} \times 5.5 \times 10^{11}\sqrt{5.5 \times 10^{11}} + 1.883 \times 10^{-12} \times 5.5 \times 10^{11}$$
$$= 1.205 \times 10^{-3} + 1.036 = 1.037 \approx 1 \qquad (2.58)$$

Hence,

$$f_C = 5.5 \times 10^{11}\ \text{Hz} = 0.55\ \text{THz} \tag{2.59}$$

Equation (2.44) gives the skin depth as

$$\delta_S = \frac{5.033 \times 10^2}{\sqrt{f_C \sigma}} = \frac{5.033 \times 10^2}{\sqrt{5.5 \times 10^{11} \times 865\,08}} = \frac{5.033 \times 10^2}{2.1813 \times 10^3 \times 10^5}$$
$$= 2.31 \times 10^{2-3-5} = 2.31 \times 10^{-6}\ \text{m} = 2.31\ \mu\text{m} \tag{2.60}$$

From equation (2.50), the skin-effect increased series resistance is

$$R_S = 9.84 \times 10^{-7}\sqrt{f_C} + 6.27 \times 10^2 = 9.84 \times 10^{-7}\sqrt{5.5 \times 10^{11}} + 6.27 \times 10^2$$
$$= 72.975 \times 10^{-2} + 627 = 627.73\ \Omega \tag{2.61}$$

2.6 Range of applicability of traditional SBD model

The traditional model is criticized from the following viewpoints:

i. A fundamental parameter governing the ultimate limit to the switching speed of a Schottky diode is the relaxation time of carriers, which is the time required for a crystal lattice to return to its equilibrium condition when upset by an external disturbance. When the switching speed is expressed in terms of frequency, the commonly used term is dielectric relaxation frequency ω_D, the frequency of relaxation of mobile charges, dipoles or ions, in a dielectric material when subjected to an applied alternating electric field. But the dielectric relaxation frequency concept is generalized to conducting materials as well. For any material, there is a frequency corresponding to the relaxation time referred to as its relaxation frequency. A material reacts to an applied field through several mechanisms, electronic, ionic, dipolar, etc centered around its characteristic relaxation frequency dependent upon its electrical conductivity σ and dielectric constant ε:

$$\omega_D = \frac{\sigma}{\varepsilon_0 \varepsilon} \tag{2.62}$$

The constituent materials in a Schottky diode are the metal and the semiconductor. For a metal, the dielectric relaxation frequency is ~10^{17} Hz. For a Schottky diode, the doping concentrations of epitaxial N-layer (N_D) used varies from 10^{16} to 4×10^{18} cm^{-3}. For GaAs, the relaxation frequency for $N_D = 10^{16}$ cm^{-3} is

$$|\omega_D|_{N_D=10^{16}} = \frac{\sigma}{\varepsilon_0 \varepsilon} = \frac{10^{16} \times 10^6 \times 1.602 \times 10^{-19} \times 5000 \times 10^{-4}}{8.854 \times 10^{-12} \times 10.89}$$
$$= 83.07 \times 10^{16+6-19-4+12} = 8.31 \times 10^{12}\ \text{Hz} = 8.31\ \text{THz} \tag{2.63}$$

assuming an electron mobility of 5000 $cm^2\ (V{-}s)^{-1}$, while that for $N_D = 4 \times 10^{18}\ cm^{-3}$ is

$$|\omega_D|_{N_D=4\times 10^{18}} = \frac{\sigma}{\varepsilon_0\varepsilon} = \frac{4\times 10^{18}\times 10^{6}\times 1.602\times 10^{-19}\times 2500\times 10^{-4}}{8.854\times 10^{-12}\times 10.89} \tag{2.64}$$
$$= 166.15\times 10^{18+6-19-4+12} = 1.6615\times 10^{15}\ \text{Hz} = 1662\ \text{THz}$$

for an assumed mobility of 2500 $cm^2\ (V{-}s)^{-1}$.

ii. Another parameter relevant to high-frequency operation of SBD is the carrier scattering frequency ω_S, which governs the mass-inertial behavior of charge carriers. It is given by

$$\omega_S = \frac{q}{m^*\mu_n} \tag{2.65}$$

where q is the electronic charge, m^* is the effective mass of charge carriers (here electrons) and μ_n is the electron mobility in the semiconductor. For GaAs,

$$m^* = 0.067 m_0 \tag{2.66}$$

where m_0 is the rest mass of electrons. Taking electron mobility of 5000 $cm^2\ (V{-}s)^{-1}$ for $N_D = 10^{16}\ cm^{-3}$ N–GaAs,

$$|\,\omega_S\,|_{10^{16}} = \frac{q}{m^*\mu_n} = \frac{1.602\times 10^{-19}}{0.067\times 9.109\times 10^{-31}\times 5000\times 10^{-4}} \tag{2.67}$$
$$= 0.001\,05\times 10^{-19+31+4} = 5.25\times 10^{12} = 5.25\ \text{THz}$$

and similarly with mobility of 2500 $cm^2\ (V{-}s)^{-1}$ for $N_D = 4\times 10^{18}\ cm^{-3}$ N-GaAs, it is

$$|\,\omega_S\,|_{4\times 10^{18}} = 10.5\ \text{THz} \tag{2.68}$$

Effects of relaxation and scattering frequencies are not taken into account in the traditional model, and therefore this model remains valid as long we are dealing with frequencies ω satisfying the conditions:

$$\omega \ll \omega_D \tag{2.69}$$

$$\omega \ll \omega_S \tag{2.70}$$

Non-consideration of ω_D leads to the exclusion of the effects of displacement current. When ω_S is not taken into account, the mass-inertial behavior of charge carriers is ignored.

2.7 Extended model of SBD

To overcome the deficiencies of the elementary model incorporating the skin effect, an improved model was propounded by Champlin and Eisenstein (1978). It is an all-encompassing model including the contributions from skin effect, carrier inertia and displacement current. In this model, the two assumptions of traditional model that $\omega << \omega_S$ and $\omega << \omega_D$ are nullified by replacing the impedance Z of the SBD by an impedance Z' consisting of two terms representing a complex bulk-spreading impedance and a complex skin-effect impedance:

$$\begin{aligned} Z = & \text{ Complex bulk-spreading impedance}\left(Z_{\text{Bulk-Spreading}}\right) \\ & + \text{Complex skin-effect impedance}\left(Z_{\text{Skin effect}}\right) \end{aligned} \tag{2.71}$$

To include these impedances, the modified equivalent circuit shown in figure 2.9 contains two additional segments, the bulk-spreading impedance and the skin-effect impedance, in comparison to the traditional model of figure 2.5.

The segment for the bulk-spreading impedance contains three components:

$$\text{DC speading resistance } R_S = \frac{1}{4\sigma a} \tag{2.72}$$

$$\text{Inertial inductance } L_S = \frac{R_S}{\omega_S} \tag{2.73}$$

$$\text{Displacement capacitance } C_S = \frac{1}{R_S \omega_D} \tag{2.74}$$

The DC spreading resistance is the same as that of the original model. Parallel resonance of $Z_{\text{Bulk-spreading}}$ is related to plasma resonance taking place with the plasma resonance frequency

$$\omega_P = \frac{1}{\sqrt{L_S C_S}} \tag{2.75}$$

Substituting for L_S and C_S from equations (2.73) and (2.74) in equation (2.75)

$$\omega_P = \frac{1}{\sqrt{\left(\frac{R_S}{\omega_S}\right)\left(\frac{1}{R_S \omega_D}\right)}} = \frac{1}{\sqrt{\frac{1}{\omega_S \omega_D}}} = \sqrt{\omega_S \omega_D} \tag{2.76}$$

By equations (2.63), (2.67) and (2.76), the plasma resonance frequency for $N_D = 1 \times 10^{16}$ cm^{-3} N–GaAs

$$|\omega_P|_{N_D = 1\times 10^{16}} = \sqrt{5.25 \times 8.31} = 6.605 \text{ THz} \tag{2.77}$$

while that for $N_D = 4 \times 10^{18}$ cm^{-3} N–GaAs is

$$|\omega_P|_{N_D = 4\times 10^{18}} = \sqrt{10.5 \times 1662} = 132.1 \text{ THz} \tag{2.78}$$

from equations (2.64), (2.68) and (2.76).

The quality or Q-factor of this resonance is

$$Q = \frac{\omega_P L_S}{R_S} = \frac{\omega_P}{\frac{R_S}{L_S}} \tag{2.79}$$

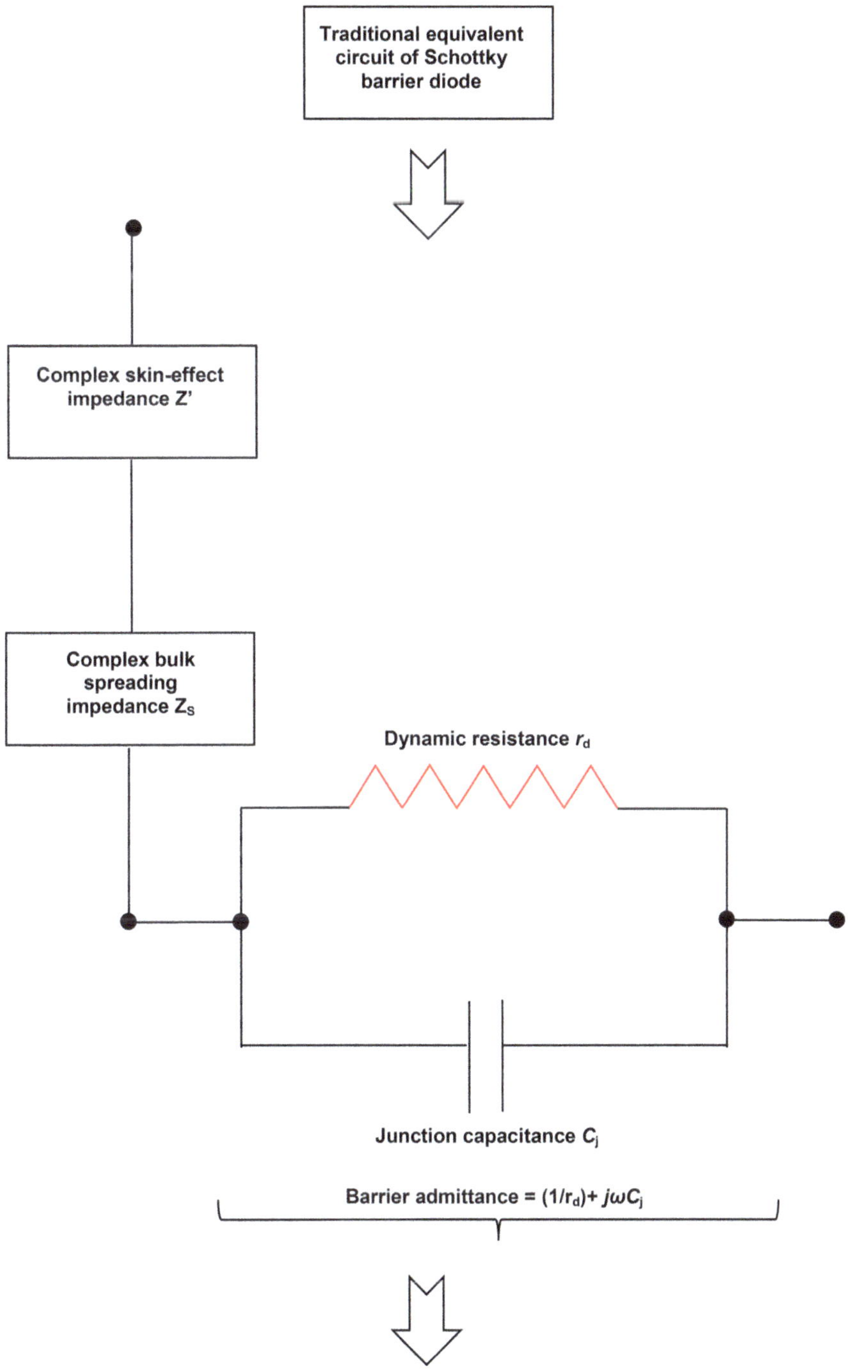

Figure 2.9. Evolution of the extended electrical equivalent circuit model of Schottky barrier diode from its traditional model.

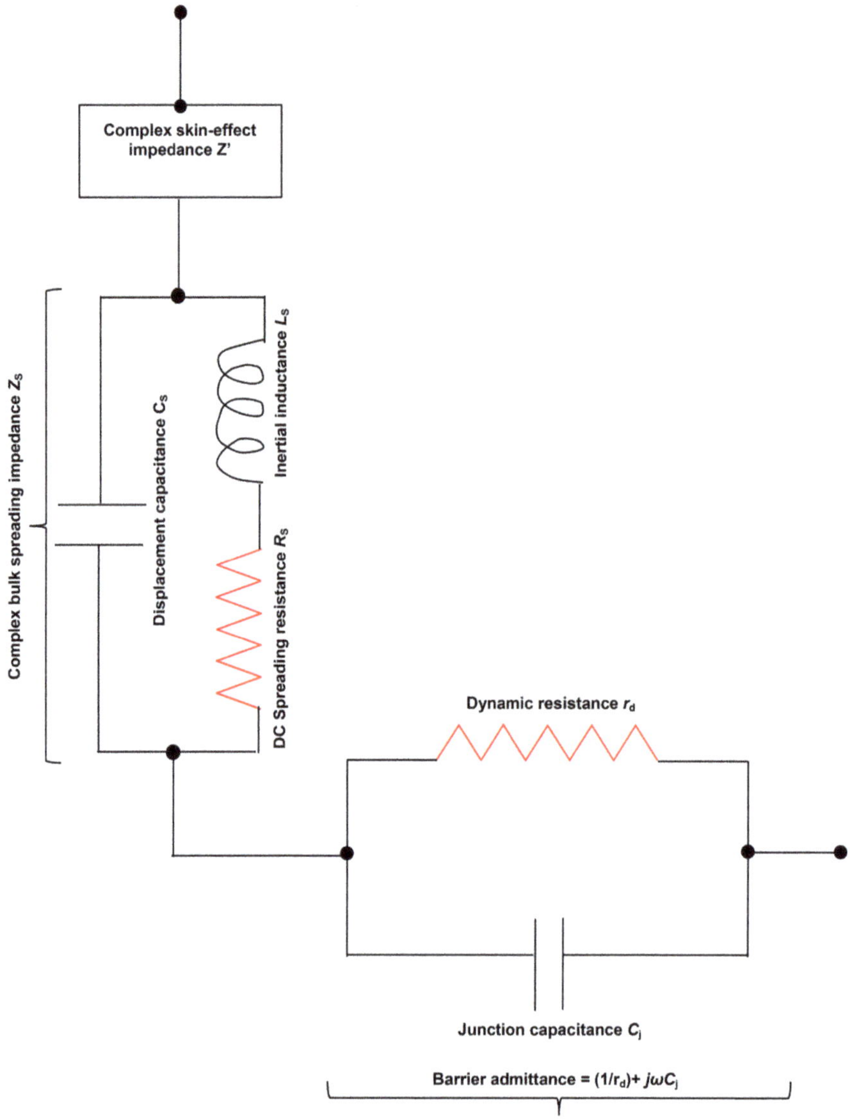

Figure 2.9. (Continued.)

Equations (2.79) and (2.73) give

$$Q = \frac{\omega_P}{\frac{R_S}{L_S}} = \frac{\omega_P}{\omega_S} \tag{2.80}$$

Then from equations (2.80) and (2.76)

$$Q = \frac{\sqrt{\omega_S \omega_D}}{\omega_S} = \sqrt{\frac{\omega_D}{\omega_S}} \tag{2.81}$$

The properties of a material at high frequencies can be described by two independent sets of parameters, either ω_S and ω_D or ω_P and Q. The parameter set (ω_S and ω_D) is convenient when $\omega_D < \omega_S$ because then $Q < 1$ leading to non-observation of true resonance. The parameter set (ω_P and Q) is used for situations in which $\omega_D > \omega_S$.

By applying the extended model, exhaustive calculations are performed by Champlin and Eisenstein (1978) for N–GaAs and N Si SBDs. These calculations reveal that for given structural parameters of the diode and when working at frequencies < plasma frequency, the cut-off frequency estimated from the revised model differs from that given by the traditional model for N–GaAs but yields almost identical results for N–Si diode. For the N–GaAs diode, the frequency depreciation factor is 1/6 while for the N–Si diode, the discrepancy is comparatively less. The lower value of cut-off frequency is due to the larger series resistance obtained in the case of GaAs. Evidently, the plasma frequencies being well above the terahertz gap, working with the traditional model including the skin-effect correction is a useful approximation.

2.8 Schottky diodes with terahertz operational frequencies

2.8.1 High-frequency performance degrading factors

Factors which limit the high frequency performance of the Schottky diodes are:

i. Junction Capacitance: As we have seen, junction capacitance is the primary factor limiting the operational frequency of a Schottky diode. For raising the frequency to terahertz range, the junction area must be taken as small as practically easily realizable by reducing the anode size. Electron beam lithography enables us to define such small geometries.

ii. Parasitic Capacitance: Even if junction capacitance has been reduced by size downscaling, the parasitic capacitances between the contact electrodes always remain annoying.

A strategy frequently adopted is decoupling of the contact pads using a micrometric air bridge (Giovine *et al* 2011, Bagni *et al* 2013). Looking at figure 2.10, we see that the diode is built with a submicron T-shaped finger joined to a bridge hanging in the air and connecting two mesas separated by a gap of several microns. The purpose is to connect the two contact pads through a closed path as: Ohmic contact 1 (anode)/gold T-shaped air-bridge finger/Schottky metal/N–GaAs epilayer/Ohmic contact 2 (cathode)/N–GaAs epilayer/N^+GaAs buffer layer/semi-insulating (SI)–GaAs substrate/N^+GaAs buffer layer/N–GaAs epilayer/Ohmic contact 1 (anode). Consequently, the parasitic fringe capacitance between the Schottky metal and contact pad 1 has a capacitance connected in series with it. This series capacitor has a thick film of air and SI substrate as the dielectric medium. It has therefore a very small capacitance, and its placement in series with the parasitic fringe capacitance renders the resultant parasitic capacitance almost negligible.

A novel design of AlGaN/GaN planar Schottky diode with air bridge and a half-through hole at the center of a circular anode was found to be very effective in

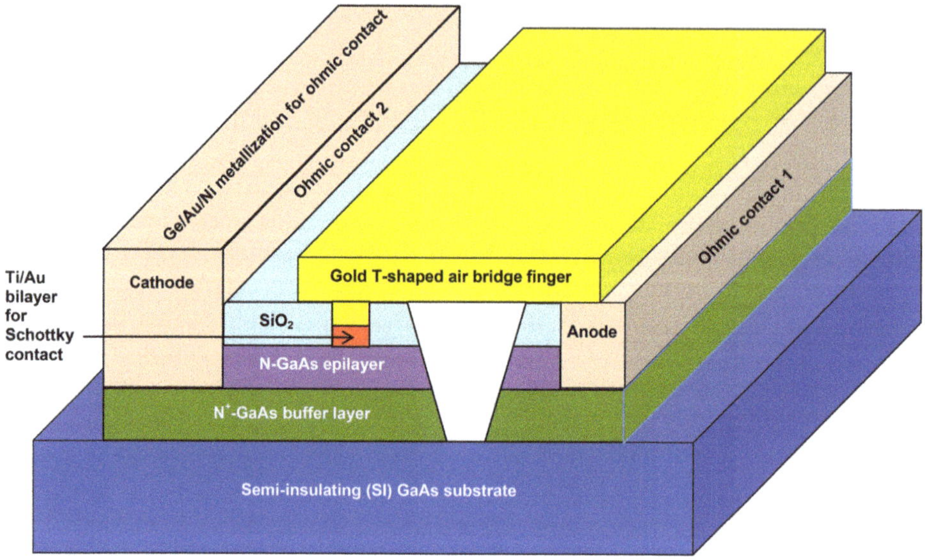

Figure 2.10. Micrometric air-bridged Schottky barrier diode.

decreasing the capacitance between the anode metal and the 2DEG formed at the AlGaN/GaN interface (Yang *et al* 2020).

Series Resistance: Reduction of anode dimension for decreasing the junction area is accompanied by an increase in series resistance. The series resistance is decreased by fabricating the Schottky diode in a material with high carrier mobility such as gallium arsenide or exploiting the 2DEG formed at a heterojunction such as AlGaN/GaN. The path length between anode and ohmic contact must be kept small. The doping level in this region must be high.

2.8.2 Terahertz Schottky barrier diodes

Giovine *et al* (2011) fabricated two series of Schottky diodes with T-anode and air bridge structure:

- Large foot print diode: 10 μm length × 0.2 μm breadth, and
- Small foot print diode: 2 μm length × 0.2 μm breadth.

Owing to fringing field effects, the junction capacitances do not scale linearly with device area, and the cut-off frequencies of these diodes are estimated as 1.1 THz and 7.2 THz. Wang *et al* (2018) fabricated Schottky diodes with footprints 3.15 μm^2 and 0.385 μm^2, for which the cut-off frequencies were estimated as 4.82 and 8.87 THz.

2.8.3 Schottky diodes in terahertz detection

A detector circuit using planar GaAs Schottky diode as the core component was fabricated (Schiselski *et al* 2016). The circuit contains a Schottky diode connected to an on-chip patch antenna. It works at 0.3 THz.

Uncooled InGaAs Schottky diodes were used for detection of terahertz waves from tens of GHz to more than 3 THz (Daghestani *et al* 2019). The noise equivalent power (NEP) is 50 pW $Hz^{-1/2}$ at 0.3 THz. NEP, a measure of photodetector sensitivity, is the minimum optical power that yields an output signal-to-noise ratio of unity.

2.8.4 Terahertz sources utilizing Schottky diode frequency-multiplier chains

A broadband chain made from four cascaded Schottky barrier frequency doublers was demonstrated at 1.5 THz (Chattopadhyay *et al* 2004). Driven by an input power of 100 mW in the 0.088–0.099 THz range, the peak efficiency at room temperature is 30% at 0.19 THz and falls to 4% at 1.5 THz. The peak output power is 40 μW at 120 K.

Maestrini *et al* (2004, 2005) reported chains consisting of three cascaded ×2×3×3 frequency multipliers fabricated using GaAs Schottky diodes to cover the 1.55–1.75 THz band. The first multiplier is a balanced frequency doubler while the second and third multipliers are balanced triplers. Pumped with 100 mW input power, the all solid-state local oscillator chain delivers power output of more than 4 μW across the 1.55–1.75 THz band at 120 K with peak power of 100 μW at 1.665 THz. At room temperature, the peak power output drops to 21 μW.

The Schottky diode is widely used for making multiplier chains because of easy reduction of active area to nanoscale dimensions providing junction capacitance in a range of fraction of a femtoFarad. It can be easily manufactured in standard cost-effective processes using materials such as GaAs, InP or SiGe. However, there exists a practical upper limit to building chains for frequency multiplication. The order of multiplication for chains built from Schottky diodes is usually up to $N = 3$ and infrequently $N = 4$. One reason from which this restriction arises is the saturation effect due to which the efficiency of conversion falls as the number segments in the chain increases. Another reason is the increased vulnerability of diodes to thermal or reverse breakdown. These effects drastically reduce the lifetimes of chains beyond a certain stage (Maestrini *et al* 2008). The output power achieved at room temperature (120 K) with frequency multipliers is typically 100 μW (200 μW) at 1.2 THz, decreasing to 3 μW (30 μW) at 1.9 THz (Maestrini *et al* 2010).

2.8.5 Frequency mixing using Schottky diodes

Room temperature mixers are necessary both for downconverting and upconverting terahertz frequencies in conjunction with laser local oscillators. At frequencies less than 2.5 THz, fundamental/subharmonic mixers are used for downconversion of terahertz RF signals. But at higher frequencies, harmonic mixers are required for downconverting these signals. A harmonic mixer is developed by Bulcha *et al* (2016) using lower harmonic factors. Low conversion loss obtained fulfils an essential requirement for phase locking at terahertz frequencies. Phase-locked quantum cascade laser is useful as a local oscillator source. Multipixel receivers and spectrometers providing high resolution can be built with this source.

GaAs Schottky diode is the critical mixer element used in the circuit. The mixer comprises a GaAs-based THz circuit containing the Schottky diode along with local oscillator and intermediate frequency circuits. It is characterized in the frequency range 1.8–3.2 THz. For third harmonic mixing, the conversion loss is 27 dB and increases to 30 dB for fourth harmonic mixing.

2.9 Non-PN junction diodes

2.9.1 Metal–insulator–metal (MIM) diode

2.9.1.1 Construction and principle

This consists of a thin insulating film of thickness < 10 nm sandwiched between two metallic electrodes (Moddel and Grover 2013). Figure 2.11 shows the MIM diode and its current–voltage characteristics. This diode works on the principle of quantum mechanical tunneling of electrons through the insulating film, governed by the thickness and material of the insulating film, and the work functions of the metals used for the two electrodes. The tunneling efficiency increases exponentially as the film thickness decreases. The capability of these diodes to operate at terahertz and up to optical frequencies is a consequence of the extremely small time required for tunneling ~ 10^{-15} s.

2.9.1.2 Origin of asymmetric current–voltage characteristic

In order that a diode is used for rectification, it should have an asymmetric current–voltage characteristic. The asymmetry is defined as the ratio of currents flowing through the diode in two opposite directions, namely the ratio of forward and reverse currents. How does this asymmetry arise in an MIM diode? The difference between the work function ϕ_M of a metal and the electron affinity χ of the insulator equals the barrier height φ_B at the metal–insulator interface. Because two metals of different work functions ϕ_{M1}, ϕ_{M2} are used, the barrier heights φ_{B1}, φ_{B2} at the interfaces of metal M1–insulator and metal M2–insulator differ. The difference between these barrier heights φ_{B1}, φ_{B2} is responsible for asymmetry in current flow in the two directions and hence diode behavior exhibited by the MIM structure. The energy band diagrams shown in figure 2.12 clarify this reasoning.

MIM diodes of various structures have been fabricated e.g. Ni/NiO/Au, Cr/Cr_2O_3/Pd, Al/Al_2O_3/Pt, Au–ZnO–Ni, etc (Azad *et al* 2018).

2.9.1.3 Radiation detection and rectification by nano-rectenna

Gadalla *et al* (2014) fabricated a 28.3 THz rectenna using an Cu/CuO/Au MIM diode of contact area 67 nm × 67 nm patterned by electron beam lithography. The CuO film of thickness 0.7 nm is formed by atomic layer deposition. The resulting zero-bias resistance of the diode is 500 Ω which is easily impedance matched with the antenna. A high zero-bias responsivity of 4 amperes per watt of incident power, is reported.

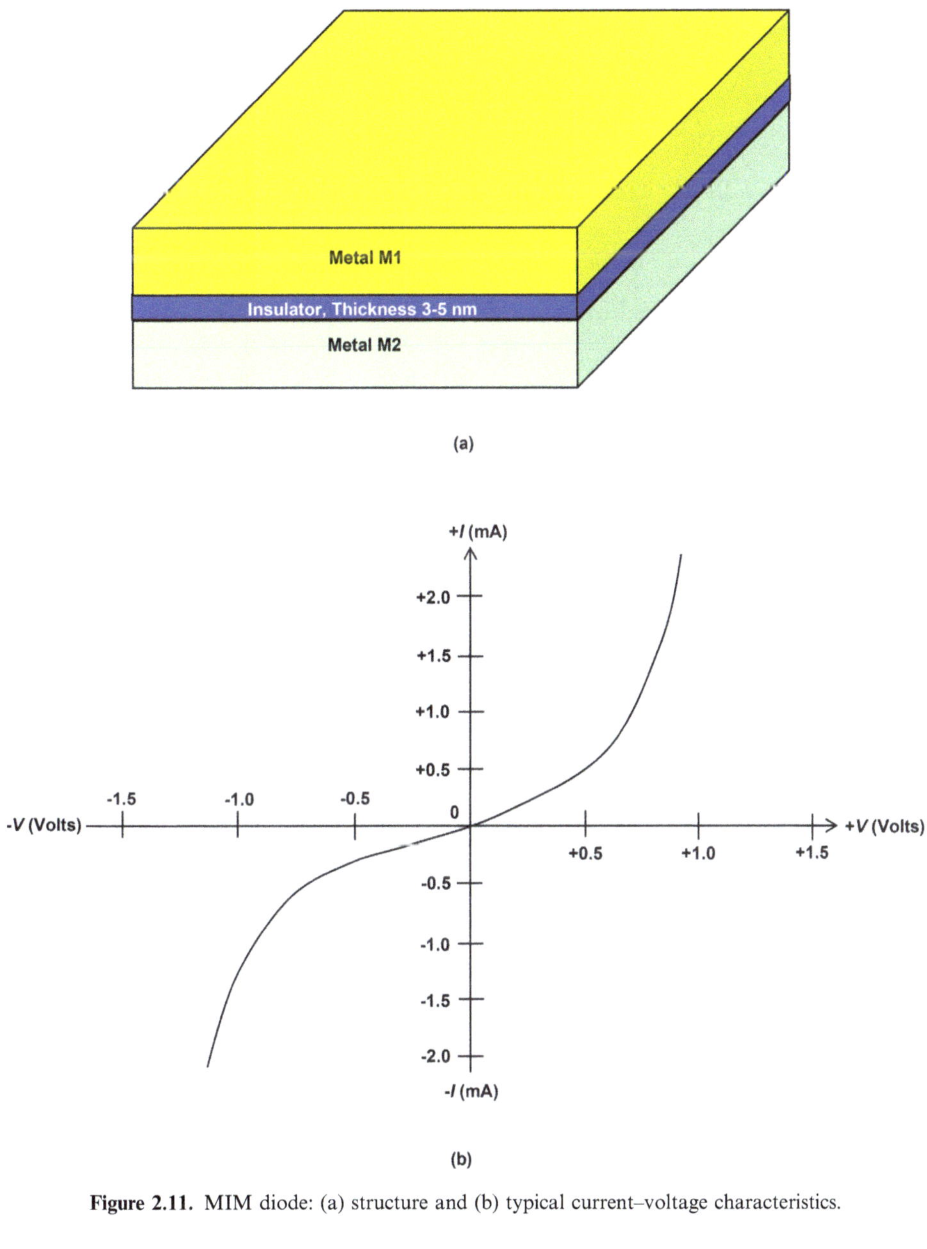

Figure 2.11. MIM diode: (a) structure and (b) typical current–voltage characteristics.

2.9.2 Self-switching diode (SSD)

2.9.2.1 Principle and simplicity of fabrication process

This is a planar nanodiode working on the field-effect principle (Song *et al* 2003). Owing to its structural simplicity (figure 2.13), it can be fabricated by a single step of nanolithography by etching insulating trenches on the surface of a semiconductor substrate. The simple fabrication process of the diode allows easy downscaling to nano dimensions. Using nanowires with intentionally broken symmetry, detection of

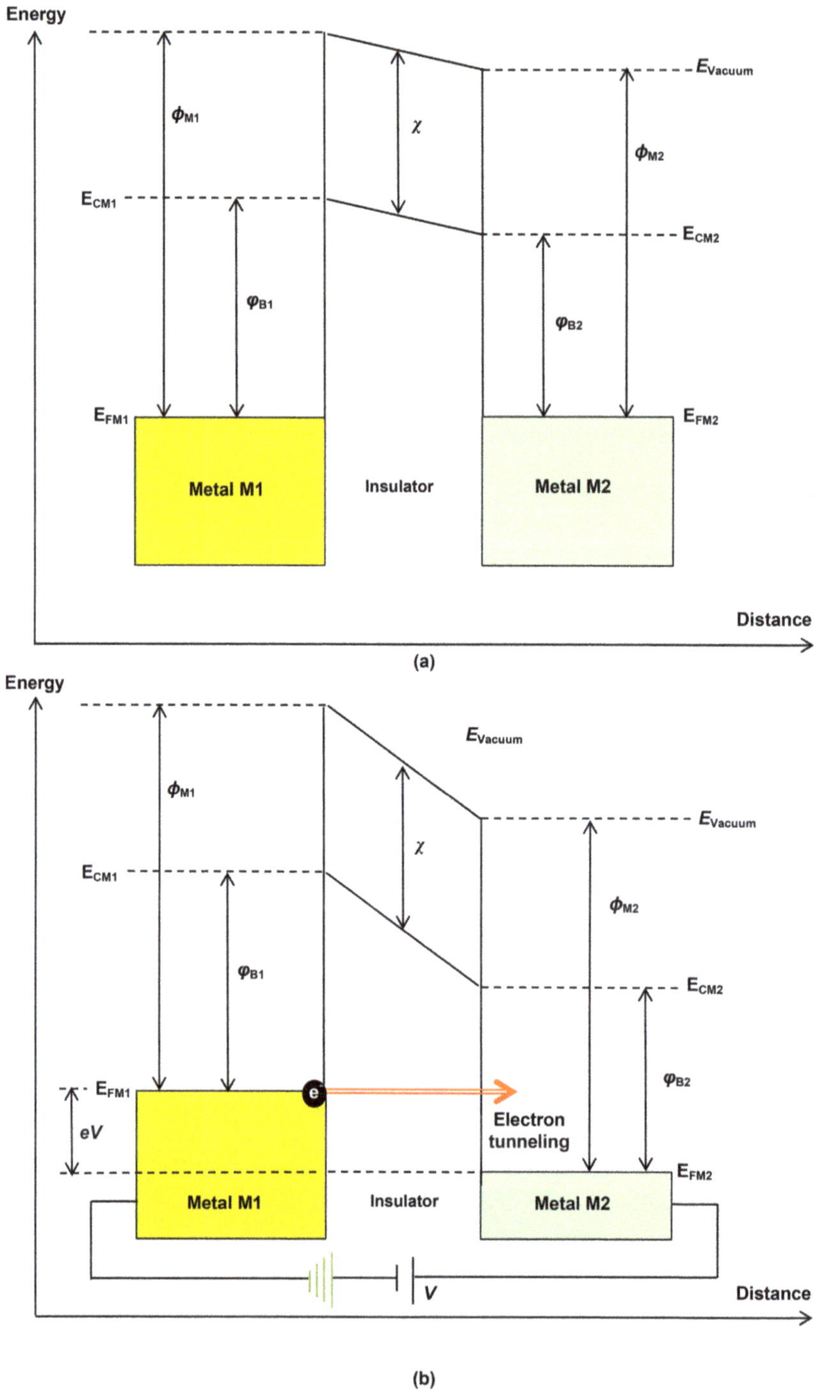

Figure 2.12. Energy-band diagrams of the metal–insulator–metal diode under three different conditions: (a) thermal equilibrium, (b) forward-biased mode and (c) reverse-biased mode, explaining the asymmetrical forward and reverse response behaviors.

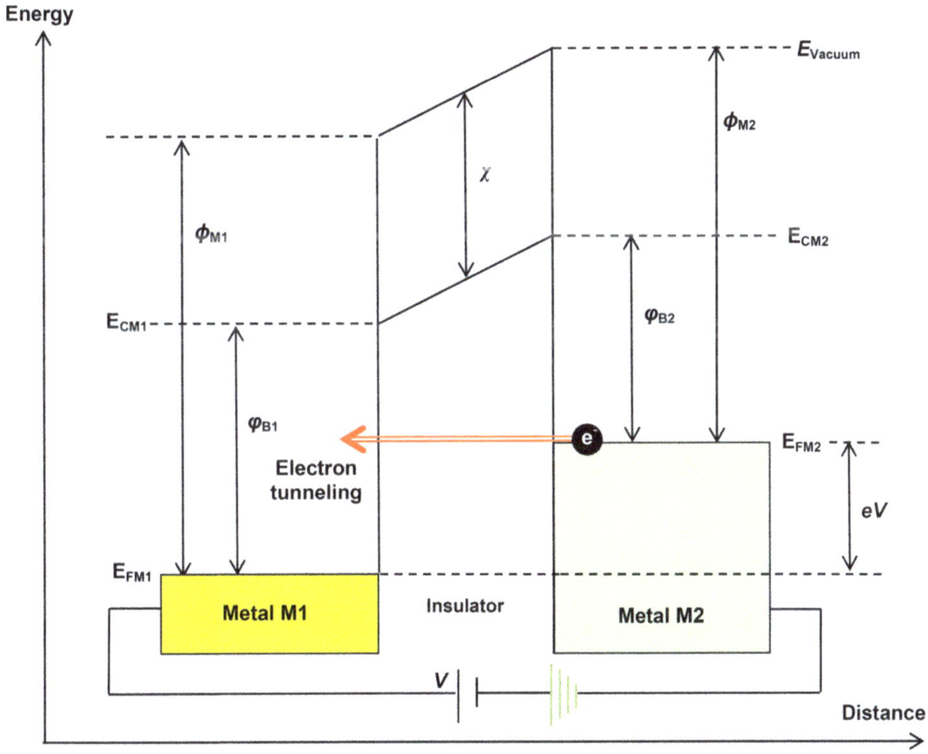

(c)

Figure 2.12. (Continued.)

electromagnetic waves at 0.11 THz was demonstrated at room temperature (Bolocco *et al* 2005). When built using high electron mobility material systems, such as InGaAs channels, the intrinsic cut-off frequencies can be raised to terahertz range by design optimization (Mateos *et al* 2005).

2.9.2.2 The channel, and source and drain terminals

On a two-dimensional plane, as shown in figure 2.14(a), one can see a semiconductor bar called the channel and marked as A, separated by trenches from two flanges, marked B. The two ends of the semiconductor are labelled as source and drain terminals.

2.9.2.3 Self-switching operation

Consider an N-type semiconductor substrate. The source is kept grounded (figure 2.14(b)). When a negative voltage is applied to the drain, the negative potential acts across the insulating trench on the electrons in the channel. It repels the electrons from regions close to the etched areas deeper into the bar, thereby forming depletion regions next to the etched areas. The higher the negative voltage,

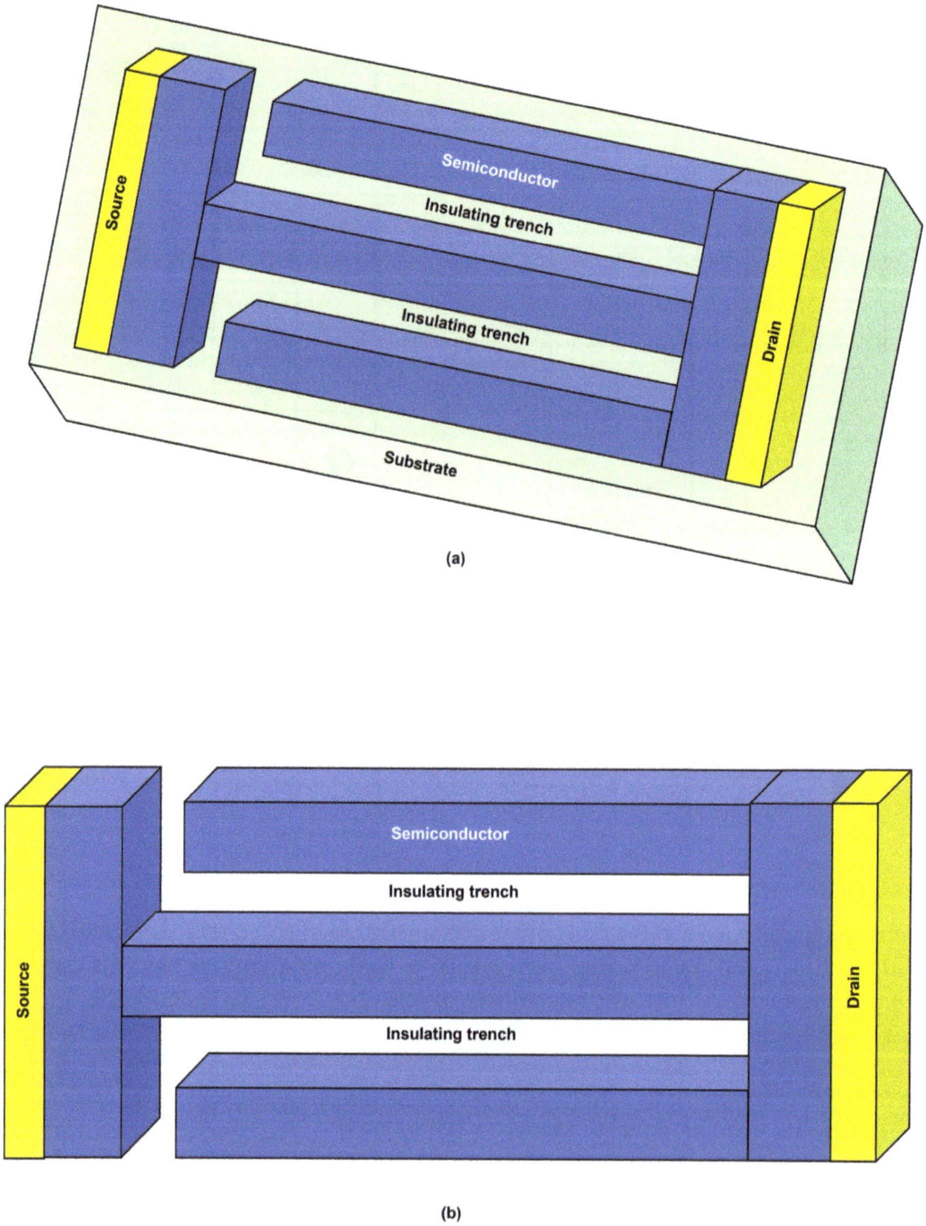

Figure 2.13. Self-switching diode: (a) 3D view on substrate and (b) top view without substrate.

the deeper the depletion regions penetrate into the channel, leaving a narrow path for current flow, which can be looked upon as closure of the channel As the channel width decreases, the electron current flowing from source to drain falls, resulting in the reversed-bias condition.

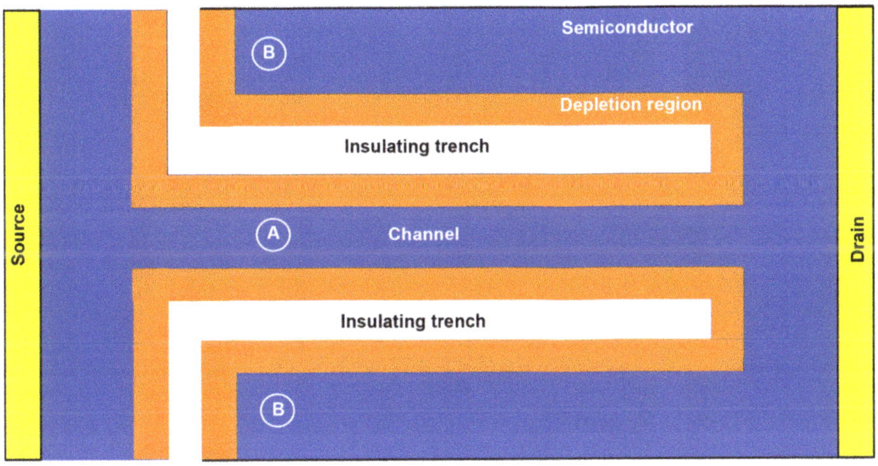

(a)

Figure 2.14. Explanation of the current–voltage characteristics of a self-switching diode from the depletion layer spreading and contraction: (a) diagram showing the native depletion region which is formed in the as-fabricated structure close to the etched boundaries due to surface charges, (b) reverse-biased diode with the depletion region spreading to constrict the channel and hence prevent current flow and (c) forward-biased diode with the depletion region shrinking and thereby allowing current to flow easily. The spreading and shrinkage of the depletion region and the consequent decrease and increase of channel width leads to diode-like characteristics.

Next, consider the reversal of polarity of applied voltage. On applying a positive potential to the drain (figure 2.14(c)), the electrons in the channel are attracted to the regions near the etched areas. The depletion regions shrink, thereby allowing a wider path for current flow. This is the forward biased condition during which a large current flows between drain and source through the broader channel.

Thus the structure provides a non-linear current–voltage characteristic. The nonlinearity arises from the decrease in channel width when a negative voltage is applied to the drain (reverse bias) and increase in channel width with opposite polarity of applied voltage (forward bias). Since the diode switches between the two biasing conditions by the action of applied voltage, it is referred to as a self-switching device. Further, the ability to rectify AC signals and the participation of a single type of carrier makes it a unipolar rectifying diode.

2.9.2.4 Use of InAs self-switching diodes for terahertz detection

Westlund *et al* (2013) described zero-bias terahertz detectors employing InAs self-switching diodes. The detectors were characterized in the frequency range 0.002–0.315 THz. A responsivity of 0.7 V W^{-1} is achieved at 0.6 THz by coupling a self-switching diode to an antenna.

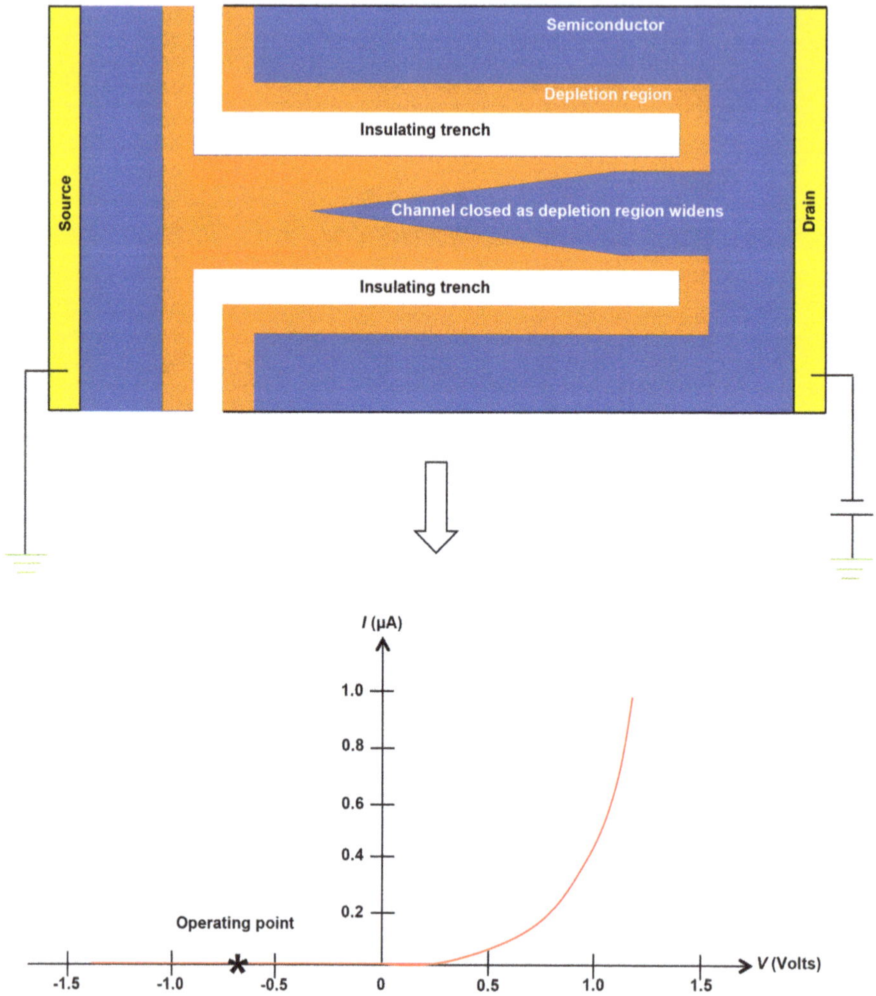

(b)

Figure 2.14. (Continued.)

2.9.3 Geometric diode

2.9.3.1 The common funnel-shaped design

It is a diode in which asymmetry between carrier transport in forward and reverse directions is created by utilizing the shape of the structure through which the carriers propagate (Moddel *et al* 2012). The style of architectural design guides electrons to flow naturally in the preferential direction. A common shape is in the form of a

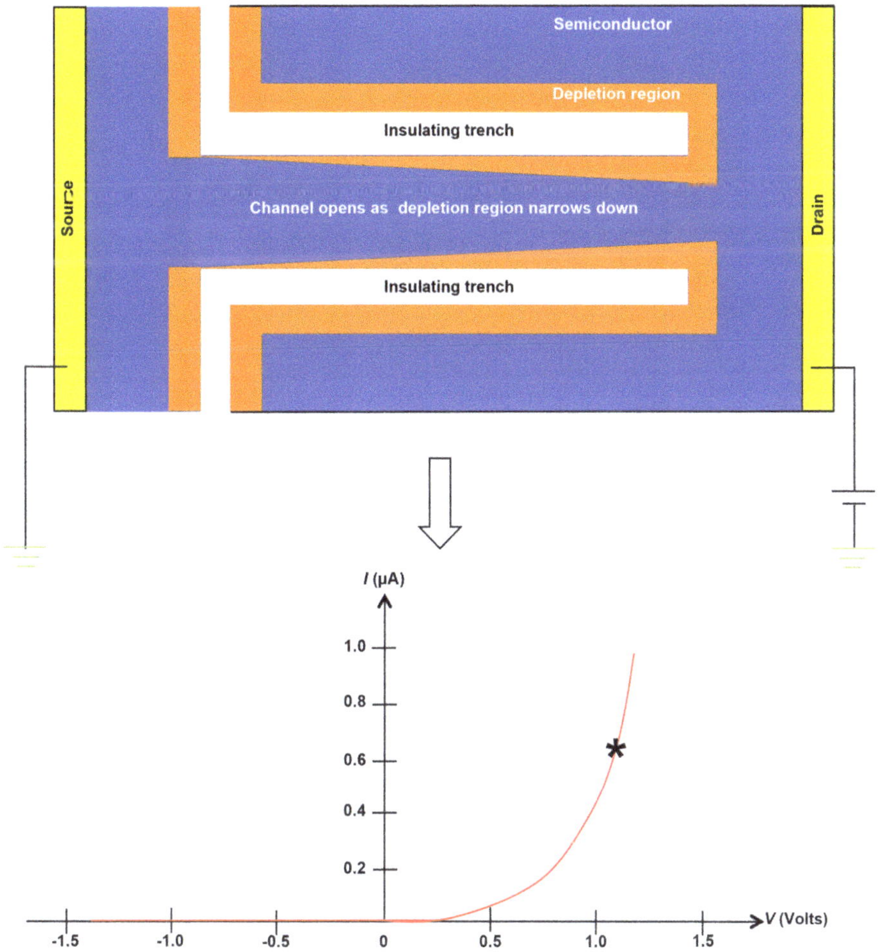

(c)

Figure 2.14. (Continued.)

conical funnel (figure 2.15). The electrons flow easily from the broad side to the narrow constriction producing forward-bias state. However, their motion is impeded in the opposite direction because electrons experience difficulty in squeezing through the funnel hole to enter the broader part of the funnel cone. Hence the reverse bias state is produced.

2.9.3.2 Ballistic transport mechanism, and mean free path consideration

The diode works on ballistic or quasi-ballistic transport of carriers by making the shape dissimilarity for forward and reverse bias conduction of the comparable

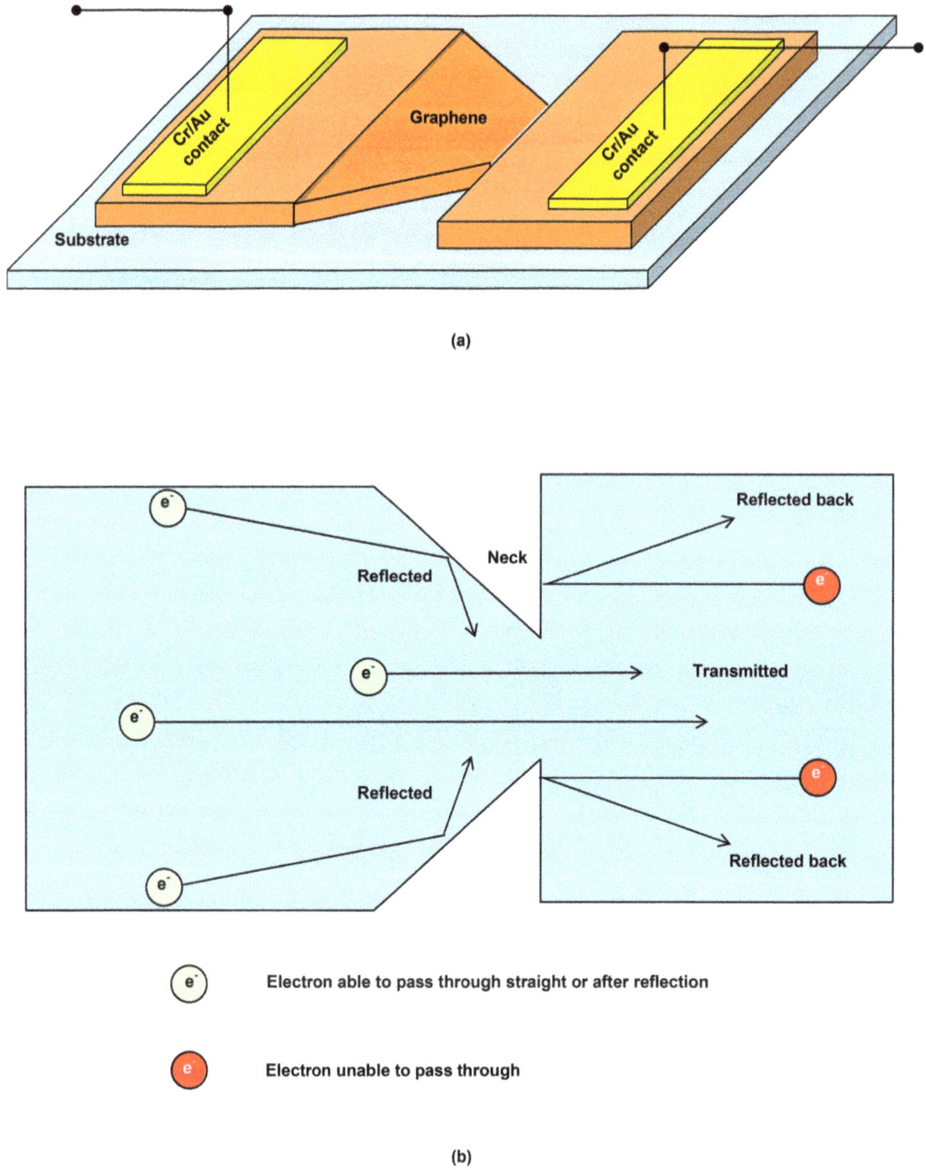

Figure 2.15. Geometric diode: (a) topological layout and (b) permitted and restricted electron paths.

dimensions to the mean free path of the carriers. The mean free path of electrons is a few nms in metals but can be several tens or even hundreds of nms in semiconductors, or may be as long as a micron in some systems. So, the diode can be fabricated easily by using a material with a long mean free path. When using materials with shorter mean free paths, nanolithographic techniques must be resorted to.

2.9.3.3 Low power consumption but high leakage current

An advantage of this diode is its low power requirement. As the diode does not contain any PN junction, there is no depletion region or potential barrier. Hence there is no built-in potential or threshold for conduction. So the diode can turn on at nearly zero bias. As a result, it operates with little power consumption. However, the advantage is offset by the high leakage current during reverse bias.

2.9.3.4 Fast switching behavior

The absence of potential barrier eliminates the capacitance associated with it. The effect of time constant for charging/discharging of this capacitance is therefore mitigated. Together with the fact that this diode is a majority carrier device makes it fast in operation, attaining terahertz switching frequencies because transit time of carriers across the structure is the only speed limiting factor.

2.9.3.5 Using graphene geometric diodes in terahertz rectennas

The rectenna formed by coupling graphene geometric diode to optical antenna converts the incident solar radiation signal at 28 THz into direct current (Zhu *et al* 2011, Joshi *et al* 2012). The zero-bias responsivity is 0.12 A W^{-1}. Under bias, the responsivity is 0.2 A W^{-1} (Zhu *et al* 2013).

Example 2.5. An MIM diode is fabricated with nickel oxide dielectric of relative permittivity 10, and film thickness $d = 3.5$ nm. The contact area is 110 nm × 110 nm. If the zero-bias resistance of the diode is 340 Ω, find its time constant and cut-off frequency.

Solution: Capacitance of the diode is

$$\begin{aligned} C &= \frac{\varepsilon_0 \varepsilon_{\text{Insulator}} A}{d} = \frac{8.854 \times 10^{-12} \times 10 \times 110 \times 10^{-9} \times 110 \times 10^{-9}}{3.5 \times 10^{-9}} \\ &= \frac{8.854 \times 10^{-12-9+2+1} \times 11 \times 11}{3.5} = 306.095 \times 10^{-18}\ \text{F} \\ &= 3.061 \times 10^{-16}\ \text{F} \end{aligned} \tag{2.82}$$

Here $R_0 = 340\ \Omega$. The time constant of the MIM diode is

$$\begin{aligned} \tau &= R_0 C = 340 \times 3.061 \times 10^{-16}\ \text{s} = 1040.74 \times 10^{-16}\ \text{s} \\ &= 104.074 \times 10^{-15}\ \text{s} \approx 104\ \text{fs} \end{aligned} \tag{2.83}$$

The cut-off frequency of the diode is

$$\begin{aligned} |\nu_{\text{Cut-off}}|_{\text{MIM diode}} &= \frac{1}{2\pi C R_0} = \frac{1}{2 \times 3.141\,59 \times 3.061 \times 10^{-16} \times 340} \\ &= \frac{10^{16}}{2 \times 3.141\,59 \times 3.061 \times 340} = 1.529 \times 10^{12}\ \text{Hz} \\ &= 1.53\ \text{THz} \end{aligned} \tag{2.84}$$

Example 2.6. Calculate the cut-off frequency of a metal1/CuO/metal2 MIM diode of contact area 35 nm × 35 nm. The CuO film has a relative permittivity 2.4. The CuO film thickness is 0.5 nm, and the zero-bias resistance of the diode is 139 Ω.

Solution: The cut-off frequency of the diode is

$$\begin{aligned}|\nu_{\text{Cut-off}}|_{\text{MIM diode}} &= \frac{1}{2\pi CR_0} = \frac{1}{2\pi\frac{\varepsilon_0\varepsilon_{\text{Insulator}}A}{d}R_0} = \frac{d}{2\pi\varepsilon_0\varepsilon_{\text{Insulator}}AR_0} \\ &= \frac{0.5\times10^{-9}}{2\times3.141\,59\times8.854\times10^{-12}\times2.4\times35\times10^{-9}\times35\times10^{-9}\times139} \\ &= \frac{0.5\times10^{-9+12+9+9}}{2\times3.141\,59\times8.854\times2.4\times35\text{x}35\times139} = \frac{0.5\times10^{21}}{22\,734\,277.136} \\ &= 219.93\times10^{11}\text{Hz} = 21.993\times10^{12}\text{ Hz} \approx 22\text{ THz}\end{aligned} \tag{2.85}$$

Example 2.7. Find the time constant and cut-off frequency of a geometric diode for the given equivalent circuit parameters: capacitance $C = 0.025$ fF and zero-bias resistance $R_0 = 200\ \Omega$.

Solution: The time constant is

$$\tau = R_0C = 200\times0.025\times10^{-15}\text{ s} = 5\times10^{-15}\text{ s} \tag{2.86}$$

The cut-off frequency is given by

$$\begin{aligned}|\nu_{\text{Cut-off}}|_{\text{Geometric diode}} &= \frac{1}{2\pi CR_0} = \frac{1}{2\times3.141\,59\times0.025\times10^{-15}\times200} \\ &= \frac{10^{15}}{2\times3.141\,59\times2.5\times2} = \frac{10^{15}}{31.4159} \\ &= 3.1831\times10^{13}\text{ Hz} = 31.83\times10^{12}\text{ Hz} = 31.83\text{ THz}\end{aligned} \tag{2.87}$$

2.10 Discussion and conclusions

The Schottky diode structure is examined and parameters affecting the high frequency performance of the diode are identified. From this analysis, design and fabrication techniques are devised to realize Schottky diodes to comply with the targeted frequency specifications.

Planar Schottky diode architectural design is of great significance in the development of diodes for terahertz frequency operation. Diodes with air-bridge anodes fabricated in high-mobility materials are widely used in multiplier chains and mixer circuits. However, Schottky diode operation remains satisfactory only at lower terahertz frequencies. For extending operation to higher terahertz frequencies up to the optical range for efficient solar energy harvesting, one must adopt the help of non-PN junction diodes. Using these diodes, high frequency electromagnetic radiation can be rectified by coupling to integrated antennas.

References

Azad I, Ram M K, Goswami D Y and Stefanakos E 2018 Fabrication and characterization of NiO based metal–insulator–metal diode using Langmuir–Blodgett method for high frequency rectification *AIP Adv.* **8** 045219-1–15

Bagni R, Giovine E, Carta S, Di Gaspare A, Casini R, Ortolani M, Foglietti V, Evangelisti F and Notargiacomo A 2013 Fabrication of air-bridge sub-micron Schottky junctions on Ge/SOI for THz detection *Microelectron. Eng.* **110** 470–3

Bolocco C *et al* 2005 Microwave detection at 110 GHz by nanowires with broken symmetry *Nano Lett.* **5** 1423–27

Bulcha B T, Hesler J L, Drakinskiy V, Stake J, Valavanis A, Dean P, Li L H and Barker N S 2016 Design and characterization of 1.8–3.2 THz Schottky-based harmonic mixers *IEEE Trans. Terahertz Sci. Technol.* **6** 737–46

Champlin K S and Eisenstein G 1978 Cut-off frequency of Schottky submillimeter diodes *IEEE Trans. Microw. Theory Tech. MTT* **26** 31–4

Chattopadhyay G, Schlecht E, Ward J S, Gill J J, Javadi H H S, Maiwald F and Mehdi I 2004 An all-solid-state broad-band frequency multiplier chain at 1500 GHz *IEEE Trans. Microw. Theory Tech.* **52** 1538–47

Daghestani N, Parow-Souchon K, Pardo D, Liu H, Brewster N, Frogley M, Cinque G, Alderman B and Huggard P G 2019 Room temperature ultrafast InGaAs Schottky diode based detectors for terahertz spectroscopy *Infrared Phys. Technol.* **99** 240–47

Dickens L E 1967 Spreading resistance as a function of frequency *IEEE Trans. Microw. Theory Tech. MTT* **15** 101–9

Gadalla M N, Abdel-Rahman M and Shamim A 2014 Design, optimization and fabrication of a 28.3 THz nano-rectenna for infrared detection and rectification *Sci. Rep.* **4** 4270

Giovine E, Casini R, Dominijanni D, Notargiacomo A, Ortolani M and Foglietti V 2011 Fabrication of Schottky diodes for terahertz imaging *Microelectron. Eng.* **88** 2544–6

Hudait M K, Venkateswarlu P and Krupanidhi S B 2001 Electrical transport characteristics of Au/n-GaAs Schottky diodes on n-Ge at low temperatures *Solid-State Electron.* **45** 133–41

Joshi S, Zhu Z, Grover S and Moddel G 2012 Infrared optical response of geometric diode rectenna solar cells *2012 38th IEEE Photovoltaic Specialists Conf.* 3–8 June 2012 *(Austin, TX)* pp 002976–8

Maestrini A, Ward J, Gill J, Javadi H, Schlecht E, Chattopadhyay G, Maiwald F, Erickson N R and Mehdi I 2004 A 1.7–1.9 THz local oscillator source *IEEE Microw. Wirel. Compon. Lett.* **14** 253–5

Maestrini A, Ward J S, Javadi H, Tripon-Canseliet C, Gill J, Chattopadhyay G, Schlecht E and Mehdi I 2005 Local oscillator chain for 1.55 to 1.75 THz with 100-μW peak power *IEEE Microw. Wirel. Compon. Lett.* **15** 871–3

Maestrini A, Ward J, Chattopadhyay G, Schlecht E and Mehdi I 2008 Terahertz sources based on frequency multiplication and their applications Frequenz *J. RF-Eng. Telecom.* **5/6** 118–22

Maestrini A, Thomas B, Wang H, Jung C, Treuttel J, Jin Y, Chattopadhyay G, Mehdi I and Beaudin G 2010 Schottky diode based terahertz frequency multipliers and mixers *Compt. Rend. Phys* **11** 480–95

Mateos J, Vasallo B G, Pardo D and González T 2005 Operation and high-frequency performance of nanoscale unipolar rectifying diodes *Appl. Phys. Lett.* **86** 212103

Moddel G and Grover S (ed) 2013 Metal single-insulator and multi-insulator diodes for rectenna solar cells *Rectenna solar cells* (New York: Springer) pp 89–109

Moddel G, Zhu Z, Grover S and Joshi S 2012 Ultrahigh speed graphene diode with reversible polarity *Solid State Commun.* **152** 1842–5

Schiselski M, Laabs M, Neumann N, Kovalev S, Green B, Gensch M and Plettemeier D 2016 A planar Schottky diode based integrated THz detector for fast electron pulse diagnostics *2016 IEEE MTT-S Int. Microw. Symp. (IMS) (San Francisco, CA, 22–27 May)* (Piscataway, NJ: IEEE)

Sehgal B K, Balakrishnan V R, Gulati R and Tewari S P 2003 Analysis of electrical properties of Ti/Pt/Au Schottky contacts on (n)GaAs formed by electron beam deposition and RF sputtering *J. Semicond. Technol. Sci.* **3** 1–12

Song A M, Missous M, Omling P, Peaker A R, Samuelson L and Seifert W 2003 Unidirectional electron flow in a nanometer-scale semiconductor channel: a self-switching device *Appl. Phys. Lett.* **83** 1881–83

Sze S M and Ng K K 2007 Physics of seiconductor devices *Metal–Semiconductor Contacts* (Hoboken, NJ: Wiley) pp 134–96

Tyagi M S 1984 Physics of Schottky barrier junctions *Metal-semiconductor Schottky barrier junctions and their applications* ed B L Sharma (New York: Springer) pp 1–60

Wang W, An N, Tong X and Zeng J 2018 Fabrication and characterization of 8.87 THz schottky barrier mixer diodes for mixer *AIP Conf. Proc.* **1955** 040177

Westlund A, Sangare P, Ducournau G, Nilsson P-A, Gaquière C, Desplanque L, Wallart X and Grahn J 2013 Terahertz detection in zero-bias InAs self-switching diodes at room temperature *Appl. Phys. Lett.* **103** 133504-1–4

Yang L, Yao W, Liu Y, Wang L, Dai Y, Liu H, Wang F, Ren Y, Wu Z, Liu Y and Zhang B 2020 Low capacitance AlGaN/GaN based air-bridge structure planar Schottky diode with a half through-hole *AIP Adv* **10** 045219-1–5

Zhu Z, Grover S, Krueger K and Moddel G 2011 Optical rectenna solar cells using graphene geometric diodes *2011 37th IEEE Photovoltaic Specialists Conf.* 19–24 June, 2011 *(Seattle, WA)* pp 002120–2

Zhu Z, Joshi S, Grover S and Moddel G 2013 Graphene geometric diodes for terahertz rectennas *J. Phys. D: Appl. Phys* **46** 185101

IOP Publishing

Practical Terahertz Electronics: Devices and Applications, Volume 1

Solid-state devices and vacuum tubes

Vinod Kumar Khanna

Chapter 3

Resonant tunneling diodes

The resonant tunneling diode is represented by an equivalent circuit consisting of parallely-connected conductance and capacitance in series with a resistance, followed by derivation of the formula for its oscillation frequency. The maximum power delivered to the load is expressed in terms of the difference between peak and valley currents, the oscillation and maximum frequencies, and diode capacitance. The small-signal transit-time equivalent circuit model reveals the parameters which control the transit-time cut-off frequency. A physics-based small-signal equivalent circuit model is introduced, and the concepts of quantum conductance and inductance are defined. The advantages offered by InAs/AlSb material combination over GaAs/AlAs for fabricating resonant tunneling diodes are pointed out. Several resonant diode structures providing oscillation frequencies up to 1.98 THz are described, and the necessary material, process and design improvements implemented are discussed.

In this chapter, equivalent circuits of resonant tunneling diodes are studied for analyzing its high-frequency behavior, and the diode structures fabricated for terahertz frequencies are surveyed.

3.1 Resonant tunneling diode working and high-frequency capability

3.1.1 Structure and operation of the resonant tunneling diode

The configuration of a resonant tunneling diode is a double barrier quantum well (DBQW) structure (figure 3.1). In this DBQW structure, the barrier layers are made of a large bandgap material such as $Al_xGa_{1-x}As$ (bandgap increasing from 1.422 eV to 2.16 as x varies from 0 to 1) and the well is made of a small bandgap material like GaAs (bandgap = 1.422 eV). The DBQW structure is sandwiched between heavily doped GaAs layers on its two sides which are called the emitter and the collector.

doi:10.1088/978-0-7503-3171-5ch3

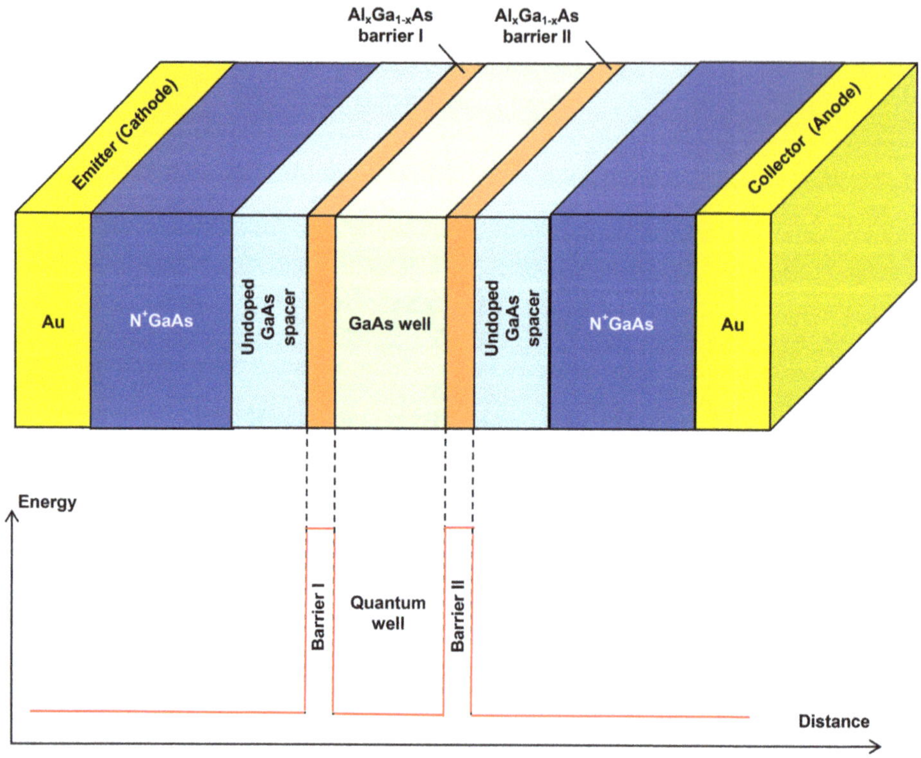

Figure 3.1. GaAs/AlGaAs resonant tunneling diode structure showing the sequential arrangement of the constituent layers. The barrier and quantum well layers are clearly marked in the energy band diagram drawn below the sketch.

In a quantum well, the electrons which have freedom of movement in three dimensions are restricted to two dimensions. This quantum confinement takes place when the thickness of the well layer becomes comparable to the De Broglie wavelength of electrons. It leads to discretization of energy levels and produces energy subbands. Hence, in a resonant tunneling diode there are dimensionality changes at the emitter/DBQW and DBQW/collector interfaces because of the existence of three-dimensional density of states in the emitter and collector and two-dimensional density of states in the quantum well (Sun *et al* 1998).

An electron in the emitter suffers multiple reflections at the emitter/DBQW interface due to its wave nature. Depending on its frequency, either constructive or destructive interference occurs. The transmission probability of a particular electron having a specific frequency becomes unity if constructive interference is favored at the energy of this frequency. Stated in other words, resonant tunneling takes place. The corresponding energy levels are known as resonant energy levels. Broadening mechanisms impart finite widths to these resonant levels.

Figure 3.2 displays the current–voltage characteristics of a resonant tunneling diode. We shall try to explain the reason for the peculiar shape of its characteristics and also the cause for the negative resistance shown by the diode.

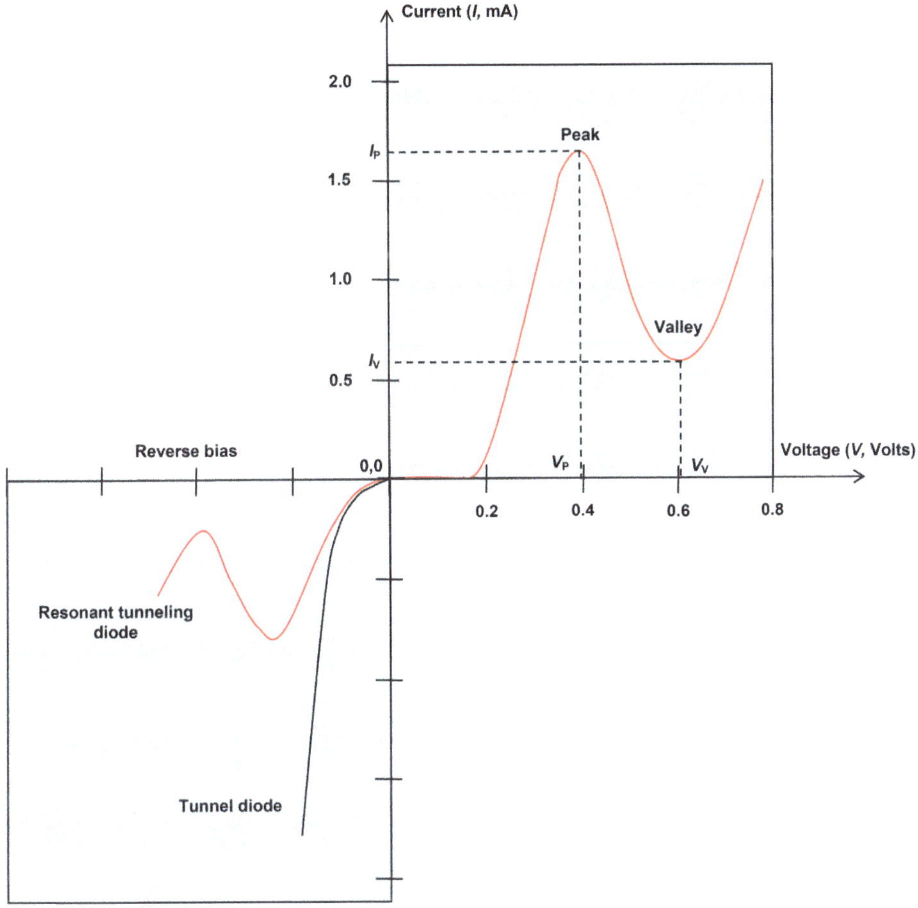

Figure 3.2. The current–voltage characteristics of resonant tunneling diode under forward and reverse biasing; the reverse-bias characteristic of a tunnel diode is also shown for comparison. The reverse current of a resonant tunneling diode is less than that of a tunnel diode.

In figure 3.3, a resonant tunneling diode is considered. Across this diode an external voltage is applied. At low levels of applied bias, the Fermi level in the emitter is below the resonant level in the well region. So the diode allows only a small current to flow through it. As the applied bias is increased, the Fermi level in the emitter rises up, moves closer to and aligns with the resonant energy level in the well. Consequently, the number of electrons tunneling across the well increases and so does the current. The current continues to rise and reaches the maximum value when the edge of the conduction band in the emitter aligns with the resonant energy level in the well. This is the peak current in the diode. Subsequently, further increase of applied voltage raises the conduction band edge in the emitter above the resonant energy level in the well so that its alignment with the resonant energy level is disturbed to greater extents. Now the current flowing in the diode registers a decrease, i.e., the current decreases as the applied bias increases. Thus a negative resistance region is observed. The minimum current obtained is known as the valley current. The ratio of peak and valley currents is referred to as the peak-to-valley ratio. It is an important figure of merit serving as a benchmark for diode performance (Sugiyama *et al* 2011).

Beyond the valley current, further upliftment of applied bias raises the edge of conduction band still more until alignment is achieved with the higher-located resonant energy level in the well. Then the current once again increases with applied bias.

Besides the GaAs/AlGaAs structure discussed above, two other examples of material systems used for fabrication of resonant tunneling diodes include InGaAs/AlAs and InAs/AlSb (figure 3.4). In the former, the quantum wells are made of InGaAs ($E_g = 0.75$ eV) and the barrier layers from AlAs ($E_g = 2.16$ eV). The latter has InAs ($E_g = 0.43$ eV) wells with AlSb ($E_g = 1.615$ eV) cladding layers for barrier formation.

3.1.2 Reason for high switching speed of resonant tunneling diode

The resonant tunneling diode (RTD) owes its fast speed to the fact that it is a device based on quantum-mechanical tunneling and tunneling is an intrinsically fast process taking place in a time interval ~10^{-15} s (Feiginov 2019). The time spent in tunneling through the double barrier structure of a resonant tunneling diode is in the range of 10^{-12} s or less. Apart from the tunneling time, the time required for charging of the extremely small capacitance of this diode at high current density $>1 \times 10^5$ A cm^{-2} is also short, enabling faster switching. This behavior clearly distinguishes the resonant tunneling diode from the Esaki tunnel diode. In the latter

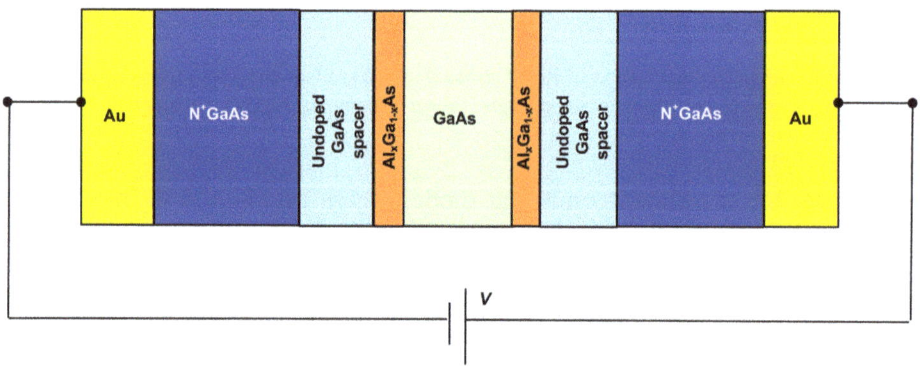

(a)

Figure 3.3. Explanation of the origin of current–voltage characteristics of a resonant tunneling diode in forward-biased mode in terms of energy-band model: (a) a forward-biased GaAs/AlGaAs resonant tunneling diode with negative voltage applied to emitter and positive voltage to collector; (b), (c), (d) and (e): the shifting of conduction band edge with increasing applied bias to create different regions of the *I–V* curve:(b) when the Fermi level in the emitter $E_{F(Emitter)}$ is below the resonant energy level E_0, (c) when $E_{F(Emitter)}$ aligns with E_0, (d) when $E_{C(Emitter)}$ aligns with E_0, and (e) when $E_{C(Emitter)}$ moves upward beyond E_0.

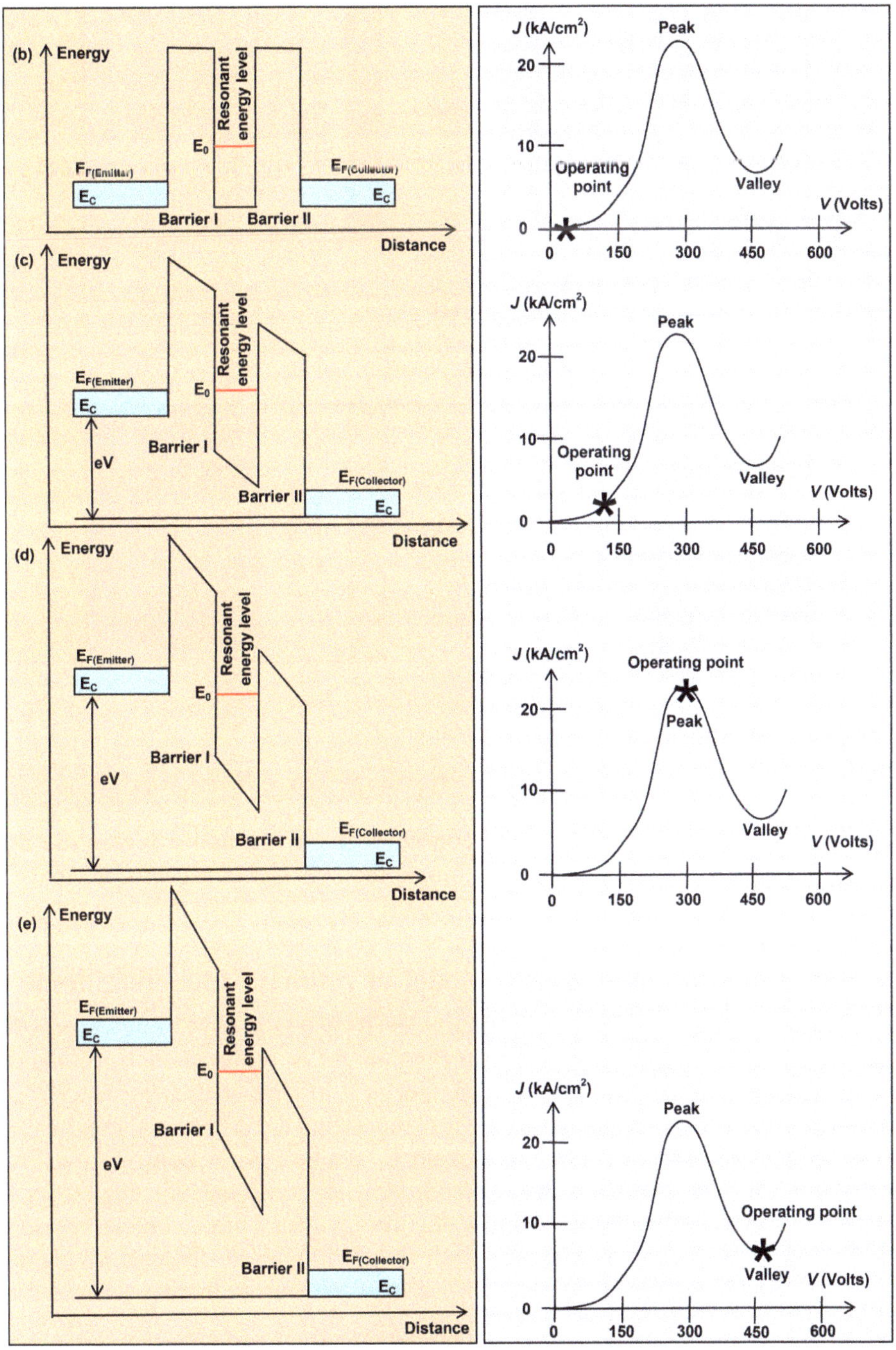

Figure 3.3. (Continued.)

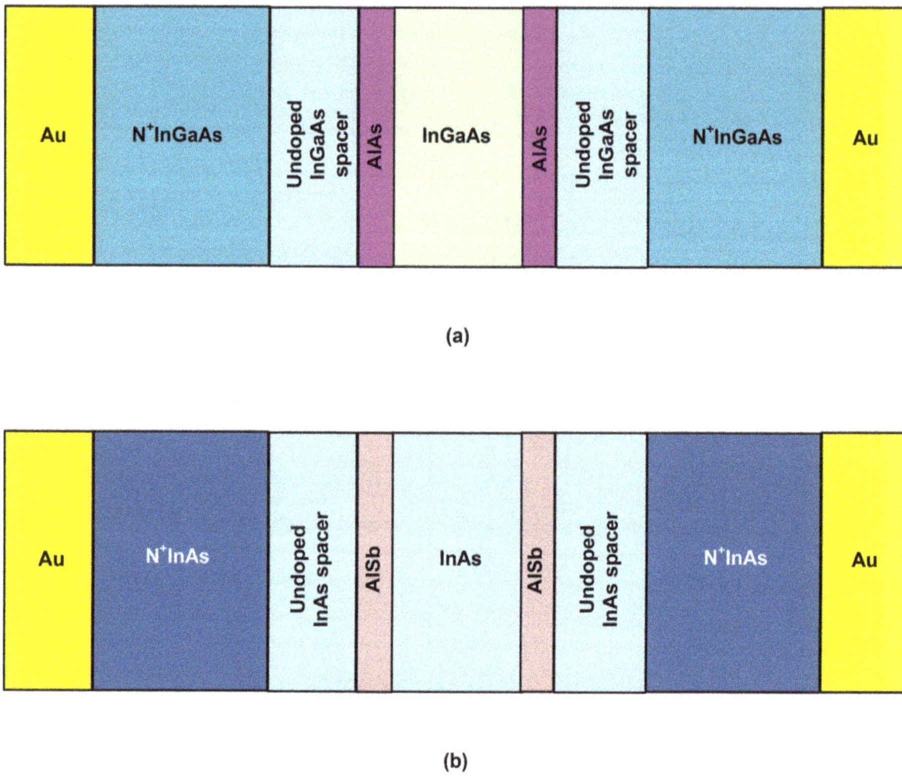

Figure 3.4. Partitions in the structures of resonant tunneling diodes: (a) InGaAs/AlAs and (b) InAs/AlSb.

diode, the current density and doping concentration in the active region are both high. As a result the capacitance of the tunnel diode is high and hence its charging time is long.

3.2 Simplest equivalent circuit model of resonant tunneling diode

It is a lumped element equivalent circuit model (Brown *et al* 1991, Yugvesson 1991, Alkeev *et al* 2004). The resonant tunneling diode is represented by a parallel combination of conductance $-G$ and capacitance C_0, in series with a resistance R_S, as shown in figure 3.5(a). The circuit element G represents the differential conductance across the active region of RTD, the element C_0 symbolizes the contributions from the accumulation layer on the emitter (cathode) side and depletion region on the collector (anode) side while the element R_S accounts for ohmic contact resistance together with the spreading resistance from the RTD mesa to the ground plane of the device. In the circuit of figure 3.5(b), a load with an inductive component is included.

3.2.1 Oscillation frequency formula

Reactance of capacitor C_0 is

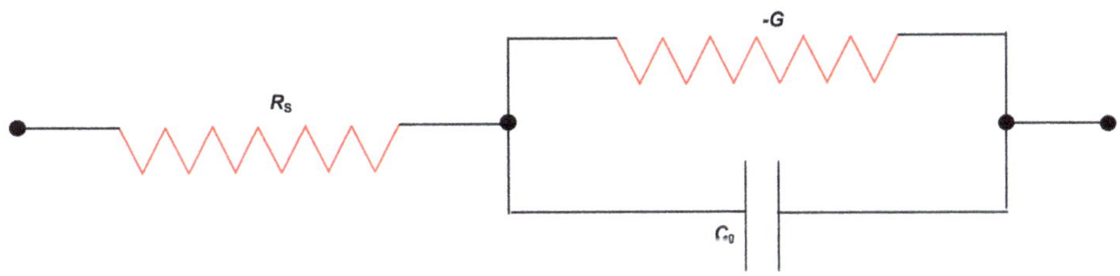

(a)

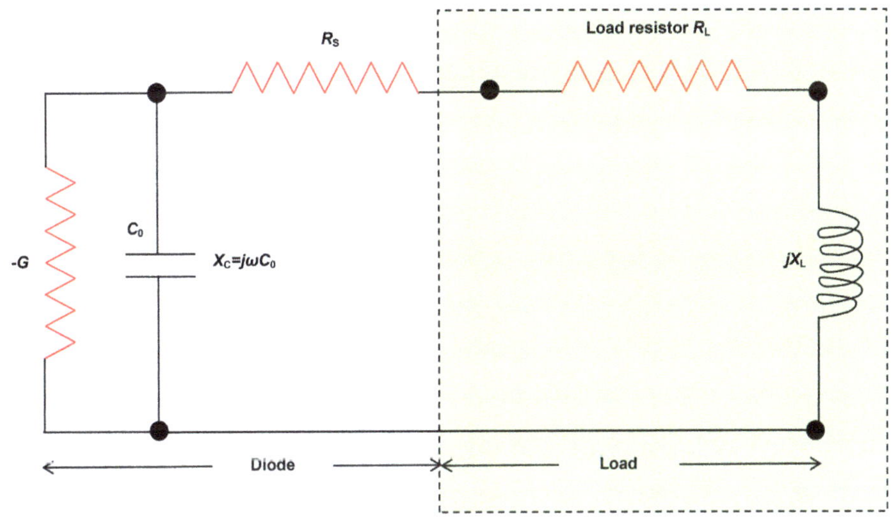

(b)

Figure 3.5. Electrical equivalent circuit of a resonant tunneling diode: (a) elementary equivalent circuit and (b) equivalent circuit including a load with an inductive component.

$$X_C = \frac{1}{j\omega C_0} = \frac{j}{j^2\omega C_0} = -\frac{j}{\omega C_0} \tag{3.1}$$

Since the impedance of components $-1/G$ and $-j/\omega C_0$ connected in parallel is given by

$$\frac{1}{Z} = -G - \frac{\omega C_0}{j} = -\frac{jG + \omega C_0}{j} \tag{3.2}$$

or,

$$Z = -\frac{j}{jG + \omega C_0} \tag{3.3}$$

the complex input impedance of resonant tunneling diode is expressed as

$$\begin{aligned} Z_{in} &= R_L + R_S - Z = R_L + R_S - \frac{j}{jG + \omega C_0} = R_L + R_S - \frac{j(-jG + \omega C_0)}{(jG + \omega C_0)(-jG + \omega C_0)} \\ &= R_L + R_S - \frac{G + j\omega C_0}{G^2 + \omega^2 C_0^2} = R_L + R_S - \frac{G}{G^2 + \omega^2 C_0^2} - \frac{j\omega C_0}{G^2 + \omega^2 C_0^2} \end{aligned} \tag{3.4}$$

At the oscillation frequency $\omega = \omega_{osc}$, the real part of complex input impedance is zero. Hence,

$$R_L + R_S - \frac{G}{G^2 + \omega_{Osc}^2 C_0^2} = 0 \tag{3.5}$$

or,

$$R_L + R_S = \frac{G}{G^2 + \omega_{Osc}^2 C_0^2} \tag{3.6}$$

or,

$$(R_L + R_S)\left(G^2 + \omega_{Osc}^2 C_0^2\right) = G \tag{3.7}$$

or,

$$G^2 + \omega_{Osc}^2 C_0^2 = \frac{G}{R_L + R_S} \tag{3.8}$$

or,

$$\omega_{Osc}^2 C_0^2 = \frac{G}{R_L + R_S} - G^2 \tag{3.9}$$

or,

$$\omega_{Osc}^2 = \frac{1}{C_0^2}\left(\frac{G}{R_L + R_S} - G^2\right) \tag{3.10}$$

or,

$$\omega_{Osc} = \frac{1}{C_0}\sqrt{\frac{G}{R_L + R_S} - G^2} \tag{3.11}$$

or,

$$2\pi f_{Osc} = \frac{1}{C_0}\sqrt{\frac{G}{R_L + R_S} - G^2} \tag{3.12}$$

or,

$$f_{Osc} = \frac{1}{2\pi C_0}\sqrt{\frac{G}{R_L + R_S} - G^2} \tag{3.13}$$

The oscillation frequency will increase in the extreme case

$$R_L = 0 \tag{3.14}$$

Then we get

$$f_{Osc} = \frac{1}{2\pi C_0}\sqrt{\frac{G}{R_S} - G^2} \tag{3.15}$$

3.2.2 Maximum oscillation frequency formula

To get maximum oscillation frequency we set the derivative of f_{osc} with respect to G to zero getting

$$\frac{df_{Osc}}{dG} = \frac{1}{2\pi C_0} \times \frac{1}{2}\frac{1}{\sqrt{\frac{G}{R_S} - G^2}} \times \left(\frac{1}{R_S} - 2G\right) = 0 \tag{3.16}$$

It is only possible that

$$\frac{1}{R_S} - 2G = 0 \tag{3.17}$$

because remaining factors $\neq 0$. Hence

$$\frac{1}{R_S} = 2G \tag{3.18}$$

or,

$$G = \frac{1}{2R_S} \tag{3.19}$$

Substituting for G from equation (3.19) in equation (3.15) we have

$$\begin{aligned} f_{max} = \left|f_{Osc}\right|_{max} &= \frac{1}{2\pi C_0}\sqrt{\frac{\frac{1}{2R_S}}{R_S} - \left(\frac{1}{2R_S}\right)^2} = \frac{1}{2\pi C_0}\sqrt{\frac{1}{2R_S^2} - \frac{1}{4R_S^2}} \\ &= \frac{1}{2\pi C_0}\sqrt{\frac{2-1}{4R_S^2}} = \frac{1}{2\pi C_0}\sqrt{\frac{1}{4R_S^2}} = \frac{1}{2\pi C_0} \times \frac{1}{2R_S} = \frac{1}{4\pi C_0 R_S} \end{aligned} \tag{3.20}$$

Alternatively, equation (3.10) is rearranged as

$$G^2 - \frac{G}{R_L + R_S} + \omega_{Osc}^2 C_0^2 = 0 \tag{3.21}$$

which is a quadratic equation in G with the solution

$$\begin{aligned} G &= \frac{\frac{1}{R_L + R_S} \pm \sqrt{\left(-\frac{1}{R_L + R_S}\right)^2 - 4 \times 1 \times \omega_{Osc}^2 C_0^2}}{2 \times 1} \\ &= \frac{1}{2(R_L + R_S)} \pm \frac{1}{2}\sqrt{\frac{1}{(R_L + R_S)^2} - 4\omega_{Osc}^2 C_0^2} \\ &= \frac{1}{2(R_L + R_S)} \pm \sqrt{\frac{1}{4(R_L + R_S)^2} - \frac{4}{4}\omega_{Osc}^2 C_0^2} \\ &= \frac{1}{2(R_L + R_S)} \pm \sqrt{\frac{1}{4(R_L + R_S)^2} - \omega_{Osc}^2 C_0^2} \\ &= \frac{1}{2(R_L + R_S)} \pm \sqrt{\frac{1 - 4\omega_{Osc}^2 C_0^2 (R_L + R_S)^2}{4(R_L + R_S)^2}} \\ &= \frac{1}{2(R_L + R_S)} \pm \frac{1}{2(R_L + R_S)}\sqrt{1 - 4(R_L + R_S)^2 \omega_{Osc}^2 C_0^2} \end{aligned} \tag{3.22}$$

Choosing the negative sign to obtain negative G,

$$\begin{aligned} G &= \frac{1}{2(R_L + R_S)}\left(1 - \sqrt{1 - 4(R_L + R_S)^2 \omega_{Osc}^2 C_0^2}\right) \\ &= \frac{1}{2(R_L + R_S)}\left[1 - \sqrt{1 - \{2(R_L + R_S)\omega_{Osc} C_0\}^2}\right] \end{aligned} \tag{3.23}$$

In order that G does not acquire an imaginary value it is necessary that

$$1 - \{2(R_L + R_S)\omega_{Osc} C_0\}^2 > 0 \tag{3.24}$$

or,

$$\{2(R_L + R_S)\omega_{Osc} C_0\}^2 < 1 \tag{3.25}$$

or,

$$2(R_L + R_S)\omega_{Osc} C_0 < 1 \tag{3.26}$$

or,

$$\omega_{Osc} < \frac{1}{2(R_L + R_S)C_0} < \frac{1}{2R_S C_0} \tag{3.27}$$

or,

$$f_{Osc} < \frac{1}{2\pi} \times \frac{1}{2R_S C_0} < \frac{1}{4\pi R_S C_0} \tag{3.28}$$

So, the maximum frequency must be

$$f_{\text{Max}} = \frac{1}{4\pi R_S C_0} \tag{3.29}$$

Example 3.1

(a) Find the oscillation frequency of a resonant tunneling diode of area 1.25×10^{-7} cm^2, thickness 700 Å and negative conductance 50 mS. The specific contact resistance is 2.5×10^{-7} Ω-cm^2. The spreading resistance for the circular contact may be approximated with the resistivity 0.0007 Ω-cm. A uniform dielectric constant of 12.9 may be assumed for all the layers.

(b) Describe the circuit by an RC time constant and find the value of this time constant.

(c) What is the maximum oscillation frequency?

Solution:

(a)

$$\text{Contact resistance} = \frac{\text{Specific contact resistance}}{\text{Area}} = \frac{2.5 \times 10^{-7}}{1.25 \times 10^{-7}} = 2\,\Omega \tag{3.30}$$

$$\text{Area} = \pi r^2 = 1.25 \times 10^{-7}\text{cm}^2 \tag{3.31}$$

or,

$$r^2 = \frac{1.25 \times 10^{-7}}{3.141\,59} \tag{3.32}$$

or,

$$r = \sqrt{\frac{1.25 \times 10^{-7}}{3.141\,59}} = 1.995 \times 10^{-4}\text{ cm} \tag{3.33}$$

Spreading resistance

$$R_{\text{Spr}} = \frac{\text{Resistivity}}{2 \times \text{Radius of contact}} = \frac{\rho}{2r} = \frac{0.0007\,\Omega - \text{cm}}{2 \times 1.995 \times 10^{-4}\text{ cm}} = 1.754\,\Omega \tag{3.34}$$

$$R_S = (2 + 1.754)\Omega = 3.754\,\Omega \tag{3.35}$$

$$C_0 = \frac{\epsilon_0 \epsilon A}{t} = \frac{8.854 \times 10^{-12} \times 12.9 \times 1.25 \times 10^{-7} \times 10^{-4}}{700 \times 10^{-10}} = 2.0396 \times 10^{-14}\text{ F} = 20.396\text{ fF} \tag{3.36}$$

$$f_{\text{Osc}} = \frac{1}{2\pi C_0}\sqrt{\frac{G}{R_S} - G^2} = \frac{1}{2 \times 3.141\,59 \times 2.0396 \times 10^{-14}}$$
$$\sqrt{\frac{50 \times 10^{-3}}{3.754} - (50 \times 10^{-3})^2} = 0.078 \times 10^{14} \quad (3.37)$$
$$\times \sqrt{1.332 \times 10^{-2} - 2.5 \times 10^{-3}} = 7.8 \times 10^{12} \times 0.104 = 0.8112$$
$$\times 10^{12}\ \text{Hz} = 0.81\ \text{THz}$$

(b) Equation (3.15) can be written as

$$f_{\text{Osc}} = \frac{1}{2\pi C_0 R_0} \quad (3.38)$$

where

$$R_0 = \frac{1}{\sqrt{\frac{G}{R_S} - G^2}} \quad (3.39)$$

The circuit is described by the time constant

$$\tau = R_0 C_0 = \frac{C_0}{\sqrt{\frac{G}{R_S} - G^2}} = \frac{2.0396 \times 10^{-14}}{\sqrt{\frac{50 \times 10^{-3}}{3.754} - (50 \times 10^{-3})^2}} = \frac{2.0396 \times 10^{-14}}{0.104}$$
$$= 19.6 \times 10^{-14}\ \text{s} = \frac{19.6 \times 10^{-14}}{10^{-15}}\text{fs} = 196\ \text{fs} \quad (3.40)$$

(c) Applying equation (3.29)

$$f_{\text{Max}} = \frac{1}{4\pi R_S C_0} = \frac{1}{4 \times 3.141\,59 \times 3.754 \times 2.0396 \times 10^{-14}} = 1.039 \times 10^{12}\ \text{Hz} \quad (3.41)$$

Example 3.2 From the condition for peak value of frequency find the series resistance of an RTD having a negative conductance of 70 mS.

Solution: By equation (3.19), the condition for maximum frequency is

$$G = \frac{1}{2R_S} \quad (3.42)$$

from which

$$R_S = \frac{1}{2G} = \frac{1}{2 \times 70 \times 10^{-3}} = 7.14\ \Omega \quad (3.43)$$

3.3 Maximum output power conveyed to the load resistor R_L

The power delivered to load resistance R_L is calculated as

$$
\begin{aligned}
P_{Del} &= \text{RMS Current} \times \text{RMS Voltage} \\
&= \frac{\frac{\Delta I}{2}}{\sqrt{2}}\frac{\frac{\Delta V}{2}}{\sqrt{2}}\cos 0^\circ \times \text{Factor representing load resistance effect} \\
&\quad \times \text{Factor representing negative conductance effect} \\
&= \frac{1}{2}\frac{\Delta I}{2} \times \frac{\Delta V}{2} \times \text{Factor representing load resistance effect} \\
&\quad \times \text{Factor representing negative conductance effect} \\
&= \frac{1}{2}\frac{\Delta I}{2} \times \frac{\Delta I}{2} R_L \\
&\quad \times \text{Fraction of total voltage which is dropped across load resistor} R_L \\
&\quad \times \text{Factor by which power is increased by negative conductance} \\
&= \frac{1}{2}\frac{\Delta I}{2} \times \frac{\Delta I}{2} R_L \times \frac{R_L}{R_L + R_S} \\
&\quad \times \text{Factor by which power is increased by negative conductance} \\
&= \frac{1}{2}\frac{\Delta I}{2} \times \frac{\Delta I}{2} R_L \times \frac{R_L}{R_L + R_S} \times \frac{1}{GR_L} = \frac{\Delta I^2}{8} \times \frac{R_L}{R_L + R_S} \times \frac{1}{G} \\
&= \frac{\Delta I^2}{8} \times \frac{R_L}{R_L + R_S} \times \frac{2(R_L + R_S)}{1 - \sqrt{1 - \{2(R_L + R_S)\omega_{Osc} C_0\}^2}} \\
&= \frac{\Delta I^2}{4} \times \frac{R_L}{1 - \sqrt{1 - \{2(R_L + R_S)\omega_{Osc} C_0\}^2}}
\end{aligned}
\tag{3.44}
$$

where

$$\Delta I = \text{Peak current} - \text{Valley current} = I_P - I_V \tag{3.45}$$

$$\Delta V = \text{Peak voltage} - \text{Valley voltage} = V_P - V_V \tag{3.46}$$

and

$$
\begin{aligned}
\frac{\Delta I}{2} &= \text{Amplitude of the high-frequency current}, \\
\frac{\Delta V}{2} &= \text{Amplitude of the high-frequency voltage}
\end{aligned}
\tag{3.47}
$$

for the diode. The phase lag $\cos\phi = \cos 0^\circ = 1$ is zero because the load is purely resistive. Equation (3.23) for G has been used. Also, for resistive load, the relation

$$P_{\mathrm{Del}} = I_{\mathrm{RMS}} V_{\mathrm{RMS}} = \frac{I_{\mathrm{Peak}}}{\sqrt{2}} \frac{V_{\mathrm{Peak}}}{\sqrt{2}} = \frac{1}{2} I_{\mathrm{Peak}} V_{\mathrm{Peak}} = \frac{1}{2} I_{\mathrm{Peak}} I_{\mathrm{Peak}} R = \frac{1}{2} I_{\mathrm{Peak}}^2 R \\ = \frac{1}{2} I_{\mathrm{Peak}}^2 R_{\mathrm{L}} \times \frac{R_{\mathrm{L}}}{R_{\mathrm{L}} + R_{\mathrm{S}}} \tag{3.48}$$

is applied remembering that $I_{\mathrm{peak}} = \Delta I/2$, $V_{\mathrm{Peak}} = \Delta V/2$. Further, for negative resistance/conductance, the formula including the factor by which the power increases due to negative conductance is

$$P_{\mathrm{Del}} = P_{\text{Without negative resistance}} \times \frac{\text{Negative resistance}}{\text{Load resistance}} \tag{3.49}$$

or,

$$P_{\mathrm{Del}} = P_{\text{Without negative conductance}} \times \frac{1}{\text{Negative conductance} \times \text{Load resistance}} \\ = P_{\text{Without negative conductance}} \times \frac{1}{GR_{\mathrm{L}}} \tag{3.50}$$

The maximum power is delivered when

$$1 - \{2(R_{\mathrm{L}} + R_{\mathrm{S}})\omega_{\mathrm{Osc}} C_0\}^2 = 0 \tag{3.51}$$

or,

$$\{2(R_{\mathrm{L}} + R_{\mathrm{S}})\omega_{\mathrm{Osc}} C_0\}^2 = 1 \tag{3.52}$$

or,

$$2(R_{\mathrm{L}} + R_{\mathrm{S}})\omega_{\mathrm{Osc}} C_0 = 1 \tag{3.53}$$

or,

$$\omega_{\mathrm{Osc}} = \frac{1}{2(R_{\mathrm{L}} + R_{\mathrm{S}})C_0} \tag{3.54}$$

or,

$$f_{\mathrm{Osc}} = \frac{1}{2 \times 2\pi(R_{\mathrm{L}} + R_{\mathrm{S}})C_0} = \frac{1}{4\pi(R_{\mathrm{L}} + R_{\mathrm{S}})C_0} \tag{3.55}$$

or,

$$R_{\mathrm{L}} + R_{\mathrm{S}} = \frac{1}{4\pi C_0 f_{\mathrm{Osc}}} \tag{3.56}$$

From equation (3.44), the maximum power is

$$P_{\mathrm{Max}} = \frac{\Delta I^2}{4} \times \frac{R_{\mathrm{L}}}{1 - \sqrt{1 - \left\{2\left(\frac{1}{4\pi C_0 f_{\mathrm{Osc}}}\right)\omega_{\mathrm{Osc}} C_0\right\}^2}}$$

$$= \frac{\Delta I^2}{4} \times \frac{R_{\mathrm{L}}}{1 - \sqrt{1 - \left\{2\left(\frac{1}{2C_0\omega_{\mathrm{Osc}}}\right)\omega_{\mathrm{Osc}} C_0\right\}^2}} \tag{3.57}$$

$$= \frac{\Delta I^2}{4} \times \frac{R_{\mathrm{L}}}{1 - \sqrt{1 - 1}} = \frac{\Delta I^2}{4} \times \frac{R_{\mathrm{L}}}{1 - 0} = \frac{\Delta I^2 R_{\mathrm{L}}}{4}$$

This equation can be put in the form

$$P_{\mathrm{Max}} = \frac{\Delta I^2}{4}\{(R_{\mathrm{L}} + R_{\mathrm{S}}) - R_{\mathrm{S}}\} = \frac{\Delta I^2}{4}\left\{\frac{1}{4\pi C_0 f_{\mathrm{Osc}}} - \frac{1}{4\pi C_0 f_{\mathrm{Max}}}\right\} \tag{3.58}$$

by applying equations (3.56) and (3.29). Equation (3.58) can be re-written as

$$P_{\mathrm{Max}} = \frac{\Delta I^2}{4 \times 4\pi C_0}\left(\frac{1}{f_{\mathrm{Osc}}} - \frac{1}{f_{\mathrm{Max}}}\right) = \frac{\Delta I^2}{16\pi C_0}\left(\frac{1}{f_{\mathrm{Osc}}} - \frac{1}{f_{\mathrm{Max}}}\right) \tag{3.59}$$

The optimum value of load resistor R_{L} to obtain maximum power is obtained from equation (3.57) as

$$R_{\mathrm{L}} = \frac{4P_{\mathrm{Max}}}{\Delta I^2} \tag{3.60}$$

Substituting for P_{Max} from equation (3.59) in equation (3.60) we have

$$R_{\mathrm{L}} = \frac{4P_{\mathrm{Max}}}{\Delta I^2} = \frac{\frac{4\Delta I^2}{16\pi C_0}\left(\frac{1}{f_{\mathrm{Osc}}} - \frac{1}{f_{\mathrm{Max}}}\right)}{\Delta I^2} = \frac{1\Delta I^2}{4\pi C_0}\left(\frac{1}{f_{\mathrm{Osc}}} - \frac{1}{f_{\mathrm{Max}}}\right) \times \frac{1}{\Delta I^2} \tag{3.61}$$

$$= \frac{1}{4\pi C_0}\left(\frac{1}{f_{\mathrm{Osc}}} - \frac{1}{f_{\mathrm{Max}}}\right)$$

Example 3.3

(a) If the series resistance of a resonant tunneling diode is 5 Ω and the geometrical capacitance is 100 fF, what is its maximum frequency of oscillation? (b) The diode has a peak current of 2.6 mA and valley current of 1.3 mA. If the current is oscillating at a frequency of 100 MHz, find the maximum power conveyed to the load? (c) Calculate the maximum power at a frequency of 3 GHz. (d) Repeat the maximum power calculation for 80 GHz.

Solution: (a) Here $R_S = 5\ \Omega$, $C_0 = 100$ fF Applying equation (3.29).

$$f_{\text{Max}} = \frac{1}{4\pi R_S C_0} = \frac{1}{4 \times 3.141\,59 \times 5 \times 100 \times 10^{-15}} = 0.0159 \times 10^{13} \\ = 0.159 \times 10^{12}\ \text{Hz} = 0.159\ \text{THz} \tag{3.62}$$

(b) Given:

$$\Delta I = I_P - I_V = (2.6\text{–}1.3) \times 10^{-3} = 1.3 \times 10^{-3}\ \text{A} \tag{3.63}$$

$$f_{\text{Osc}} = 100\ \text{MHz} = 100 \times 10^6\ \text{Hz} = 10^8\ \text{Hz} \tag{3.64}$$

Hence, by equation (3.59)

$$\begin{aligned} P_{\text{Max}} &= \frac{\Delta I^2}{16\pi C_0}\left(\frac{1}{f_{\text{Osc}}} - \frac{1}{f_{\text{Max}}}\right) \\ &= \frac{(1.3 \times 10^{-3})^2}{16 \times 3.141\,59 \times 100 \times 10^{-15}}\left(\frac{1}{1 \times 10^8} - \frac{1}{0.159 \times 10^{12}}\right) \\ &= \frac{1.69 \times 10^7}{50.265}(1 \times 10^{-8} - 6.289 \times 10^{-12}) = 3.362 \times 10^5 \\ &\quad \times 10^{-8}(1 - 6.289 \times 10^{-4}) = 3.362 \times 10^{-3} \times 0.999\,37 \\ &= 3.36 \times 10^{-3}\ \text{W} = 3.36 \times 10^{-3} \times 10^3\ \text{mW} = 3.36\ \text{mW} \end{aligned} \tag{3.65}$$

(c) Given:

$$f_{\text{Osc}} = 3\ \text{GHz} = 3 \times 10^9\ \text{Hz} \tag{3.66}$$

Hence, from equation (3.59)

$$\begin{aligned} P_{\text{Max}} &= \frac{\Delta I^2}{16\pi C_0}\left(\frac{1}{f_{\text{Osc}}} - \frac{1}{f_{\text{Max}}}\right) = 3.362 \times 10^5\left(\frac{1}{3 \times 10^9} - \frac{1}{0.159 \times 10^{12}}\right) \\ &= 3.362 \times 10^5(0.33 \times 10^{-9} - 6.289 \times 10^{-12}) = 3.362 \times 10^5 \\ &\quad \times 10^{-9}(0.33\text{–}6.289 \times 10^{-3}) = 3.362 \times 10^{-4} \times 0.3237 \\ &= 1.088 \times 10^{-4}\ \text{W} = 1.088 \times 10^{-4} \times 10^3\ \text{mW} = 0.109\ \text{mW} \end{aligned} \tag{3.67}$$

(d) Since

$$f_{\text{Osc}} = 80\ \text{GHz} = 80 \times 10^9\ \text{Hz} = 8 \times 10^{10}\ \text{Hz} \tag{3.68}$$

equation (3.59) gives

$$\therefore P_{\text{Max}} = \frac{\Delta I^2}{16\pi C_0}\left(\frac{1}{f_{\text{Osc}}} - \frac{1}{f_{\text{Max}}}\right) = 3.362 \times 10^5\left(\frac{1}{8 \times 10^{10}} - \frac{1}{0.159 \times 10^{12}}\right)$$
$$= 3.362 \times 10^5(0.125 \times 10^{-10} - 6.289 \times 10^{-12}) = 3.362 \times 10^5 \tag{3.69}$$
$$\times 10^{-10}(0.125 - 6.289 \times 10^{-2}) = 3.362 \times 10^{-5} \times 0.062 = 0.2084$$
$$\times 10^{-5}\,\text{W} = 2.084 \times 10^{-6} \times 10^6\,\mu\text{W} = 2.084\,\mu\text{W}$$

3.4 Small-signal transit-time equivalent circuit model of RTD

Figure 3.6 shows the quantum well injection transit time (QWITT) diode. Its equivalent circuit model is also shown. Substantial transit time effects are shown by this diode with an anode-side depletion layer.

Recalling the structure of the resonant tunneling diode, we note that there are two GaAs spacers, one on the cathode side and the other on the anode side of the quantum well. In the GaAs/AlAs diode structure shown (Kesan *et al* 1988), the

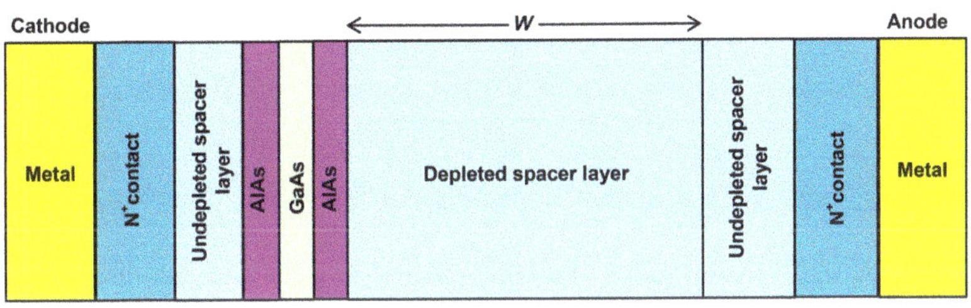

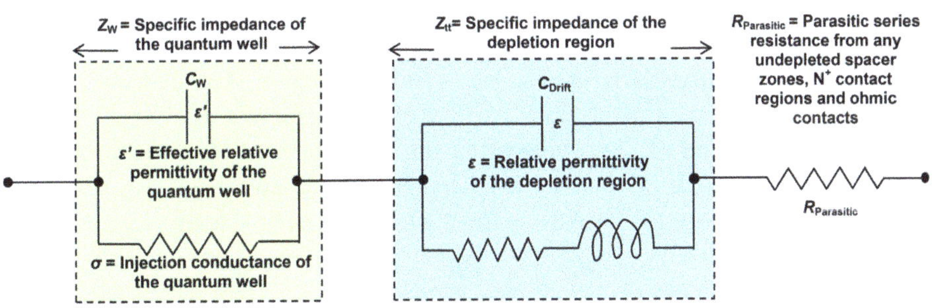

Figure 3.6. Resonant tunneling diode with significant transit-time effect: (a) structure of the diode and (b) its small-signal equivalent circuit.

GaAs spacer on the cathode side of the quantum well is made very thin. This helps to decrease any parasitic series resistance accrued from it. At the same time, the width of the GaAs spacer on the anode side of the quantum well is taken sufficiently large to form a thick depletion layer having thickness W. Then the transit time of carriers through this depletion region is much longer than through the quantum well region. Since the quantum well length (L)« thickness of depletion region (W), electron transport through the quantum well is considered instantaneous. The quantum well is treated as an electron injecting cathode. It is represented by a capacitance C_W in parallel with a normalized conductance σ defined as

$$\sigma = \left| L\left(\frac{\partial J_W}{\partial V_W}\right)\right|_{V_0} \tag{3.70}$$

where J_W is the instantaneous current density flowing through the quantum well, V_W is the instantaneous voltage across the quantum well and V_0 is the DC bias. If ε_0 is permittivity of vacuum, ε is the dielectric constant of the well region, the specific impedance of the quantum well region at an angular frequency ω is

$$Z_W = \frac{L}{\sigma + j\omega\varepsilon_0\varepsilon'} \tag{3.71}$$

whose real part gives the specific resistance of quantum well region

$$R_W = \frac{L/\sigma}{1 + \left(\frac{\omega}{\sigma/\varepsilon_0\varepsilon'}\right)^2} \tag{3.72}$$

Similarly for the depletion region, the transit-time negative resistance obtained from the real part of specific impedance Z_{tt} is

$$R_{tt} = \frac{2\left(\frac{v_S\varepsilon_0\varepsilon}{2\sigma^2}\right)}{\left(\frac{\omega}{\sigma/\varepsilon_0\varepsilon}\right)^2\left\{1 + \left(\frac{\omega}{\sigma/\varepsilon_0\varepsilon}\right)^2\right\}} \times \left\{1 - \cos\left(\frac{\omega W}{v_S}\right) + \left(\frac{\omega}{\sigma/\varepsilon_0\varepsilon}\right)\sin\left(\frac{\omega W}{v_S}\right)\right\} \tag{3.73}$$

where v_S is the saturated electron velocity in the drift region. The dielectric constant of drift region is assumed to be same as that of well region ($\varepsilon = \varepsilon'$).

Equations (3.72) and (3.73) representing the resistances of the quantum well and transit-time depleted drift regions are the governing equations of this model. Differentiating equation (3.73) with respect to W and σ, i.e., from

$$\frac{\partial R_{tt}}{\partial W} = 0, \quad \frac{\partial R_{tt}}{\partial \sigma} = 0 \tag{3.74}$$

an optimum thickness of depletion region (W_{Optimum}) and an optimum injection conductance (σ_{Optimum}) are found. These optimal quantities provide maximum small-signal negative resistance of RTD. The analysis reveals that the maximum negative resistance is inversely proportional to the square of frequency ω and directly proportional to carrier velocity v_S. It is also found that at low frequencies $\omega \ll |\sigma/\varepsilon_0\varepsilon|$, the negative resistance decreases with increasing injection conductance of the quantum well, whereas at high frequencies $\omega \gg |\sigma/\varepsilon_0\varepsilon|$, the opposite behavior is observed. Focusing on a device with negative σ, at low frequencies the transit-time negative resistance is broadband and frequency independent. The optimum depletion layer thickness too is independent of frequency. But at high frequencies, keeping W and σ fixed, the negative resistance declines as cube of frequency. Eventually, it changes sign from negative to positive at a particular frequency given by

$$f_{\text{Cut-off}} \cong \frac{v_S}{4\pi W}\left(\pi + \sqrt{\pi^2 - 8\frac{|\sigma|\,W}{\varepsilon_0\varepsilon v_S}}\right) \tag{3.75}$$

At this frequency called the cut-off frequency, the oscillations cease. Equation (3.75) shows that for an RTD, the intrinsic transit-time cut-off frequency $f_{\text{Cut-off}}$ is determined by the saturated carrier drift velocity v_S, the width of the depletion region W, and the injection conductance of the quantum well σ (Kesan *et al* 1988).

Example 3.4 Find the cut-off frequency of an RTD in which the injection conductance is $-0.3\ \Omega^{-1}\text{cm}^{-1}$. Take the high-frequency dielectric constant of RTD material as 10.89 and carrier drift velocity as $6 \times 10^6\ \text{cm s}^{-1}$. The depletion region is 70 nm thick.

Solution: From equation (3.75)

$$\begin{aligned} f_{\text{Cut-off}} &\cong \frac{v_S}{4\pi W}\left(\pi + \sqrt{\pi^2 - 8\frac{|\sigma|\,W}{\varepsilon_0\varepsilon v_S}}\right) = \frac{6 \times 10^6 \times 10^{-2}}{4 \times 3.141\,59 \times 70 \times 10^{-9}} \\ &\times \left(3.141\,59 + \sqrt{(3.14159)^2 - 8\frac{0.3 \times 100 \times 70 \times 10^{-9}}{8..854 \times 10^{-12} \times 10.89\ \times\ 6 \times 10^6 \times 10^{-2}}}\right) \\ &= 6.8 \times 10^{10} \times \left(3.141\,59 + \sqrt{(3.14159)^2 - 8\frac{0.3 \times 100 \times 70 \times 10^{-1}}{8..854 \times 10.89 \times 6}}\right) \\ &= 6.8 \times 10^{10} \times (3.141\,59 + \sqrt{9.87 - 2.903\,96}) = 6.8 \times 10^{10} \\ &\times (3.141\,59 + 2.6393) = 39.31 \times 10^{10} = 0.39 \times 10^{12}\ \text{Hz} = 0.39\ \text{THz} \end{aligned} \tag{3.76}$$

3.5 Physics-based small-signal equivalent circuit model

This model is proposed by proposed by Liu *et al* (2004). The model is shown in figure 3.7.

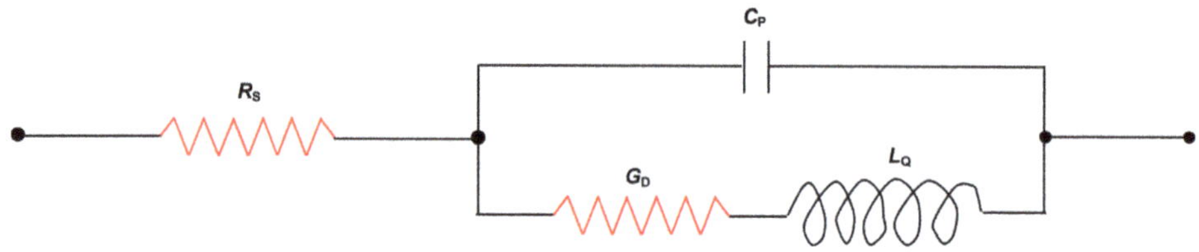

Figure 3.7. Physics-based small-signal parallel inductance model of a resonant tunneling diode.

In this model

$$\begin{aligned} R_S &= \text{Series resistance} \\ &= \text{Contact resistance between metal and semiconductor} \\ &\quad + \text{Sheet resistance of heavily} \\ &\quad - \text{doped semiconductor contact region} \end{aligned} \tag{3.77}$$

$$G_D = \text{Differential conductance} = \frac{dI}{dV} \tag{3.78}$$

where I and V are current and potential difference between neighboring points

$$\begin{aligned} C_P &= \text{Total parallel capacitance} \\ &= \text{Geometrical depletion capacitance } (C_0) \\ &\quad + \text{Quantum capacitance } (C_Q) \end{aligned} \tag{3.79}$$

or,

$$C_P = C_0 + C_Q \tag{3.80}$$

where C_0 comprises components from quantum well, barrier and depletion region:

$$C_0 = \frac{A}{\frac{L_W}{\varepsilon_0\varepsilon_W} + \frac{2L_B}{\varepsilon_0\varepsilon_B} + \frac{L_D}{\varepsilon_0\varepsilon_D}} \tag{3.81}$$

where ε_0 is permittivity of free space

$$L_W, \varepsilon_W = \text{Width and relative permittivity of quantum well region} \tag{3.82}$$

$$L_B, \varepsilon_B = \text{Width and relative permittivity of barrier} \tag{3.83}$$

and

$$L_D, \varepsilon_D = \text{Width and relative permittivity of depletion region} \tag{3.84}$$

The factor 2 in the barrier term accounts for the two barriers in the RTD structure. The quantum capacitance arises from the charging/discharging of the quantum well. It is given by

$$C_Q = -\frac{G_D}{v_C} = -\tau_C G_D \tag{3.85}$$

where v_C is the rate of escape of electrons from the quantum well to the collector with an associated electron lifetime

$$\tau_C = \frac{1}{v_C} \tag{3.86}$$

$$L_Q = \text{Quantum inductance} = \frac{\tau}{G_D} \tag{3.87}$$

where τ is the electron lifetime in the quantum well given by

$$\tau = \frac{1}{v_E + v_C} \tag{3.88}$$

v_E, v_C are electron escape rates from the quantum well into the emitter and collector, respectively.

The electron lifetime associated with v_E is

$$\tau_E = \frac{1}{v_E} \tag{3.89}$$

The quantum inductance describes the lagging behavior of quantum well charge with respect to change of bias. It depends both on device geometry and the scattering mechanism in the device.

Typical tunneling rates for an InGaAs–AlAs–InGaAs–InAs–InGaAs–AlAs–InGaAs RTD are (Liu *et al* 2004): $\tau = 2.58$ ps, $\tau_E = -1.14$ ps, $\tau_C = 0.79$ ps. The capacitance $C_0 = 29.3$ fF for $L_W = 4$ nm, $L_B = 1.5$ nm, $L_D = 4$ nm; RTD area = 1.6 μm ×1.6 μm.

Example 3.5 In an InAs/AlSb RTD, the width of the depletion region is 80 nm. What is the electron transit time through the depletion region if the drift velocity is 8×10^7 cm s^{-1}?

Solution:

$$\begin{aligned}\text{Depletion region transit time} &= \frac{\text{Thickness of depletion region}}{\text{Drift velocity of electrons}}\\ &= \frac{80 \times 10^{-9}\ \text{m}}{8 \times 10^7 \times 10^{-2}} = 10 \times 10^{-14}\ \text{s} \\ &= \frac{10 \times 10^{-14}}{10^{-15}}\text{fs} = 100\ \text{fs}\end{aligned} \tag{3.90}$$

Example 3.6 In a GaInAs (Emitter)/AlAs(Barrier)/GaInAs (Well)/AlAs(Barrier)/GaInAs (Collector) RTD of area 1.2 μm ×1.2 μm, width of the barrier layer = 2 nm, width of the quantum well = 5 nm, and depletion region thickness = 60 nm.

Calculate the geometric capacitance of the diode taking the high-frequency dielectric constant of AlAs in accordance with the equation for $Al_xGa_{1-x}As$ ($x = 1$)

$$\varepsilon_{\text{AlGaAs}} = 10.89 - 2.73x \tag{3.91}$$

and the dielectric constant of GaInAs according to the equation for $Ga_xIn_{1-x}As$ ($x = 0.47$)

$$\varepsilon_{\text{InGaAs}} = 12.3 - 1.4x \tag{3.92}$$

Solution: From equation (3.91).

Dielectric constant of AlAs is

$$\varepsilon_{\text{AlAs}} = 10.89 - 2.73x = 10.89 - 2.73 \times 1 = 10.89 - 2.73 = 8.16 \tag{3.93}$$

Applying equation (3.92).

Dielectric constant of $Ga_xIn_{1-x}As$ is

$$\varepsilon_{\text{GaInAs}} = 12.3 - 1.4 \times 0.47 = 12.3 - 0.658 = 11.642 \tag{3.94}$$

Equation (3.81) gives

$$\begin{aligned} C_0 &= \frac{A}{\frac{1}{\varepsilon_0}\left\{\frac{L_{\text{W}}}{\varepsilon_{\text{W}}} + \frac{2L_{\text{B}}}{\varepsilon_{\text{B}}} + \frac{L_{\text{D}}}{\varepsilon_{\text{D}}}\right\}} = \frac{\varepsilon_0 A}{\frac{L_{\text{GaInAs}}}{\varepsilon_{\text{GaInAs}}} + \frac{2L_{\text{AlAs}}}{\varepsilon_{\text{AlAs}}} + \frac{L_{\text{GaInAs}}}{\varepsilon_{\text{GaInAs}}}} \\ &= \frac{8.854 \times 10^{-12} \times 1.2 \times 1.2 \times (10^{-6})^2}{\frac{5 \times 10^{-9}}{11.642} + \frac{2 \times 2 \times 10^{-9}}{8.16} + \frac{60 \times 10^{-9}}{11.642}} = \frac{10^{-24}}{10^{-9}} \\ &\quad \times \frac{8.854 \times 1.2 \times 1.2}{\frac{5}{11.642} + \frac{2 \times 2}{8.16} + \frac{60}{11.642}} = 10^{-15} \times \frac{12.75}{0.429 + 0.4902 + 5.154} \\ &= 10^{-15} \times \frac{12.75}{6.0732} = 2.0994 \times 10^{-15}\ \text{F} = 2.1\ \text{fF} \end{aligned} \tag{3.95}$$

Example 3.7 An RTD has a differential conductance of 10 mS. Calculate its: (a) quantum conductance and (b) quantum inductance. Electron lifetime in the collector (τ_{C}) = 0.8 ps, and electron lifetime in the quantum well (τ_{W}) = 2.6 ps.

(c) If the geometrical capacitance of the RTD is 13 fF and its series resistance is 10 Ω, estimate from the simple G-C_0-R_S circuit, the maximum oscillation frequencies taking the quantum capacitance into account and also by excluding its effect.

Solution:

(a) Applying equation (3.85)

$$C_Q = -\tau_C G_D = -0.8 \times 10^{-12} \times 10 \times 10^{-3} = -8 \times 10^{-15}\,\text{F} = -8\,\text{fF} \quad (3.96)$$

(b) Equation (3.87) gives

$$L_Q = \frac{\tau}{G_D} = \frac{2.6 \times 10^{-12}}{10 \times 10^{-3}} = 0.26 \times 10^{-9} = 0.26\,\text{nH} \quad (3.97)$$

(c)

$$C_P = C_0 + C_Q = 13 - 8 = 5\,\text{fF} \quad (3.98)$$

By equations (3.29) and (3.98)

$$\left|f_{\text{Max}}\right|_{C_P} = \frac{1}{4\pi R_S C_P} = \frac{1}{4 \times 3.141\,59 \times 10 \times 5 \times 10^{-15}} = 1.59 \times 10^{12}\,\text{Hz} \quad (3.99)$$
$$= 1.6\,\text{THz}$$

Using equation (3.29) and excluding C_Q from equation (3.98)

$$\left|f_{\text{Max}}\right|_{C_0} = \frac{1}{4\pi R_S C_0} = \frac{1}{4 \times 3.141\,59 \times 10 \times 13 \times 10^{-15}} \quad (3.100)$$
$$= 0.612 \times 10^{12}\,\text{Hz} = 0.6\,\text{THz}$$

This exercise shows the effect of quantum capacitance on the maximum oscillation frequency.

3.6 Terahertz resonant tunneling diodes

Diodes containing GaAs quantum wells with two ~1.1 nm AlAs barrier layers oscillated up to 0.420 THz but the peak-to-valley ratio is only 1.5:1 (Brown *et al* 1989). The frequency limitation arises from the RC time delay in the negative differential resistance region.

InAs/AlSb material combination is preferred over GaAs/AlAs system because:

i. it provides a higher current density for the same barrier thickness, thereby decreasing the *RC* time delay; the larger current arises from the band offset at the InAs/AlSb interface allowing electron tunneling through the AlAs barrier with little diminution;
ii. the transit time through the depletion layer is smaller for InAs than GaAs; and
iii. the electron mobility is higher in InAs than GaAs at a given dopant concentration resulting in lower series resistance. Therefore performance

improvement was observed in double barrier diodes using the InAs/AlSb material system. The power density of 90 W cm^{-2} obtained with InAs/AlSb diodes was 50 times that with GaAs/AlAs devices at 0.36 THz. The InAs/AlSb diode oscillated up to 0.712 THz giving 0.3 μW power, which is consistent with theoretical maximum oscillation frequency f_{max} of 1.24 THz (Brown *et al* 1991).

A GaInAs/AlAs RTD oscillating at 1.02 THz (harmonic content) was realized with 0.342 THz fundamental frequency (Orihashi *et al* 2005). The output power was 23 μW for the fundamental mode and decreased to 2.6% of the fundamental mode for the harmonic component.

The electron transit time through the depletion region at the collector can be shortened by designing a graded emitter. This design also decreases the operating bias voltage of the device. Additionally, thin barrier layers can be used to reduce the tunneling time. Following the emitter grading approach and making barriers of thickness 1.2 nm, an InGaAs/AlAs RTD was fabricated on InP substrate. Fundamental oscillation frequency of this diode was 1.04 THz at room temperature (Sugiyama *et al* 2011).

The RTD oscillator is coupled to a planar slot resonator with the RTD placed at the center of a slot in a metal layer (Asada and Suzuki 2011). The slot acts as an antenna and oscillations occur if the absolute magnitude of negative resistance is greater than the radiation loss of the antenna. An output power ~400 μW is obtained at 0.55 THz by the offset fed slot antenna (Shiraishi *et al* 2011).

The frequency of oscillation of RTDs integrated with slot antennas varies with length of the antenna. The highest frequency was found to increase to 1.55 THz on decreasing the antenna length to an optimum 16 μm value (Maekawa *et al* 2014). The frequency is also increased by using a thick antenna electrode to decrease the conduction loss. A 0.15 THz increment in frequency was observed for a 9 μm long antenna with 0.22 μm^2 area RTD by deposition of 1 μm thick gold layer on antenna electrode of thickness 0.1 μm. The highest frequency obtained was 1.63 THz (Maekawa *et al* 2016).

RTD oscillation was achieved at 1.92 THz frequency by improving the fabrication process to decrease the skin-effect induced conduction loss caused by the N^+InGaAs layer under the air bridge electrode joined to the RTD (Maekawa *et al* 2016). The oscillation frequency was increased to 1.98 THz in an RTD with slot antenna by taking the thickness of 2 μm for antenna electrode (Izumi *et al* 2017).

A drawback of slot-antenna integrated RTD oscillators was requirement of carefully placing a bulky silicon hemispherical lens of millimeter dimensions beneath the oscillator for extraction of output power. The lens overly increases the size of the oscillator. The oscillator can be miniaturized by devising a method in which the Si lens is not necessary. In pursuance of this goal, a patch antenna was fabricated on a benzocyclobutene layer stacked on the RTD. The RTD oscillator was combined with a slot-coupled patch antenna to extract power and output power radiated upwards from RTD oscillations was ~40 μW at 0.51 THz without any lens (Okada *et al* 2015). In a further study, a dipole array antenna was used on a dielectric layer stacked on the RTD. An output power of 0.73 mW was obtained from an 89 element RTD array at 1 THz in a pulse mode without the Si lens (Kasagi *et al* 2019).

3.7 Discussion and conclusions

The use of graded emitters and thin barrier layers is applied for increasing the switching speed of resonant tunneling diodes. The frequency of resonant tunneling diodes integrated with slot antennas depends on antenna length as well as thickness of antenna electrode. Optimum length of antenna and thickness of electrode must be chosen. Avoidance of the use of a hemispherical silicon lens reduces the RTD oscillator size.

References

Alkeev N V, Lyubchenko V E, Velling P, Khorenko E, Prost W and Tegude F J 2004 Equivalent circuit of a resonant tunnel InGaAs/InAlAs diode operating at millimetric waves *J. Commun. Technol. Electron.* **49** 833–8

Asada M and Suzuki S 2011 Terahertz oscillators using electron devices – an approach with resonant tunneling diodes *IEICE Electron. Express* **8** 1110–26

Brown E R, Söderström J R, Parker C D, Mahoney L J, Molvar K M and McGill T C 1991 Oscillations up to 712 GHz in InAs/AlSb resonant-tunneling diodes *Appl. Phys. Lett.* **58** 2291–3

Brown E R, Sollner T C L G, Parker C D, Goodhue W D and Chen C L 1989 Oscillations up to 420 GHz in GaAs/AlAs resonant tunneling diodes *Appl. Phys. Lett.* **55** 1777

Feiginov M 2019 Frequency limitations of resonant-tunnelling diodes in sub-THz and THz oscillators and detectors *Int. J. Infrared Millimeter Waves* **40** 365–94

Izumi R, Suzuki S and Asada M 2017 1.98 THz resonant-tunneling-diode oscillator with reduced conduction loss by thick antenna electrode *2017 42nd Int. Conf. on Infrared, Millimeter, and Terahertz Waves (IRMMW-THz) (Cancun, 27 Aug.–1 Sept. 2017)* pp 1–2

Kasagi K, Suzuki S and Asada M 2019 Large-scale array of resonant-tunneling-diode terahertz oscillators for high output power at 1 THz *J. Appl. Phys.* **125** 151601

Kesan V P, Neikirk D P, Blakey P A, Streetman B G and Linton T D 1988 The influence of transit-time effects on the optimum design and maximum oscillation frequency of quantum well oscillators *IEEE Trans. Electron Devices* **35** 405–13

Liu Q, Seabaugh A, Chahal P and Morris F J 2004 Unified AC model for the resonant tunneling diode *IEEE Trans. Electron Devices* **51** 653–7

Maekawa T, Kanaya H, Suzuki S and Asada M 2014 Frequency increase in terahertz oscillation of resonant tunnelling diode up to 1.55 THz by reduced slot-antenna length *Electron. Lett.* **50** 1214–6

Maekawa T, Kanaya H, Suzuki S and Asada M 2016 Frequency increase in resonant-tunneling-diode terahertz oscillator by reduction in conduction loss with thick antenna electrode *2016 Compound Semiconductor Week (CSW) [Includes 28th Int. Conf. on Indium Phosphide & Related Materials (IPRM) & 43rd Int. Symp. on Compound Semiconductors (ISCS)] (Toyama, Japan, 26–30 June)* pp 1–2

Maekawa T, Kanaya H, Suzuki S and Asada M 2016 Oscillation up to 1.92 THz in resonant tunneling diode by reduced conduction loss *Appl. Phys. Express* **9** 024101

Okada K, Kasagi K, Oshima N, Suzuki S and Asada M 2015 Resonant-tunneling-diode terahertz oscillator using patch antenna integrated on slot resonator for power radiation *IEEE Trans. Terahertz Sci. Technol.* **5** 613–8

Orihashi N, Suzuki S and Asada M 2005 One THz harmonic oscillation of resonant tunneling diodes *Appl. Phys. Lett.* **87** 233501-1–3

Shiraishi M, Shibayama H, Ishigaki K, Suzuki S, Asada M, Sugiyama H and Yokoyama H 2011 High output power (~400 μW) oscillators at around 550 GHz using resonant tunneling diodes with graded emitter and thin barriers *Appl. Phys. Express* **4** 064101

Sugiyama H, Suzuki S and Asada M 2011 Room-temperature resonant-tunneling-diode terahertz oscillator based on precisely controlled semiconductor epitaxial growth technology *NTT Tech. Rev.* **9** 1–7

Sun J P, Haddad G I, Mazumder P and Schulman J N 1998 Resonant tunneling diodes: models and properties *Proc. IEEE* **86** 641–61

Yugvesson S 1991 *Microwave Semiconductor Devices* (New York: Springer) pp 107–26

IOP Publishing

Practical Terahertz Electronics: Devices and Applications, Volume 1

Solid-state devices and vacuum tubes

Vinod Kumar Khanna

Chapter 4

Avalanche transit-time and transferred-electron diodes

The avalanche transit time effect-based IMPATT diodes and transferred-electron effect-based Gunn diodes represent two key milestones in the annals of two-terminal devices used to generate subterahertz/terahertz waves. While the former uses a specially designed semiconductor layered structure, the latter is made from bulk semiconductor material. The strikingly different mechanisms of production of negative resistance in these devices are explained. Silicon and GaAs IMPATT diodes have been successful, and improved performance is anticipated from diodes made from wide bandgap materials. Frequency and power levels achieved with GaAs and InGaAs Gunn diodes have steadily increased with advancements in fabrication technology and thermal management. IMPATT and Gunn diodes serve as the input RF generators feeding frequency multiplier circuits for supplying terahertz waves.

Diodes working on two different principles of producing negative resistance in a semiconductor will be described in this chapter.

4.1 Mechanisms of creation of negative resistance

4.1.1 Avalanche transit time effect

A cooperative effort involving impact ionization and time delay of carriers in travelling through the drift region of the device results in negative resistance. An example of diode utilizing this effect is IMPATT diode.

4.1.2 Transferred electron effect

The transference of electrons from a low energy, smaller effective mass, hence higher mobility valley to a high energy, larger effective mass, hence lower mobility valley in

doi:10.1088/978-0-7503-3171-5ch4

the energy–momentum profile of multiple energy valley compound semiconductors leads to negative resistance. The electron transfer occurs when the applied electric field is increased to a level above a critical value. The Gunn diode uses this effect.

The IMPATT and Gunn diodes are used to build high frequency oscillator circuits.

4.2 Frequency and power capabilities of IMPATT diode

IMPATT or impact-ionization-avalanche-transit-time diode, also called Read diode because it was first proposed by Read (1958), is a high power diode capable of producing microwaves with frequencies in the range 1 to 300 GHz at power levels of several watts in continuous operation but higher power in kilowatt range in pulse mode (Misawa 1971, Roy and Mitra 2006). It has generally a four-layer structure P^+NIN^+ made from GaAs, Si or InP; the P^+ layer is the cathode and the N^+ layer is the anode.

GaAs diodes outperform those made from other materials but are expensive due to complex fabrication technology. Diodes made from wide bandgap materials such as silicon carbide have also been demonstrated with oscillations at 7.75 GHz, 1 mW (Yuan *et al* 2001), 8.2–12.4 GHz, 300 mW pulsed power (Vassilevski *et al* 2002). These materials offer advantages of high critical field along with high electron saturation velocity and high thermal conductivity.

4.3 Diode structure and dynamic negative resistance behavior

The structure contains two distinct regions (figure 4.1): a high-doped P^+N avalanche region and an intrinsic (I) drift region. The P^+N avalanche region is biased by applying a DC voltage on the threshold of reverse breakdown. The diode can be dynamically pulled into and out of the breakdown regime by applying an AC field of peak voltage lower than the DC bias. The resulting high electric field creates a large electron–hole pair population by a multiplicative process. The electrons move across the drift region to the anode terminal while the holes migrate to the cathode terminal.

A negative resistance results from the large free carrier concentration obtained by avalanching together with the finite transit time of carriers through the intrinsic region. Hence the current in the diode increases or the resistance of the diode decreases with rise in reverse voltage. The diode works as an oscillator by utilizing this negative resistance property. But impact ionization is a noisy process with statistical nature. Consequently, the IMPATT diode suffers from a major disadvantage of high phase noise level, typically 30 dB. This kind of noise arises from random fluctuations of phase in a signal.

4.3.1 Diode operation as a combination of avalanche and transit time effects

Suppose a DC bias V_{DC} is applied to the diode (figure 4.2). The DC bias value is selected a little below that required for voltage breakdown. Over the DC bias is superimposed an AC field $V_0 \sin\omega t$ so that the total applied voltage is the sum of AC and DC components = $V_{DC} + V_0 \sin\omega t$.

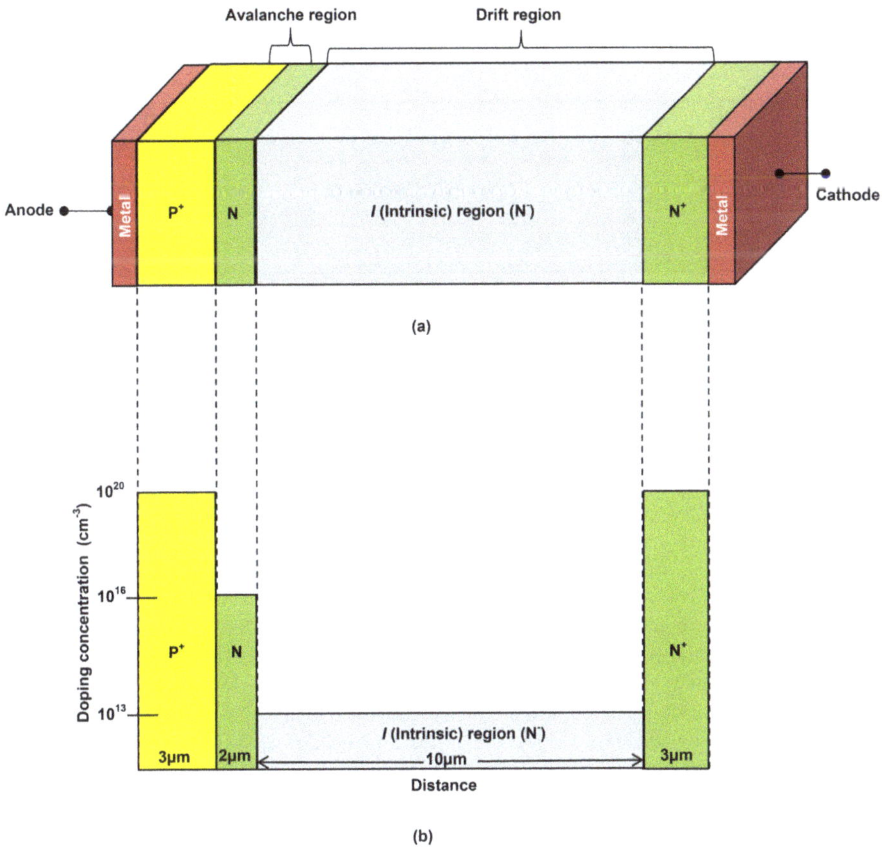

Figure 4.1. IMPATT diode: (a) $P^+NN^-N^+$ structure and (b) typical doping profile. Similar device can be fabricated in $N^+PP^-P^+$ configuration.

The AC field can change the reverse voltage across the diode with time. During at least a part of the positive half cycle of AC voltage, the total applied bias exceeds the breakdown voltage, and electrons and holes are produced by the avalanche effect. Figures 4.3–4.4 show in detail how the electron pulse sprouts, burgeons and travels across the length of the diode.

The carriers grow exponentially with the formation of secondary electron–hole pairs by collisions of accelerated electrons with covalent bonded electrons in the semiconductor and their consequent dislodging from the bonds. The liberated carriers move in opposite directions with a drift velocity equal to the saturation velocity. A time interval equal to half the AC cycle has elapsed by now.

The AC voltage falls and becomes zero just at the instant when the electrons cross over to the drift region. Also, the direction of AC voltage is reversed. Even though the electrons and holes experience deceleration, they continue drifting in the same direction. The AC voltage attains its maximum negative value and rises again towards zero. The thickness of the drift region is selected in such a manner that the carriers reach the respective terminals just at the instant that the AC voltage

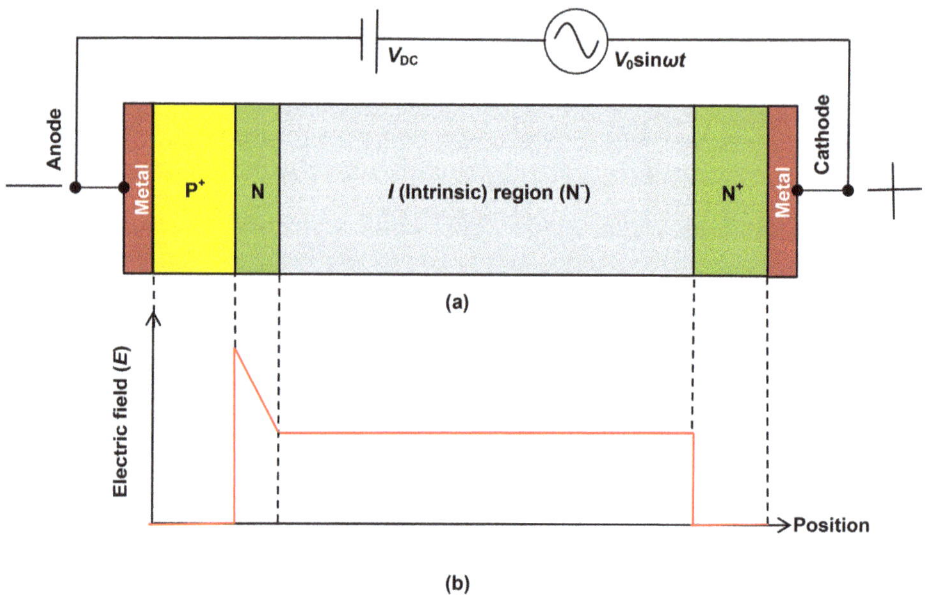

Figure 4.2. IMPATT diode under a DC +AC field with the zero level of AC voltage cycle equal to the DC voltage: (a) the biasing arrangement and (b) the electric field distribution across the diode with zero electric field across the low resistance N^+ and P^+ end regions, peak electric field at the metallurgical P^+N junction and uniform field across the drift region but lower than the peak field at P^+N junction. The DC voltage is the breakdown voltage. If the breakdown voltage is x V, the total voltage varies from x+AC voltage to x-AC voltage so that it is more than breakdown voltage during positive half cycle and less than this voltage during negative half cycle.

becomes zero again. This means that the transit time of electrons from the avalanche region to the anode terminal = ½ the period of the AC field.

The diode operation is a combination of avalanche and transit time effects. The avalanche effect produces a 90° phase lag in current behind the voltage. The transit time effect introduces an additional 90° phase shift. Thus, the current in the external circuit lags behind the voltage by a margin of 90°+ 90° = 180°, which is necessary for oscillations.

4.3.2 Resonant frequency and power formulae

Distance travelled by electrons during half cycle of AC signal is the drift length L

$$L = \text{Electron drift velocity} \times \text{Time} = v_d \tau \tag{4.1}$$

Time for half cycle of AC signal is

$$\tau = \frac{L}{v_d} \tag{4.2}$$

Time for full cycle of AC signal or periodic time is

$$T = 2\tau \tag{4.3}$$

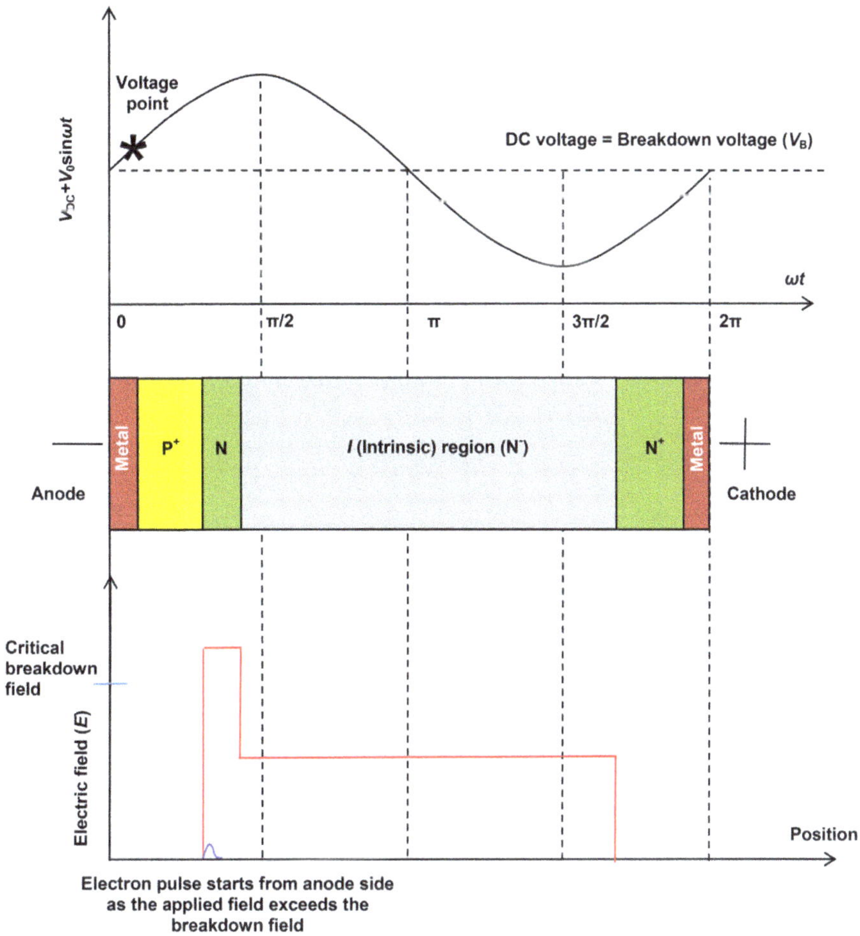

(a)

Figure 4.3. The start, growth and drift of electron pulse across the IMPATT diode structure during a cycle of the applied DC+AC voltage when: (a) ωt is slightly > 0, (b) ωt =π/2, (c) ωt =π, (d) ωt =3π/2, and (e) ωt =2π. Because the transit time of electrons is taken to be one-half of the oscillation cycle during diode design, the electron pulse is collected at the N^+contact at the instant when the voltage cycle is completed, and the cycle is repeated.

Frequency of AC signal is

$$f = \frac{1}{T} = \frac{1}{2\tau} \tag{4.4}$$

Substituting for τ from equation (4.2) in equation (4.4) we get

$$f = \frac{1}{2\tau} = \frac{1}{2\frac{L}{v_d}} = \frac{v_d}{2L} \tag{4.5}$$

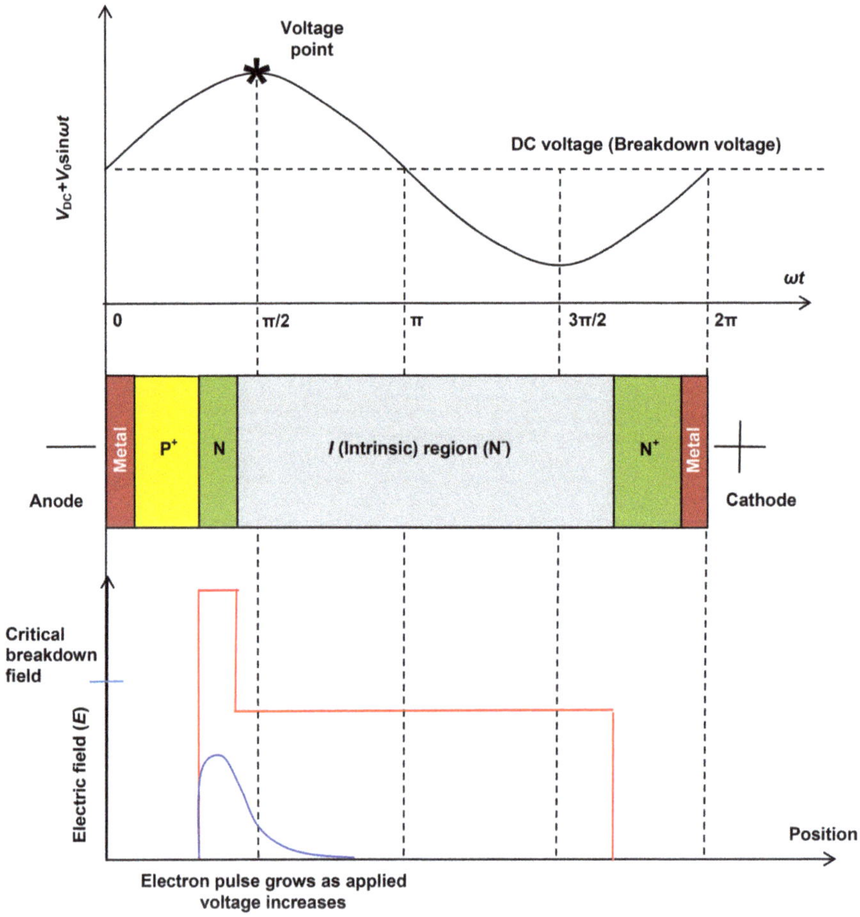

(b)

Figure 4.3. (Continued.)

Hence the resonant frequency of IMPATT diode is

$$f_r = \frac{v_d}{2L} \tag{4.6}$$

Continuous-wave output power of an IMPATT diode is given by the equation

$$P_{CW} = \eta P_{DC} \tag{4.7}$$

where η is the numerical value of efficiency, and the DC operating power P_{DC} is

$$P_{DC} = V_{Max} I_{Max} \tag{4.8}$$

where V_{Max} is the maximum DC operating voltage and I_{Max} is the maximum DC operating current.

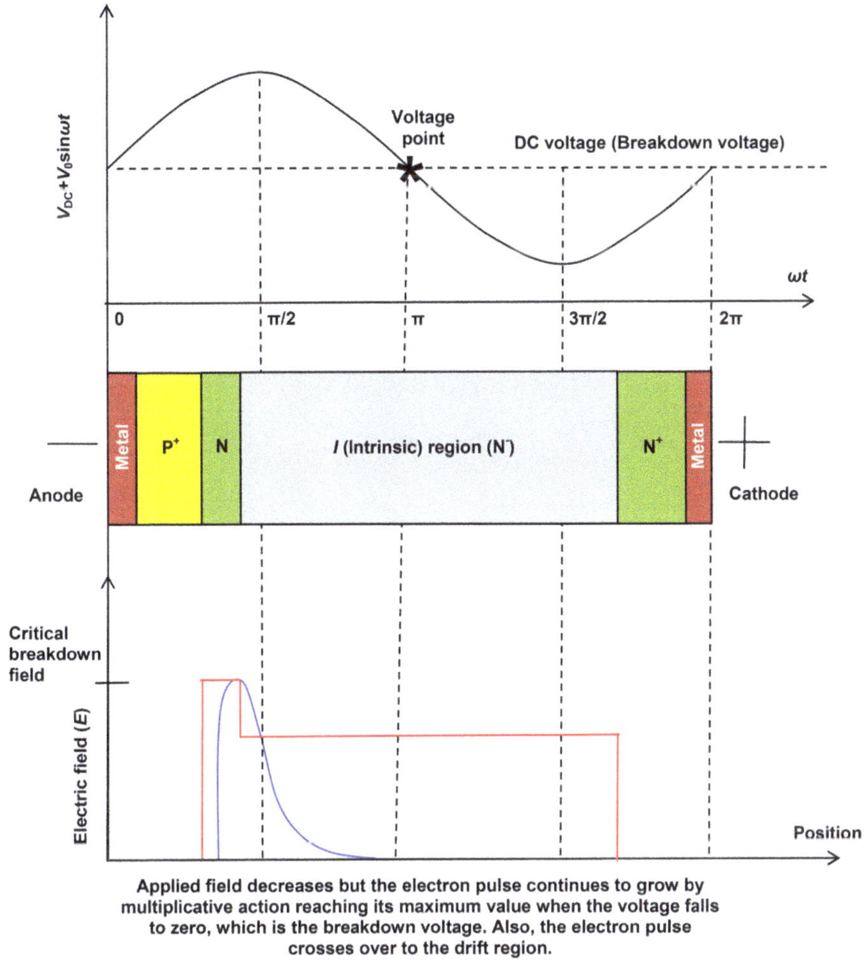

(c)

Figure 4.3. (Continued.)

Example 4.1 Derive the formula of resonant frequency of an IMPATT diode from angular frequency ω of the sinusoidal AC signal $V = V_0 \sin \omega t$ and the angular rate of transit of carriers through the drift region = angle/time, in terms of carrier drift velocity v_d and thickness of drift region L.
Solution: Angular frequency of the AC signal is

$$\omega_1 = 2\pi f \tag{4.9}$$

where f is the linear frequency.

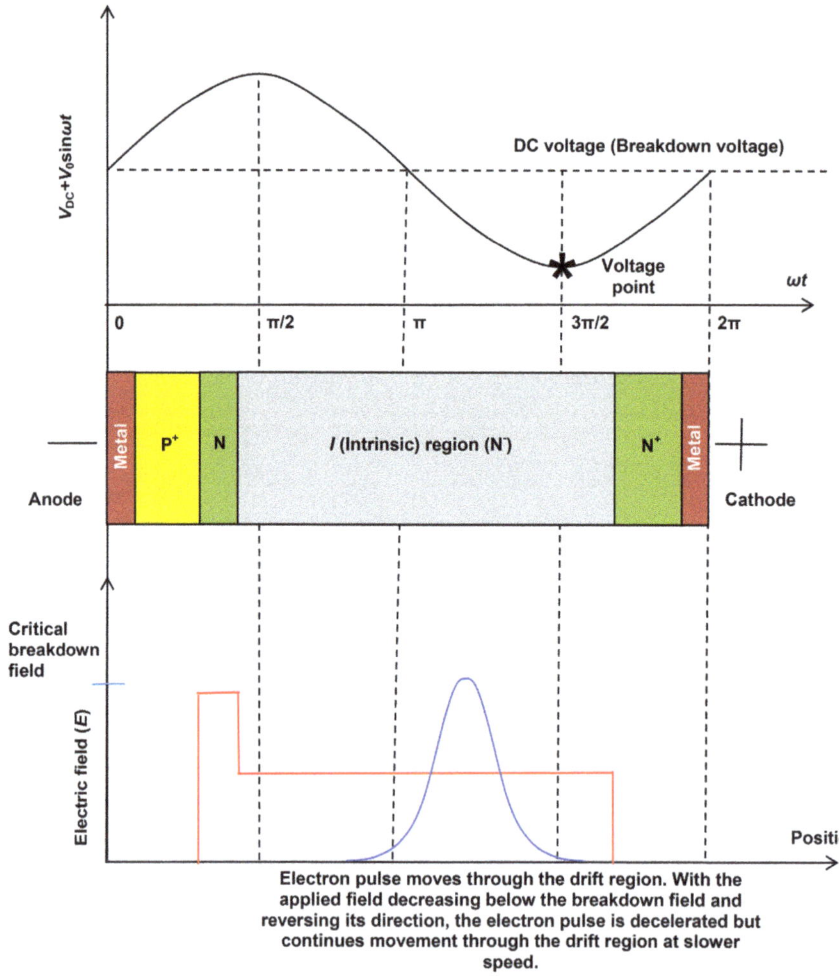

(d)

Figure 4.3. (Continued.)

Angular rate of transit of carriers through the drift region is

$$\omega_2 = \frac{\text{Angle}}{\text{Time}} = \frac{\pi}{\tau} \tag{4.10}$$

where the carriers move through the drift length L for half cycle (π radians) in time τ. At resonance,

$$\omega_1 = \omega_2 \tag{4.11}$$

so that from equations (4.9)–(4.11), we get

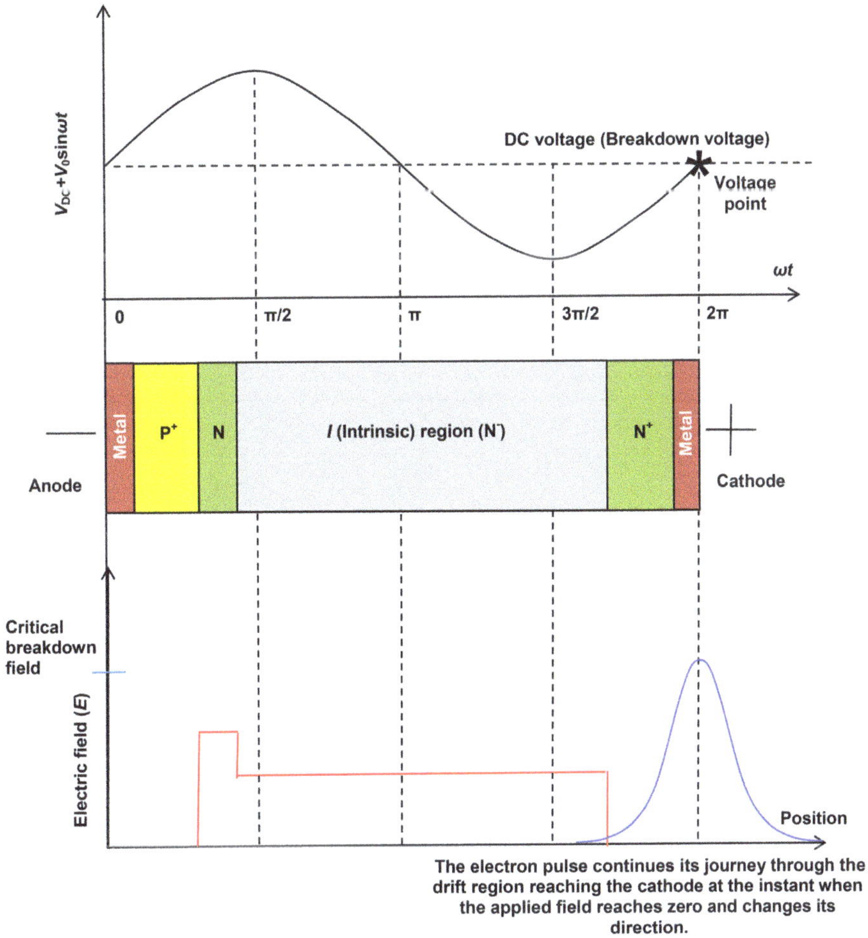

(e)

Figure 4.3. (Continued.)

$$2\pi f = \frac{\pi}{\tau} \tag{4.12}$$

or,

$$f = \frac{1}{2\tau} \tag{4.13}$$

If v_d is the drift velocity of carriers,

$$v_d = \frac{L}{\tau} \tag{4.14}$$

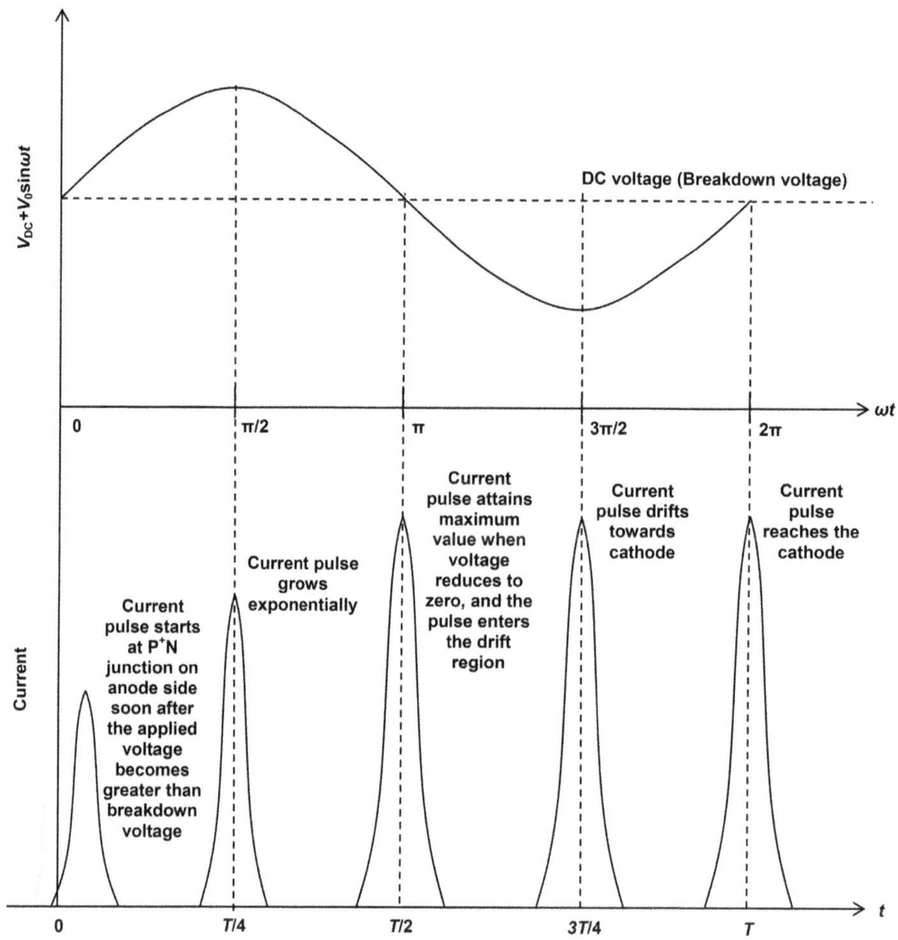

Figure 4.4. Diagram showing the sequential events taking place in the development and propagation of electron pulse through the diode structure, as shown in figure 4.3(a)–(e).

or,

$$\tau = \frac{L}{v_d} \tag{4.15}$$

Substituting for τ from equation (4.15) in equation (4.13), we have

$$f = \frac{1}{2\tau} = \frac{1}{2\frac{L}{v_d}} = \frac{v_d}{2L} \tag{4.16}$$

or,

$$f_r = \frac{v_d}{2L} \tag{4.17}$$

Example 4.2 A silicon IMPATT diode has a drift length of 0.8 μm. What is its resonant frequency? Electron saturation velocity in silicon is 1×10^5 ms^{-1}.
Solution: From equation (4.17),

$$f_r = \frac{v_\mathrm{d}}{2L} = \frac{1 \times 10^5}{0.8 \times 10^{-6}} = \frac{1 \times 10^{5+6}}{0.8} = 1.25 \times 10^{11} = 0.125 \times 10^{12}\ \mathrm{Hz} \\ = 0.125\ \mathrm{THz} \tag{4.18}$$

Example 4.3 The maximum DC operating voltage of an IMPATT diode is 300 V and the current is 100 mA. If the efficiency is 13%, find the continuous wave output power.
Solution: Applying equation (4.8),

$$P_\mathrm{DC} = V_\mathrm{Max} I_\mathrm{Max} = 300 \times 100 \times 10^{-3} = 30\ \mathrm{W} \tag{4.19}$$

From equation (4.7),

$$P_\mathrm{CW} = \eta P_\mathrm{DC} = \frac{13}{100} \times 30 = 0.13 \times 30 = 3.9\ \mathrm{W} \tag{4.20}$$

Example 4.4 If the RMS current flowing in an IMPATT diode is 500 mA, the load resistance is 2 Ω, and efficiency is 17%, find the input DC power.
Solution: Output AC power is

$$P_\mathrm{AC} = I_\mathrm{RMS} V_\mathrm{RMS} = I_\mathrm{RMS}(I_\mathrm{RMS} R) = I_\mathrm{RMS}^2 R = \left(\frac{500}{1000}\right)^2 \times 2 \\ = 0.25 \times 2 = 0.5\ \mathrm{W} \tag{4.21}$$

Since

$$\eta = \frac{P_\mathrm{AC}}{P_\mathrm{DC}} \tag{4.22}$$

$$\therefore P_\mathrm{DC} = \frac{P_\mathrm{AC}}{\eta} = \frac{0.5}{\frac{17}{100}} = \frac{0.5 \times 100}{17} = 2.941\ \mathrm{W} \tag{4.23}$$

4.4 Terahertz GaAs IMPATT diodes

Maximum output power of 30 mW is achieved at 0.10505 THz (efficiency 0.73%) and 8 mW at 0.1102 THz (efficiency 0.2%) in GaAs IMPATT diodes (Eisele *et al* 1991). The RF output power of GaAs IMPATT diodes is increased to 75 mW at 0.120 THz, 17 mW at 0.139 THz (efficiency 3.9%) and 8 mW at 0.144 THz with maximum oscillation frequency of 0.15 THz (Tschernitz *et al* 1994).

Owing to fundamental limitations in material properties, IMPATTdiodes in terahertz regime have not been possible with Si, Ge or GaAs. Nevertheless, computer simulations have shown feasibility of attaining 2.81 W output power at 0.3 THz (efficiency 18.4%) in InP IMPATT diode, and 6.23 W output power at the same frequency (efficiency 15.47%) in wurtzite-GaN IMPATT diode (Mukerjee *et al* 2010). Simulation studies have shown that 191.7–202.9 mW peak RF power at 1.0 THz (efficiency 8.48%–6.41%) is delivered by wurtzite-GaN IMPATT diode (Biswas *et al* 2018).

4.5 Transferred-electron diode

4.5.1 Operational frequency of Gunn diode

The Gunn diode is a DC-to-high frequency converter (Yngvesson 1991). It has a simplistic structure of a semiconductor bar with heavily-doped contacting regions (figure 4.5). Its operational frequency is decided in layman terms. It is mainly determined by the anode-to-cathode distance and the velocity of Gunn domain (explained in section 4.6.4) in traversing this gap. The Gunn domain is formed after the electrons are accelerated to sufficient kinetic energies at which they are promoted to a low-velocity conduction band valley of the semiconductor (explained in section 4.1.2). Obviously, the electrons cannot be accelerated over zero distance to high velocities enabling them to transfer to the aforementioned valley. Hence, the anode-to-cathode distance cannot be indefinitely reduced to increase the oscillation frequency. Some portion of the semiconductor bar is wasted for electron acceleration. This portion is known as dead space and leads to power consumption in the form of heat. It is therefore necessary to minimize the dead space.

4.5.2 Why is a Gunn diode called a 'diode'?

Accustomed to think about diodes in the context of semiconductor P–N junction diodes, one may be a little confused when told that a Gunn diode does not contain any P–N junction or metal–semiconductor Schottky contact, rather is made of a lightly doped N-type semiconductor such as gallium arsenide (GaAs), indium phosphide (InP) or cadmium telluride (CdTe) sandwiched between two heavily doped N^+ layers of the same semiconductor material having the structure : $N^+/N^-/N^+$. Equally surprising is the fact that unlike a P–N junction diode it does not possess any rectifying property. Then why is it called a diode? Simply, because it is a two-terminal device with two active electrodes, the positive end or anode and negative end or cathode, and has a non-linear current–voltage characteristic like a diode although its behavior is symmetric unlike the unsymmetric behavior of a diode.

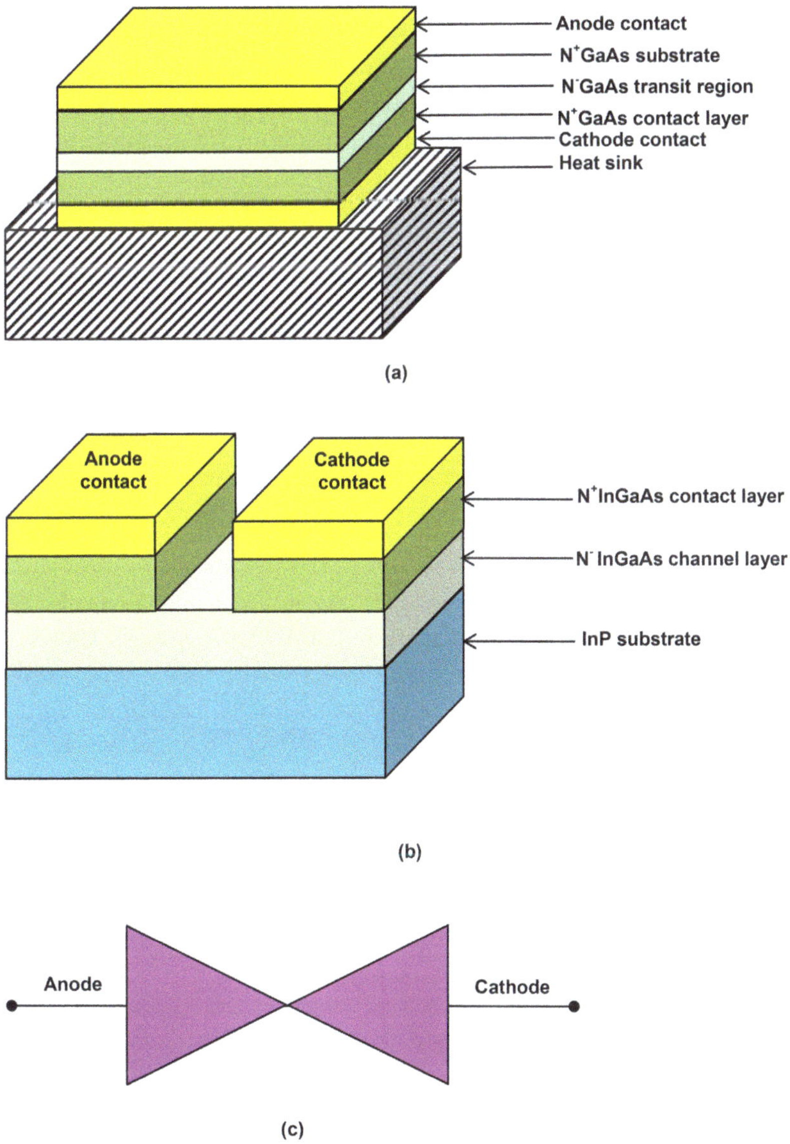

Figure 4.5. Gunn diode structure and symbol: (a) vertical GaAs diode structure, (b) planar InGaAs diode structure and (c) circuit diagram symbol.

4.6 Physics of Gunn diode operation

4.6.1 The Gunn effect

Gunn effect is the appearance of extremely high current oscillations in certain III–V semiconductors when the applied voltage exceeds a threshold level (Gunn 1963). The effect can be understood with reference to the energy–momentum (*E–k*) diagram of gallium arsenide.

4.6.2 Electron transfer mechanism and negative differential resistance

GaAs is a direct bandgap semiconductor in which the minimum energy of conduction band and the maximum energy of valence band occur at the same value of wave vector (figure 4.6). The *E–k* diagram contains three conduction subbands known as the Γ, L and X valleys. Electrons in the lower narrower, sharply curved subband or Γ valley have a small effective mass $0.0632m_0$ where m_0 is the rest mass of electron. Consequently, they have a larger mobility~ 8000 $cm^2\ V^{-1}\ s^{-1}$. On the other hand, electrons in the upper broader, gently curved subbbands or L, X satellite valleys have a large effective mass 0.55 m_0. Naturally, they have a smaller mobility ~2500 $cm^2\ V^{-1}\ s^{-1}$.

At room temperature and under low or moderate electric fields, most of the electrons reside at the bottom of the lower Γ valley. But on applying an intense electric field $\geqslant E_{Th} = 3.2$ kV m^{-1} for GaAs, the electrons are accelerated to high velocities. So, they migrate from the Γ to the L, X satellite valleys by gaining sufficient energies to overcome the intervalley separations, e.g., the Γ and L valleys are separated by a small energy gap $\Delta E = 0.300$ meV. The scattering of electrons from the high mobility Γ to the low mobility L, X satellite valleys is accompanied by a decrease in their velocities. The decrease in electron velocity (or current) with increase in electric field or (voltage) is contrary to the normal increase in current with voltage. It gives rise to a

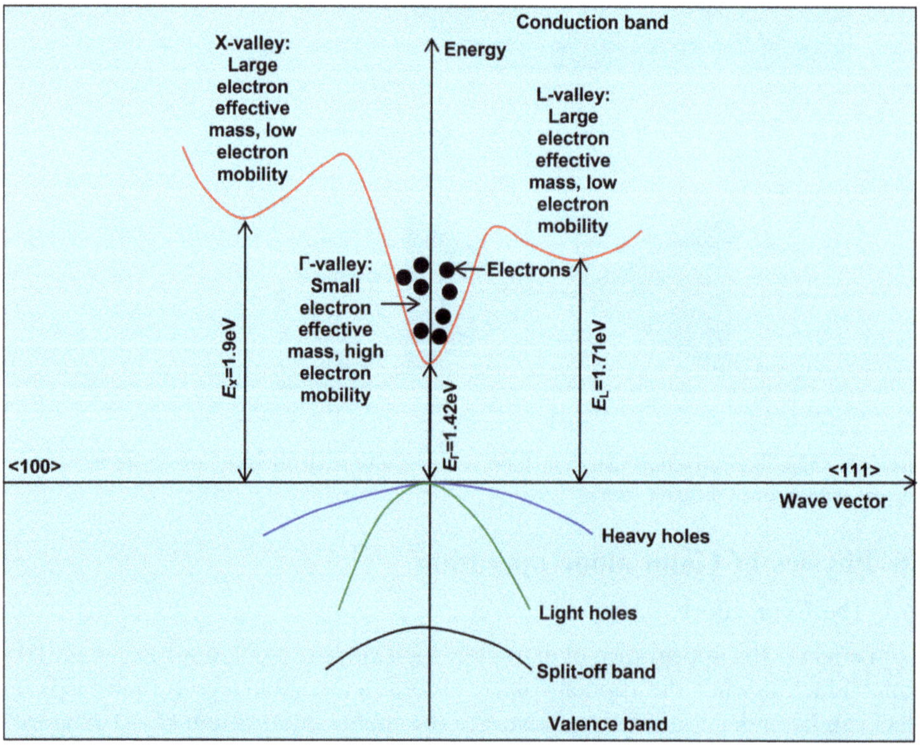

Figure 4.6. Energy-band diagram of gallium arsenide showing the Γ, L and X valleys.

negative differential resistance (NDR) effect, which is observed in the current–voltage characteristic as a portion of the curve with negative slope. Keeping this electron transfer mechanism in view, the Gunn diode is often referred to as a transferred electron device (TED). But the question may be posed, how this NDR effect leads to current oscillations?

4.6.3 Current oscillations arising from NDR

Consider a Gunn diode connected to a voltage V from a battery (figure 4.7). An external resistor R is included in the circuit. The battery voltage V is subdivided between the diode and external resistor. Let the voltage drop across the diode be V_D and that across the external resistor R be V_R:

$$V = V_D + V_R \tag{4.24}$$

As long as $V_D < V_P$, the peak voltage corresponding to point A in the I–V characteristic of Gunn diode, the current increases with voltage along OA.

When $V_D > V_P$, the diode is driven into the negative resistance region AB. Its resistance increases, the current decreases and V_D increases until it reaches the value V_v, the valley voltage at point B in the I–V characteristic of Gunn diode.

When $V_D > V_V$, the diode quickly moves to the positive resistance region BC. Its resistance decreases, the current increases and V_D decreases until it reaches the value V_P at point A.

The above sequence of events is repeated. As the diode voltage V_D moves back and forth between V_P and V_V pertaining to points A and B of the I–V characteristic of the Gunn diode, the current continuously decreases and increases. The self-repeating fall and rise of current produces an oscillatory current across the resistor R.

4.6.4 Explanation of current oscillations from the perspective of Gunn domain formation

Let us look at figures 4.7–4.8. Up to and immediately before the voltage across the Gunn diode reaches the voltage at point A on its current–voltage characteristic, the electrons are travelling fast (van Zyl *et al* 1998). But immediately after the voltage exceeds the voltage at point A and up to the point B on the current–voltage characteristic, the electrons move slowly. So there is a region of high electron concentration PQ (electron accumulation) followed by a region of low electron concentration QR (electron depletion). This dipolar region sets up a field, which opposes the externally applied electric field. It is known as a Gunn domain. The Gunn domain progressively grows until a stable domain is formed. Due to the Gunn domain formation, the increased voltage across the diode decreases. By this time the dipolar region has completed its journey from the cathode to the anode. The decrease of voltage across the diode brings it back to point A. The voltage across the diode increases again and the creation of a dipolar field due to inhomogenous electron distribution is repeated. Thus absorption of a Gunn domain at the anode contact raises the average electric field in the Gunn diode so that the Gunn domain

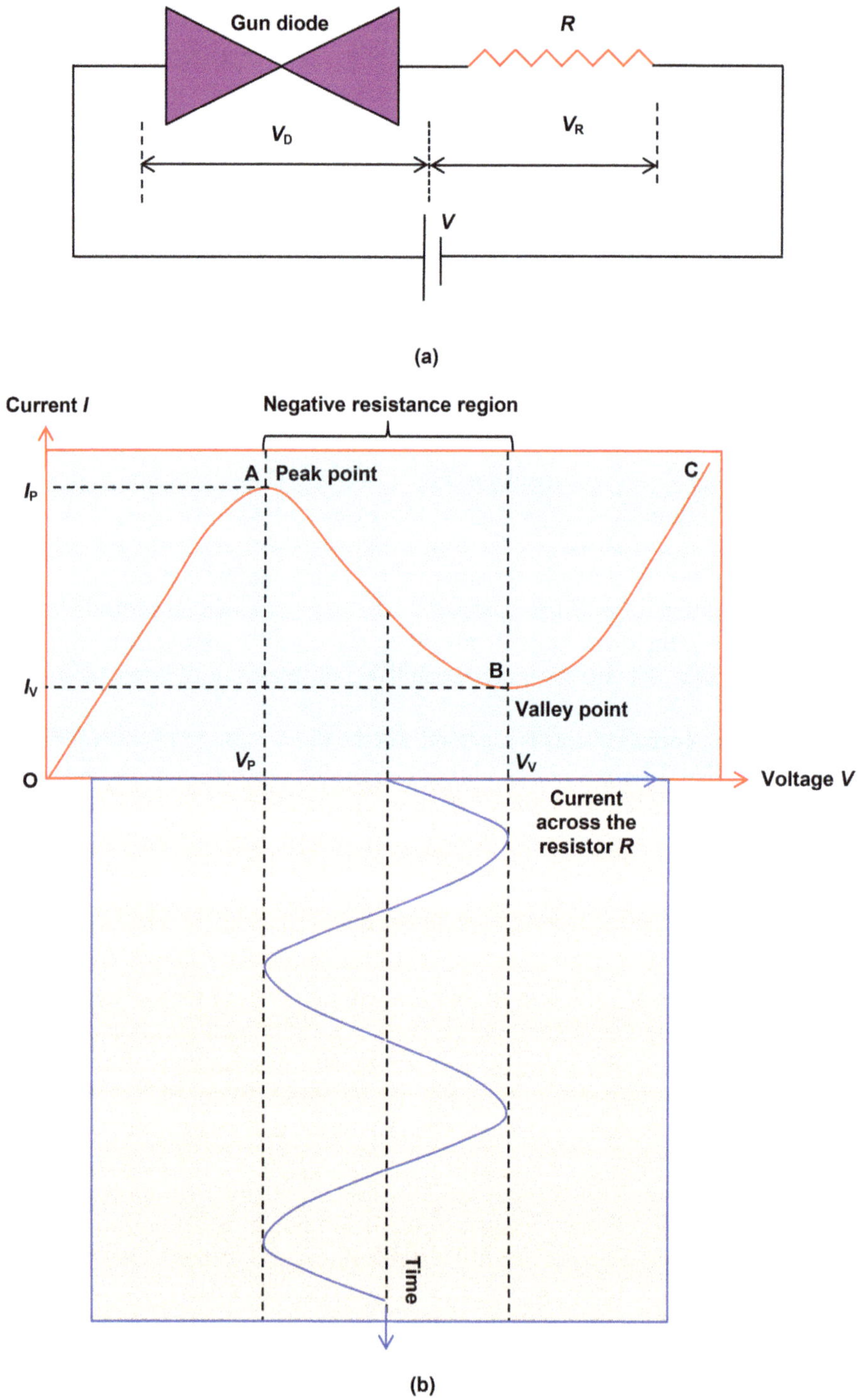

Figure 4.7. Biasing and current–voltage characteristics of a Gunn diode: (a) connections of the diode with a battery and resistor *R* and (b) the mechanism of generation of an oscillatory current from a DC source as a consequence of the negative resistance region in the characteristics.

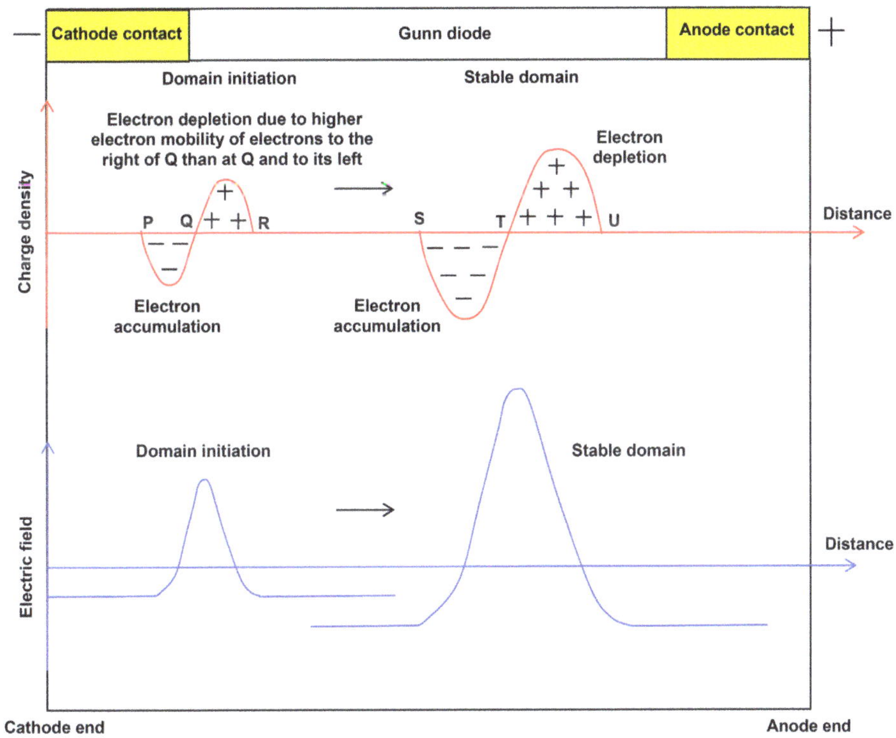

Figure 4.8. Growth and movement of Gunn domains through Gunn diode.

can form again. The successively forming and drifting Gunn domains through the diode lead to alternating current output at its contacts.

Example 4.5 A Gunn diode of length 1.4 μm is working at a frequency of 0.09 THz under an applied electric potential difference of 1 V exceeding the threshold field of 5.0×10^5 V m^{-1}. If the dead space is 0.2 μm, find the mobility of electrons.
Solution: Applied electric field is

$$E = \frac{\text{Potential difference}}{\text{Device length}} = \frac{V}{L} = \frac{1}{1.4 \times 10^{-6}} = 7.14 \times 10^5 \text{ V m}^{-1} \tag{4.25}$$

$$\text{Electron drift velocity} = \text{High-field domain velocity} \tag{4.26}$$

which is obtained from the frequency

$$f = \frac{v_{\text{High field domain}}}{\text{Effective channel length}} = \frac{v_{\text{High field domain}}}{\text{Total device length} - \text{Dead space}} \tag{4.27}$$

giving

$$v_{\text{High field domain}} = f \times (\text{Total device length} - \text{Dead space}) = 0.09 \times 10^{12} \times (1.4 - 0.2) \times 10^{-6}\,\text{m} = 0.09 \times 1.2 \times 10^{6} = 0.108 \times 10^{6} = 1.08 \times 10^{5}\,\text{ms}^{-1} \quad (4.28)$$

The electron mobility is negative because the applied electric field 7.14×10^5 V m^{-1} exceeds the threshold field (5.2×10^5 V m^{-1})

$$\mu_{\text{n}} = -\frac{\text{Electron drift velocity}}{\text{Applied electric field}} \quad (4.29)$$

Combining equations (4.25), (4.26), (4.28) and (4.29), we can write

$$\mu_{\text{n}} = -\frac{v_{\text{High field domain}}}{\text{Applied electric field}} = -\frac{1.08 \times 10^{5}}{7.14 \times 10^{5}} = -0.1513\,\text{m}^2\,\text{V}^{-1}\,\text{s}^{-1} = -0.1513 \times (100)^2\,\text{cm}^2\,\text{V}^{-1}\,\text{s}^{-1} = -1513\,\text{cm}^2\,\text{V}^{-1}\,\text{s}^{-1} \quad (4.30)$$

Example 4.6 What is the minimum doping concentration required for a GaAs Gunn diode of length 700 nm in order that any instability grows into a high-field domain?

Solution: The critical condition for growth of instability into a high-field domain in a Gunn diode is that the nL product (carrier concentration × device length) product in the diode must be > time required for dielectric displacement or polarization to adjust with the time-varying electric field (dielectric relaxation time), i.e.

$$nL > \frac{\varepsilon_0 \varepsilon v_d}{e\,|\mu_n|} = \frac{8.854 \times 10^{-12} \times 12.9 \times 2.5 \times 10^{5}}{1.602 \times 10^{-19} \times 150 \times 10^{-4}} = \frac{8.854 \times 12.9 \times 2.5 \times 10^{-12+5+19+4}}{1.602 \times 150} = 1.188 \times 10^{16}\text{m}^{-2} \approx 1 \times 10^{16}\,\text{m}^{-2} = \frac{1 \times 10^{16}}{10^{4}}\text{cm}^{-2} = 1 \times 10^{12}\,\text{cm}^{-2} \quad (4.31)$$

for N-type GaAs where dielectric constant (ε) of GaAs is taken as 12.9, electron drift velocity (v_{d}) in GaAs is 2.5×10^5 ms^{-1} and the electron mobility in GaAs is assumed to be 150 cm^2 (V–s)$^{-1}$. The electronic charge is $\varepsilon = 1.602 \times 10^{-19}$ C. From equation (4.31),

$$n > \frac{1 \times 10^{12}}{L} = \frac{1 \times 10^{12}\,\text{cm}^{-2}}{700 \times 10^{-9} \times 10^{2}\,\text{cm}} = \frac{1}{7} \times 10^{17} = 0.143 \times 10^{17} = 1.43 \times 10^{16}\,\text{cm}^{-3} \quad (4.32)$$

Hence the doping density of the semiconductor bar must be at least 1.43×10^{16} cm^{-3}.

Example 4.7 The dead space for an $Al_{0.23}Ga_{0.77}As$-based Gunn diode is 230 nm and the saturation velocity of electrons in this material is 9.6×10^4 m s^{-1}. In an $In_{0.53}Ga_{0.47}As$-based Gunn diode, the dead space and saturation velocity are 210 nm and 1.93×10^5 m s^{-1}, respectively. Find the oscillation frequency of the diodes fabricated taking the length of the semiconductor bar to be 1 μm for both materials.
Solution: The frequency of oscillation is given by

$$f = \frac{\text{Velocity of high-field domain}}{\text{Effective channel length}} = \frac{v_{\text{Domain}}}{\text{Total channel length} - \text{Dead Space}} = \frac{v_{\text{Domain}}}{L - L_{\text{Dead}}} \tag{4.33}$$

For $Al_{0.23}Ga_{0.77}As$-based Gunn diode,

$$\begin{aligned} f_{\text{AlGaAs}} &= \left|\frac{v_{\text{Domain}}}{L - L_{\text{Dead}}}\right|_{\text{AlGaAs}} = \frac{9.6 \times 10^4}{1 \times 10^{-6} - 230 \times 10^{-9}} = \frac{9.6 \times 10^4}{10^{-6}(1 - 0.23)} \\ &= \frac{9.6 \times 10^4}{10^{-6} \times 0.77} = \frac{9.6 \times 10^4}{7.7 \times 10^{-7}} = \frac{9.6}{7.7} \times 10^{11} = 1.2467 \times 10^{11}\ \text{Hz} \\ &= 0.12467 \times 10^{12}\ \text{Hz} = 0.125\ \text{THz} \end{aligned} \tag{4.34}$$

Equation (4.26) has been applied in this calculation.
For $In_{0.53}Ga_{0.47}As$-based Gunn diode

$$\begin{aligned} f_{\text{InGaAs}} &= \left|\frac{v_{\text{Domain}}}{L - L_{\text{Dead}}}\right|_{\text{InGaAs}} = \frac{1.93 \times 10^5}{1 \times 10^{-6} - 210 \times 10^{-9}} = \frac{1.93 \times 10^5}{10^{-6}(1 - 0.21)} \\ &= \frac{1.93 \times 10^5}{10^{-6} \times 0.79} = \frac{1.93 \times 10^5}{7.9 \times 10^{-7}} = \frac{1.93}{7.9} \times 10^{12} = 0.2443 \times 10^{12}\text{Hz} \\ &= 0.244\ \text{THz} \end{aligned} \tag{4.35}$$

Example 4.8 The output power of a Gunn diode is −8.2 dBm. Another diode shows an output power of −13.7 dBm. Express these output powers in milliwatts. Which diode shows a higher power?
Solution: dBm (decibel-milliwatt) is a power level used for expressing a power ratio with reference to 1 mW. Thus

$$P_{\text{dBm}} = 10\log_{10}\frac{P_{\text{mW}}}{1\ \text{mW}} \tag{4.36}$$

or,

$$\frac{P_{\mathrm{dBm}}}{10} = \log_{10}\frac{P_{\mathrm{mW}}}{1\ \mathrm{mW}} \tag{4.37}$$

or,

$$\frac{P_{\mathrm{mW}}}{1\ \mathrm{mW}} = 10^{\frac{P_{\mathrm{dBm}}}{10}} \tag{4.38}$$

or,

$$P_{\mathrm{mW}} = 1\ \mathrm{mW} \times 10^{\frac{P_{\mathrm{dBm}}}{10}} \tag{4.39}$$

Applying this formula

$$\begin{aligned} |P_{\mathrm{mW}}|_{-8.2\ \mathrm{dBm}} &= 1\ \mathrm{mW} \times 10^{\frac{-8.2}{10}} = 1\ \mathrm{mW} \times 10^{-0.82} = 1\ \mathrm{mW} \times 0.151 \\ &= 0.151\ \mathrm{mW} \end{aligned} \tag{4.40}$$

$$\begin{aligned} |P_{\mathrm{mW}}|_{-13.7\ \mathrm{dBm}} &= 1\ \mathrm{mW} \times 10^{\frac{-13.7}{10}} = 1\ \mathrm{mW} \times 10^{-1.37} = 1\ \mathrm{mW} \times 4.2658 \times 10^{-2} \\ &= 0.04266\ \mathrm{mW} \end{aligned} \tag{4.41}$$

From equations (4.40) and (4.41) we note that

$$|P_{\mathrm{mW}}|_{-8.2\ \mathrm{dBm}} > |P_{\mathrm{mW}}|_{-13.7\ \mathrm{dBm}} \tag{4.42}$$

Example 4.9

(a) The peak current of a Gunn diode is 100 mA, the valley current is 43 mA and the DC bias voltage is 10 V. Find the maximum expected power conversion efficiency. (b) Practically a CW power output of 200 mW is obtained with 10 W DC input power. What is the practical efficiency? Why?

Solution: (a) The AC power P_{AC} is determined by finding the fundamental Fourier component of the square voltage and current waveforms (figure 4.9). It is obtained as (Sisodia 2009)

$$P_{\mathrm{AC}} = \frac{1}{2}(1 - \alpha)(\beta - 1)V_P I_P \times \frac{8}{\pi^2} \tag{4.43}$$

where

$$\alpha = \frac{I_{\mathrm{V}}}{I_{\mathrm{P}}} \tag{4.44}$$

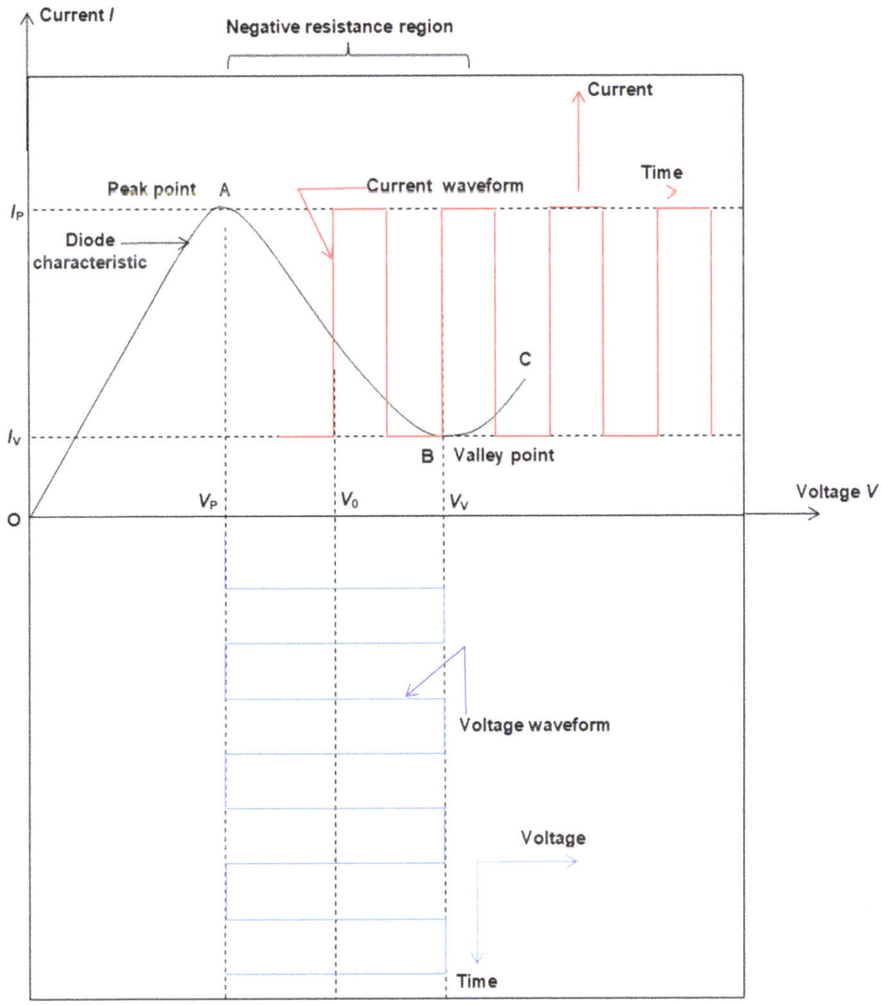

Figure 4.9. *I–V* characteristics of a Gunn diode and the square current and voltage waveforms.

$$\beta = \frac{V_0}{I_P} \tag{4.45}$$

and V_0 is the DC bias.

The DC power is

$$P_{DC} = \frac{1}{2} V_P \times \beta(I_P + I_V) \tag{4.46}$$

Dividing equation (4.43) by equation (4.46), we get the power conversion efficiency as

$$\begin{aligned}\eta &= \frac{P_{AC}}{P_{DC}} = \frac{1}{2}(1-\alpha)(\beta-1)V_P I_P \times \frac{8}{\pi^2} \times 2 \times \frac{1}{V_P \times \beta(I_P + I_V)} \\ &= (1-\alpha)(\beta-1)I_P \times \frac{8}{\pi^2} \times \frac{1}{\beta(I_P+I_V)} = \frac{(1-\alpha)(\beta-1)I_P}{\beta(I_P+I_V)} \times \frac{8}{\pi^2} \\ &= \frac{(1-\alpha)(\beta-1)}{\beta\left(\frac{I_P}{I_P}+\frac{I_V}{I_P}\right)} \times \frac{8}{\pi^2} = \frac{(1-\alpha)(\beta-1)}{\beta(1+\alpha)} \times \frac{8}{\pi^2}\end{aligned} \tag{4.47}$$

using equation (4.44).

Here,

$$\alpha = \frac{I_V}{I_P} = \frac{43 \times 10^{-3}}{100 \times 10^{-3}} = 0.43 \tag{4.48}$$

$$\beta = \frac{V_0}{I_P} = \frac{10}{100 \times 10^{-3}} = 100 \tag{4.49}$$

Calculated efficiency is

$$\begin{aligned}\eta &= \frac{(1-\alpha)(\beta-1)}{(1+\alpha)\beta} \times \frac{8}{\pi^2} = \frac{(1-0.43)(100-1)}{(1+0.43)\times 100} \times \frac{8}{(3.14159)^2} \\ &= \frac{0.57 \times 99 \times 8}{1.43 \times 100 \times 9.87} = 0.32 = 0.32 \times 100\% = 32\%\end{aligned} \tag{4.50}$$

(b) Practical efficiency is

$$\eta = \frac{P_{AC}}{P_{DC}} = \frac{200 \times 10^{-3}}{10} = 0.02 = 0.02 \times 100\% = 2\% \tag{4.51}$$

The experimentally measured efficiency is much lower than theoretical efficiency value because of inability of having a uniform electric field distribution through the device and departure of voltage and current waveforms from ideal behavior. Nonetheless the efficiency equation can be applied to compare the relative performance of Gunn diodes made from different semiconductor materials because it predicts the efficiency as a function of peak-to-valley current ratio.

4.7 Terahertz planar Gunn diodes

The planar Gunn diode is a solid-state source of terahertz waves consisting of a semiconductor film deposited on a suitable substrate. On the surface of the semiconductor film are two electrodes, the cathode and the anode. When a power supply is connected across the two electrodes, the electrons move from the cathode to the anode through the semiconductor film. The planar Gunn diode offers easy monolithic integration with supporting circuit. Compared to the vertical Gunn diode, in the planar diode, the anode-to-cathode distance can be varied by photolithography, whereas in the vertical Gunn diode, the thickness of the semiconductor film equals this distance. So, for a given film thickness, the distance becomes fixed and the flexibility of variation is lost.

4.7.1 GaAs Gunn diodes

In the first GaAs/AlGaAs heterostructure-based planar Gunn diode, the anode-to-cathode separation was 1.3 μm and the width of the device was 60 μm. The diode oscillated at 0.108 THz with maximum output power −43.5 dBm (Khalid *et al* 2007).

In an improved design which included an extra delta doping layer on each side of the active channel, planar Gunn oscillator made in GaAs/AlGaAs quantum wells on SI-GaAs substrate was found to work up to 0.1155 THz in the fundamental mode with output power −28 dBm. It exhibited negative resistance up to 0.158 THz (Li *et al* 2009).

4.7.2 InGaAs Gunn diodes

Performance of GaAs Gunn diodes is limited by the electron saturation velocity. Owing to the higher electron saturation velocity in InGaAs than GaAs, the Gunn diode fabricated in InGaAs is expected to work up to a higher frequency. Furthermore, the energy difference between the lower energy valley and the proximate energy levels is much larger in InP than GaAs. As a consequence, the disturbance caused by thermal excitation of electrons cannot decrease the peak-to-valley ratio in InP to the extent occurring in GaAs. This helps InP devices to outperform GaAs diodes.

Fundamental frequency of 0.164 THz is reported for 120 μm wide $In_{0.53}Ga_{0.47}As$ planar Gunn diode on InP substrate with active channel length 1.3 μm; the highest power is −10 dBm (Khalid *et al* 2013).

In the $In_{0.53}Ga_{0.47}As$ planar Gunn diode on InP substrate, anode-to-cathode separation was 1 μm, and the device width was 120 μm. The diode oscillated up to 0.204 THz with maximum output power −7.3 dBm (Papageorgiou *et al* 2014).

The first submicron planar Gunn $In_{0.53}Ga_{0.47}As$ diode was fabricated on InP substrate with anode-to-cathode separation 700 nm and device width 120 μm. The device oscillated above 0.265 THz giving highest output power −10 dBm (Khalid *et al* 2014a).

A subsequent improved version of a submicron planar $In_{0.53}Ga_{0.47}As$ Gunn diode on InP substrate taking anode-to-cathode separation 600 nm, and device width 120 μm, was found to oscillate at 0.307 THz yielding maximum RF power 28 μW (Khalid *et al* 2014b).

4.8 Discussion and conclusions

Positive resistances absorb power while negative resistances generate power. Two ways of generation of high frequency AC power by harnessing phenomena providing negative resistance were discussed. The oscillations from these solid-state sources at subterahertz/terahertz frequencies drive frequency multiplier chains to elevate the frequency appreciably. Planar diodes are particularly at the focus of attention as frequency sources while designing monolithic integrated circuits.

References

Biswas A, Sinha S, Acharyya A, Banerjee A, Pal S, Satoh H and Inokawa H 2018 1.0 THz GaN IMPATT source: effect of parasitic series resistance *Infrared Millim. Terahertz Waves* **39** 954–74

Eisele H, Mains R K, Haddad G I and Chen C C 1991 GaAs IMPATT diodes for frequencies above 100 GHz: technology and performance *Proc. of the Second Int. Symp. on Space Terahertz Technology (Pasadena, CA, February 26–28)* 145–53

Gunn J B 1963 Microwave oscillations of current in III–V semiconductors *Solid State Commun.* **1** 88–91

Khalid A, Li C, Papageogiou V, Dunn G M, Steer M J and Thayne I G *et al* 2013 $In_{0.53}Ga_{0.47}As$ planar Gunn diodes operating at a fundamental frequency of 164 GHz *IEEE Electron Device Lett.* **34** 39–41

Khalid A, Dunn G M, Macpherson R F, Thoms S, Macintyre D and Li C *et al* 2014b Terahertz oscillations in an $In_{0.53}Ga_{0.47}As$ submicron planar Gunn diode *J. Appl. Phys.* **115** 114502-1–6

Khalid A, Pilgrim N J, Dunn G M, Holland M C, Stanley C R, Thayne I G and Cumming D R S 2007 A planar Gunn diode operating above 100 GHz *IEEE Electron Device Lett.* **28** 849–51

Khalid A, Thoms S, Macintyre D, Thayne I G and Cumming D R S 2014a Fabrication of submicron planar Gunn diode *26th Int. Conf. on Indium Phosphide and Related Materials (IPRM2014) (Montpelier, France, 11–15 May)* (Piscataway, NJ: IEEE) 57–8

Li C, Khalid A, Pilgrim N, Holland M C, Dunn G and Cumming D S R 2009 Novel planar Gunn diode operating in fundamental mode up to 158 GHz *J. Phys.: Conf. Series* **193** 1–4

Misawa T 1971 Impatt diodes *Semiconductors and Semimetals* eds R K Willardson and A C Beer vol 7 (Amsterdam: Elsevier) pp 371–472

Mukherjee M, Banerjee S and Banerjee J P 2010 Dynamic characteristics of III–V and IV–IV semiconductor based transit time devices in the terahertz regime: a comparative analysis *Terahertz Sci. Technol.* **3** 97–108

Papageorgiou V, Khalid A, Li C and Cumming D R S 2014 Cofabrication of planar Gunn diode and HEMT on InP substrate *IEEE Trans. Electron Devices* **61** 2779–84

Read W T Jr 1958 A proposed high-frequency negative resistance diode *Bell Syst. Tech. J.* **37** 401–46

Roy S K and Mitra M 2006 *Microwave Semiconductor Devices* (New Delhi: Prentice Hall of India Pvt. Ltd) pp 48–85

Sisodia M L 2009 *Microwave Devices: Vacuum and Solid State* (New Delhi: New Age Int. (P) Limited Publishers) pp 9.20–22

Tschernitz M, Freyer J and Grothe H 1994 GaAs read-type IMPATT diodes for D band *Electron. Lett.* **30** 1070–71

van Zyl R, Perold W and Botha R 1998 The Gunn-diode: fundamentals and fabrication *Proc. of the 1998 South African Symp. on Communications and Signal Processing-COMSIG'98 (Cat. No. 98EX214) (Rondebosch, South Africa, 8 Sept. 1998)* (Piscataway, NJ: IEEE) pp 407–12

Vassilevski K V, Zorenko A V, Zekentes K, Tsagaraki K, Bano E, Banc C and Lebedev A A 2002 4H-SiC IMPATT diode fabrication and testing *Mater. Sci. Forum* **389–93** 1353–8

Yngvesson S 1991 Transferred electron devices (TED) – Gunn devices *Microwave Semiconductor Devices The Springer International Series in Engineering and Computer Science (VLSI, Computer Architectecture and Digital Signal Processing)* vol 134 (Boston, MA: Springer) pp 23 58

Yuan L, Cooper J A, Melloch M R and Webb K J 2001 Experimental demonstration of a silicon carbide IMPATT oscillator *IEEE Electron Device Lett.* **22** 266–68

IOP Publishing

Practical Terahertz Electronics: Devices and Applications, Volume 1
Solid-state devices and vacuum tubes
Vinod Kumar Khanna

Chapter 5

Heterojunction bipolar transistors

Large-signal and small-signal common emitter current gains and unilateral power gain of a transistor are defined. Various capacitances governing the frequency response characteristics of a common-emitter amplifier are explained. Amplifier bandwidth and gain bandwidth product are clarified. Two important figures of merit of high-frequency bipolar transistors, namely cut-off frequency and maximum frequency of oscillation are distinguished. The high-frequency equivalent circuit model of the heterojunction bipolar transistor (HBT) is drawn and the different terms in the cut-off frequency equation are closely examined. These terms represent the transit time of carriers through the base region, the emitter charging time, the time consumed in charging the depletion region of the collector, and the transit time of carriers across the depletion region at the collector–base junction. From this discussion, the overall meaning of the cut-off frequency equation is understood enabling one to choose appropriate values of structural parameters to make transistors suitable for high-frequency operation. Several examples of calculation of emitter and base resistances and capacitances in the transistor equivalent circuit are given to strengthen the reader's analytical reasoning. Moving to HBT fabrication aspects, the 250 and 130 nm HBT technologies are described, mentioning the HBT amplifiers and oscillators reported in the terahertz frequency range.

In this chapter, the influence of structural parameter values on cut-off frequency of the HBT will be analyzed, and HBT amplifiers and oscillators fabricated for operation at terahertz frequencies will be described.

5.1 Capability of heterojunction bipolar transistor to work at high frequencies

Let us look into the constructional details and working of the HBT vis-à-vis homojunction bipolar transistor.

doi:10.1088/978-0-7503-3171-5ch5

5.1.1 Homojunction and heterojunction bipolar transistors

The homojunction is an interface between two regions in the same semiconductor, e.g., silicon such as an N-type layer with P-type layer or a heavily doped N-type layer with low-doped N-type layer. The main point is that there is only a single semiconductor material in which are juxtaposed layers of different polarities or of the same polarity with different concentrations. There is no variation of bandgap across the junction.

A heterojunction is an interface between two dissimilar semiconductors having different bandgaps. Such junctions can be created between a wide bandgap semiconductor such as AlInAs (bandgap = 1.45 eV) or InP (bandgap = 1.35 eV) and a narrow bandgap semiconductor like GaInAs (bandgap = 0.75 eV).

A homojunction bipolar transistor is fabricated by making the emitter, base and collector regions in the same semiconductor. In a heterojunction transistor, different semiconductor materials are used for making the three constituent regions of the transistor (figure 5.1). An NPN transistor is shown. If the emitter region is made of one semiconductor and base and collector regions are made of another semiconductor, the HBT is known as a single heterojunction bipolar transistor (SHBT). The other variant of HBT is double heterojunction type DHBT in which the base material differs from emitter material and the collector material differs from base material. To be specific, a single heterojunction transistor comprises an emitter made of AlInAs or InP and both the base and collector made of GaInAs, i.e., AlInAs or InP emitter/GaInAs base/GaInAs collector (figure 5.2). If in this transistor, the collector material is InP instead of GaInAs, the transistor variant is of double heterostructure type, i.e., AlInAs or InP emitter/GaInAs base/InP collector. The advantage derived with the double heterojunction variant is its higher breakdown voltage capability attained due to the wide bandgap semiconductor used for collector which has a higher peak breakdown field thereby improving avalanche breakdown voltage rating of the device.

5.1.2 Homojunction versus heterojunction bipolar transistor

Instead of the bipolar homojunction transistor, popularly known as the bipolar junction transistor (BJT), its improved version called the heterojunction bipolar transistor (HBT) is preferred at high frequencies because it obviates several shortcomings of BJT hindering its usage at terahertz frequencies (Pulfrey 2010). These shortcomings arise from inability to make the base very thin and the helplessness to reduce the base resistance by using a high doping concentration. These two issues operate contradictorily as we see below:

(i) To get high gain in a homojunction transistor, the emitter injection efficiency must be high. To this end, the emitter is heavily doped and the base doping is taken to be comparatively lower. The idea is to set up a concentration gradient of electrons from emitter to base so that electrons can easily flow from emitter to base due to this gradient, resulting in a high emitter injection efficiency. But this has the adverse effect that the base doping must be kept low.

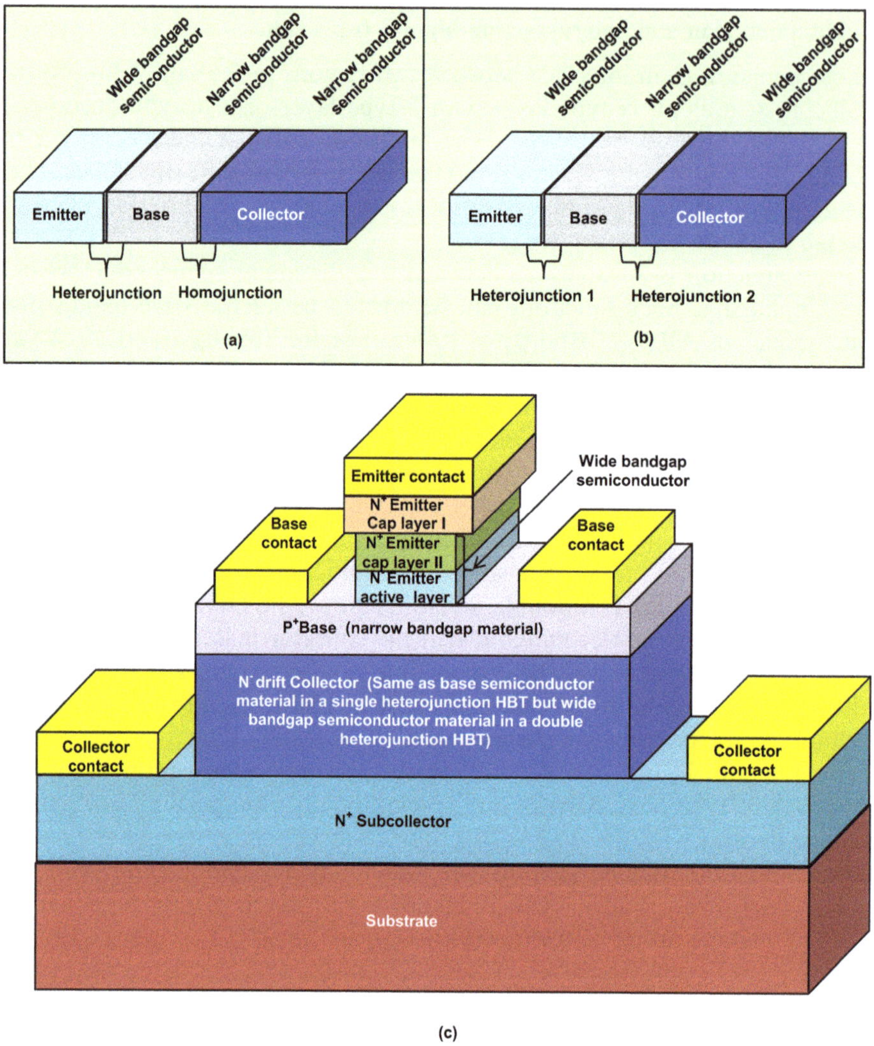

Figure 5.1. Generic construction of an NPN heterojunction bipolar transistor showing material types in single heterojunction and double heterojunction embodiments: (a) and (b) structures of SHBT and DHBT, and (c) 3D drawing.

When the base doping is low and the collector-base junction is reverse biased, the depletion region in the base spreads and touches the emitter at a low collector-base voltage causing punchthrough breakdown below the rated voltage. To avoid this breakdown, the depletion layer needs a thick base to be easily accommodated without reaching the emitter. So the base width must be large. But for a thick base, the electrons take a long time to cross the base. Due to this lengthening of the transit time of electrons through the base, the transistor takes too long to switch from off state to on state. Therefore, the wide base gives a low-cut off frequency.

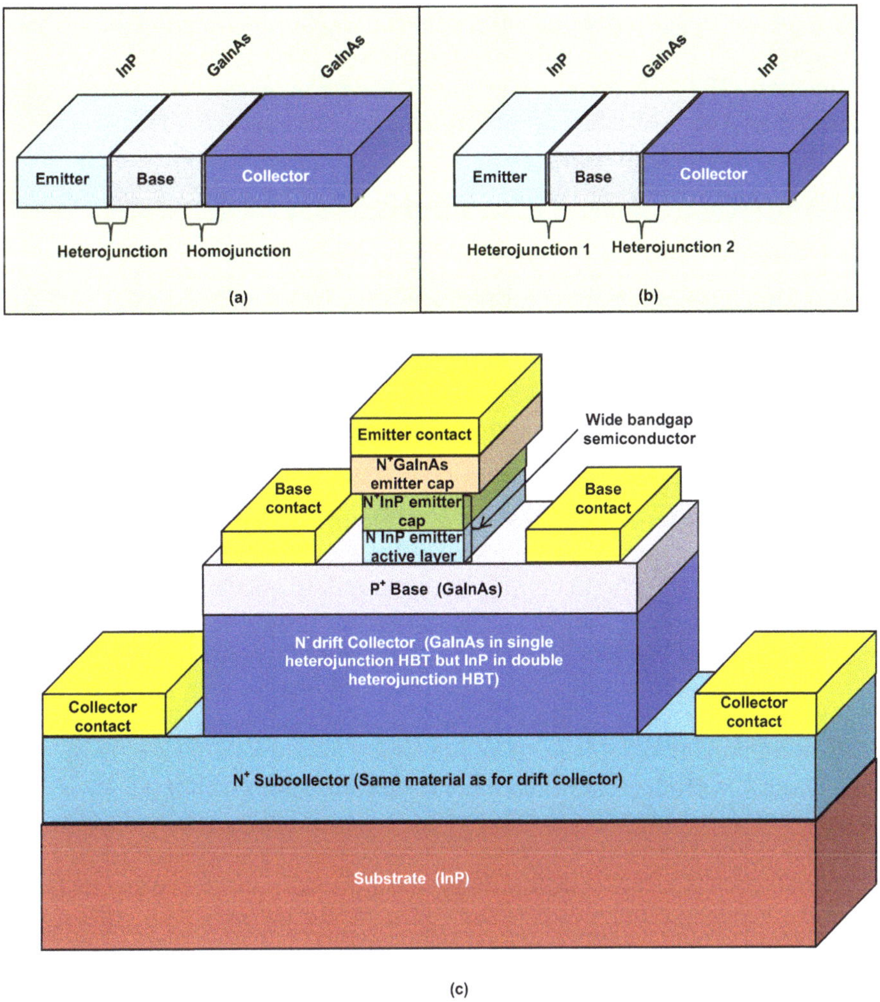

Figure 5.2. Construction of an NPN InP/GaInAs heterojunction bipolar transistor showing materials used in single heterojunction and double heterojunction embodiments: (a) and (b) structures of SHBT and DHBT, and (c) 3D drawing.

(ii) To get a high cut-off frequency, the base must be thin. If the base is thin, the depletion region enlarges and comes in contact with the emitter at a low reverse bias voltage causing premature punchthrough breakdown. Doping concentration in the base must be raised to avoid this punchthrough breakdown because the depletion region will stretch more deeply into a low-doped base at a high voltage. In opposition when the base doping is high, the current gain is lowered due to impairment of emitter injection efficiency.

Thus simultaneous attainment of high gain, high cut-off frequency and high punchthrough breakdown voltage is difficult. The base doping cannot be increased beyond a particular limit and neither can the base thickness be decreased after a certain stage, which is a perplexing situation.

HBT succeeds in getting rid of this intriguing problem by making the emitter and base regions from two semiconductor materials of different bandgaps. In a single heterojunction HBT, the emitter is made of wide bandgap semiconductor and the base is made of a narrow bandgap semiconductor. An approximate equation for the DC current gain of an HBT is

$$\beta = \frac{\begin{array}{c}\text{Doping concentration in the emitter} \times \text{Emitter width} \times \text{Diffusion} \\ \text{coefficient of minority carriers in the emitter}\end{array}}{\begin{array}{c}\text{Doping concentration in the base} \times \text{Base width} \times \text{Diffusion} \\ \text{coefficient of minority carriers in the base}\end{array}} \times \exp\left(\frac{\text{Bandgap difference between emitter and base}}{\text{Boltzmann constant} \times \text{Temperature}}\right) \tag{5.1}$$

Here also the current gain is high when the ratio of doping concentrations in emitter and base is large, which is similar to the homojunction case because it requires that the base doping be kept low. But the hope comes from the fact that the bandgap difference ΔE_G between the emitter and base affects the electron and hole transport across the heterojunction. In a graded emitter-base junction, difference between the potential barriers for hole injection and electron injection = the bandgap difference between the emitter and base semiconductors. An NPN transistor is examined. Electrons crossing from emitter to base face a smaller energy barrier, whereas holes moving from base to emitter must overcome a larger energy barrier. As a consequence, electron injection from emitter to base is facilitated while the hole injection from base to emitter is opposed.

Not only is the bandgap difference influential in carrier injection in the desired manner, but also its effect is dramatic because the current gain depends exponentially on the bandgap difference between emitter and base semiconductors. By virtue of the exponential dependence of the DC current gain β on the bandgap difference, a small bandgap difference has an appreciable effect. As a consequence, the exponential factor dominates over the doping concentration ratio and is in a commanding position for controlling the current gain. This allows us to increase the base doping concentration because the current gain will ultimately be determined by the exponential factor containing the bandgap. This means that in a heterojunction transistor, a high base doping is permitted because the electron injection depends more sensitively on the bandgap difference than on the concentration ratio. The advantage gained is that when we increase the base doping concentration we can make the base very thin without any apprehension of punchthrough breakdown of collector base junction at a low voltage. In fact in an HBT, it is possible to increase the base doping by one or two orders of magnitude as opposed to a homojunction transistor and thereby reduce the base resistance. Thus all the desirable features of high current gain, high punchthrough breakdown voltage and high cut-off frequency can be achieved in an HBT.

Now let us explain the reason why we are striving to decrease the base resistance and base thickness. It impacts the high frequency performance of the transistor in two ways by affecting the cut-off frequency and maximum oscillation frequency.

Firstly, decreasing the base resistance allows substantial thinning of the base in a heterojunction transistor. A thinner base can be used in this transistor than in a homojunction transistor because the high base doping allowed in the heterojunction version prevents punchthrough of the collector–base depletion region by the applied collector-base reverse bias. Thus the HBT has a higher cut-off frequency than a homojunction transistor because the carriers can traverse the thin base in an HBT very speedily. Secondly, the base resistance is directly related with the maximum oscillation frequency of the transistor, as we shall see below (equation 5.18) and decrease in base resistance is essential to increase this frequency. Thus not only is the cut-off frequency of an HBT higher than for the homojunction type, but also the maximum oscillation frequency is better for this case (Schwierz and Liou 2001).

Besides all these improvements, an additional merit of HBT is derived by using a graded bandgap variation ΔE_G. By engineering this graded bandgap, a built-in electric field is established across the junction which accelerates the motion of electrons making the HBT a very fast switching device.

5.2 Gain definitions

5.2.1 Large-signal common-emitter current gains (β_{DC} or h_{FE})

Consider a transistor working in the forward active region. The large-signal or DC common-emitter current gain or amplification factor is defined as the ratio of output or collector current to input or base current

$$\beta_{DC} \text{ or } h_{FE} = \frac{\text{DC collector current } (I_C)}{\text{DC base current } (I_B)} \tag{5.2}$$

or,

$$\beta_{DC} \text{ or } h_{FE} = \frac{I_C}{I_B} \tag{5.3}$$

5.2.2 Small-signal common-emitter current gains (β_{AC} or h_{fe})

Small-signal common-emitter current gain, also called incremental or AC β, is the ratio of a small change in collector current from I_C to $I_C+\Delta I_C$ produced by a small change in base current from I_B to $I_B+\Delta I_B$ at constant collector–emitter voltage (V_{CE}):

$$\begin{aligned}\beta_{AC} \text{ or } h_{fe} &= \left|\frac{\text{Change in collector current}}{\text{Change in base current}}\right|_{\text{Collector–emitter voltage = Constant}} \\ &= \left|\frac{\Delta i_C}{\Delta i_B}\right|_{v_{CE} = \text{Constant}}\end{aligned} \tag{5.4}$$

Note that the small-signal current gain is defined and measured under the condition that the collector-emitter voltage is kept at a constant value. This condition implies that there is no signal component between collector and emitter. Hence, it is often referred to as the short-circuit common-emitter gain. By 'short-circuit common emitter current gain' is meant that the current gain is measured with collector and emitter terminals of the transistor under short-circuited condition.

5.2.3 Unilateral power gain (U)

Power gain of a transistor is expressed by the ratio

$$\text{Power gain } (U) = \frac{\text{Power delivered from the output of the transistor to the load } (P_L)}{\text{Power received by the input of the transistor from signal source } (P_S)} \tag{5.5}$$

Power gain becomes maximum when power matching conditions are satisfied. The unilateral power gain (U) of a transistor is the gain obtained with a transistor having zero output-to-input feedback with output matched to the load and input matched to the signal source. The unilateral power gain is measured in decibels (dBs):

$$U = \left|\frac{P_L}{P_S}\right|_{dB} = 10\log_{10}\left(\frac{P_L}{P_S}\right) \tag{5.6}$$

5.3 Frequency response of the common-emitter transistor amplifier

Capacitive effects control the frequency response of the amplifier (Ashburn 2003). At low frequencies, the bypass/coupling capacitors of the amplifier (external to the transistor) do not act in the desired manner as short circuiting components. Due to their high reactances, the input signal is weak, which appears as low voltage gain.

The effects of capacitances inside the transistor also impact the frequency response (figure 5.3). The transistor contains two internal capacitances:

(i) emitter-base capacitance (C_{BE}) = Emitter-base diffusion capacitance $C_{Diff(BE)}$ + Emitter-base depletion capacitance $C_{Depletion(BE)}$; and
(ii) collector-base depletion capacitance $C_{Depletion(BC)}$.

The emitter–base junction being forward-biased, its depletion capacitance is negligible as compared to diffusion capacitance.

Both the internal capacitances (i) and (ii) have low values ~ a few pF producing a high reactance at low frequencies. But as frequency increases, reactances due to both these capacitances decrease. On the input side, a substantial portion of signal voltage is lost by voltage divider action between resistance of the source and reactance of C_{BE}. Hence, the voltage gain decreases. On the output side, a significant amount of output voltage is fed back out of phase to input. Due to negative feedback, the voltage gain is reduced.

Thus the frequency response of the amplifier sketched as the graph between the voltage gain

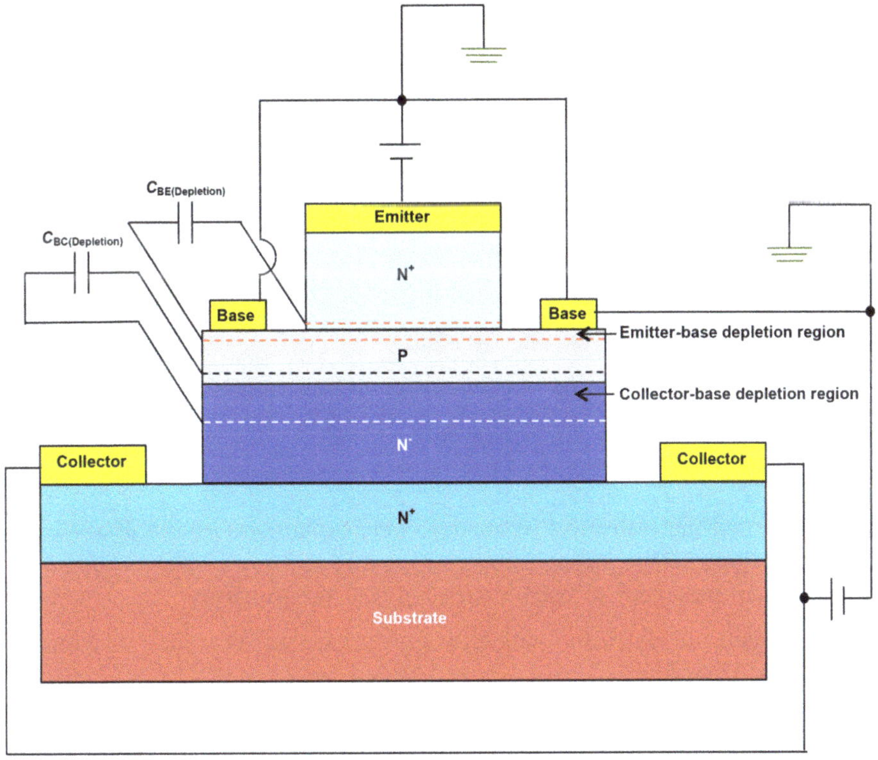

Figure 5.3. An NPN HBT biased for operation in the active region with emitter–base junction forward biased and collector–base junction reverse biased. The emitter–base depletion region is narrow because this junction is forward biased. The collector–base depletion region is broader owing to reverse biasing of this junction. The emitter–base depletion and collector–base depletion capacitances are marked. The emitter–base depletion capacitance is large because of smaller depletion region thickness but is overshadowed by the still larger diffusion capacitance of this forward-biased junction. So emitter–base capacitance is dominated by diffusion capacitance. The collector–base capacitance, however is primarily depletion capacitance because this junction is reverse-biased and there is no diffusion capacitance.

$$A = \frac{V_{\text{Output}}}{V_{\text{Intput}}} \tag{5.7}$$

with respect to frequency consists of three bands: a low frequency band, a mid-frequency band and a high frequency band. The extent of the mid-frequency band is determined by the lower and upper cut-off or corner frequencies f_{L} and f_{H} at which the voltage gain decreases by 3-dB to $1/\sqrt{2}=0.707$ times its midband value. The amplifier bandwidth BW is defined as the frequency difference between upper and lower cut-off frequencies:

$$BW \equiv f_{\mathrm{H}} - f_{\mathrm{L}} \tag{5.8}$$

Since

$$f_{\mathrm{H}} \gg f_{\mathrm{L}} \tag{5.9}$$

we can write

$$BW \cong f_{\mathrm{H}} \tag{5.10}$$

The gain-bandwidth product is defined as

$$GB = A\,|BW| \tag{5.11}$$

5.4 Figures of merit (FOMs) for high-frequency bipolar transistors

5.4.1 Cut-off frequency (f_{T}) of the transistor

It is the frequency at which the short-circuit common-emitter current gain falls to unity. Hence,

$$|h_{\mathrm{fe}}(f)|_{f=f_{\mathrm{T}}} = 1 \tag{5.12}$$

It is also called the transition frequency. This explains the symbol f_{T}.

5.4.2 Maximum frequency of oscillation (f_{Max}) of the transistor

It is the frequency at which the unilateral power gain of the transistor decreases to unity or 0 dB. Therefore,

$$|U(f)|_{f=f_{\mathrm{Max}}} = 1 \tag{5.13}$$

It represents the maximum frequency for which the transistor is capable of power amplification.

5.5 Correlation of terms in cut-off frequency equation with components of equivalent circuit of the bipolar transistor

Proceeding further, it is expedient to draw the equivalent circuit of a heterojunction bipolar transistor (figure 5.4), identify the components of the circuit in the cross-sectional diagram of the transistor, write the equation for its frequency-dependent small signal current gain and try to understand the meaning of this equation (Rücker and Heinemann 2018)

$$\begin{aligned}\frac{1}{h_{\mathrm{fe}}(f)} &= \frac{1}{\beta_{\mathrm{DC}}} + j2\pi f\Bigg[\left\{C_{\mathrm{Diff(BE)}} + C_{\mathrm{Depletion(BE)}} + C_{\mathrm{Depletion(BC)}}\right\}\frac{\partial V_{\mathrm{BE}}}{\partial I_{\mathrm{C}}} \\ &\quad + C_{\mathrm{Depletion(BC)}}\left(R_{\mathrm{Emitter}} + R_{\mathrm{Collector}}\right) + \frac{W_{\mathrm{Depletion(BC)}}}{2v_{\mathrm{Sat}}}\Bigg] = \frac{1}{\beta_{\mathrm{DC}}} \\ &\quad + j2\pi f\left\{\frac{C_{\mathrm{Diff(BE)}} + C_{\mathrm{Depletion(BE)}} + C_{\mathrm{Depletion(BC)}}}{g_{\mathrm{m}}}\right. \\ &\quad \left. + C_{\mathrm{Depletion(BC)}}\left(R_{\mathrm{Emitter}} + R_{\mathrm{Collector}}\right) + \frac{W_{\mathrm{Depletion(BC)}}}{2v_{\mathrm{Sat}}}\right\}\end{aligned} \tag{5.14}$$

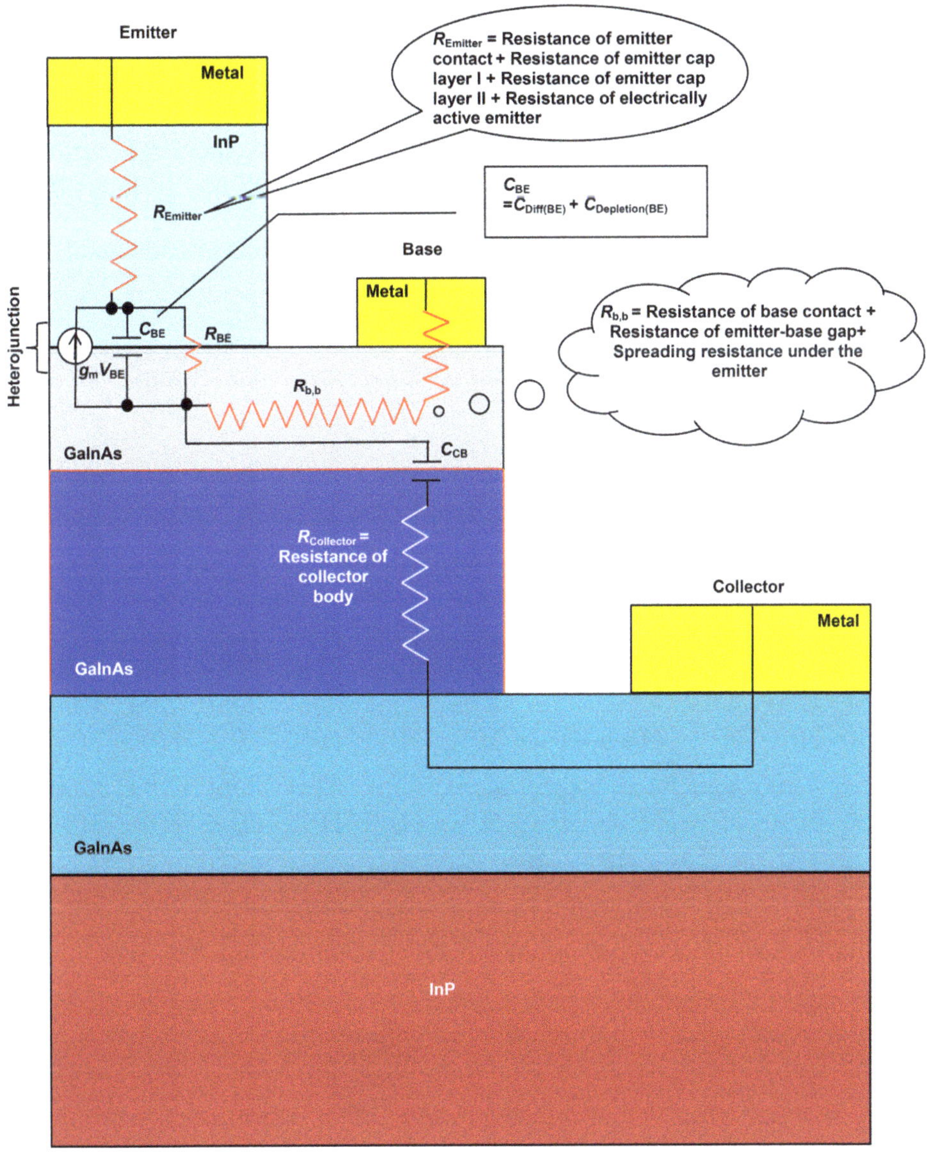

Figure 5.4. Simplified small-signal equivalent circuit model of an HBT drawn on its cross-sectional diagram.

since

$$\frac{\partial I_C}{\partial V_{BE}} = g_m \tag{5.15}$$

In equation (5.14), β_{DC} is the DC or zero-frequency beta of the transistor; $C_{Diff(BE)}$, $C_{Depletion\ (BE)}$ and $C_{Depletion(BC)}$ are the diffusion capacitance across the base-emitter junction, depletion capacitance across the emitter-base junction and depletion capacitance across the collector-base junction, respectively; g_m is the transconductance, V_{BE} is the base emitter voltage, I_C is the collector current, $R_{Emitter}$ and $R_{Collector}$ are emitter and collector resistances, $W_{Depletion\ (BC)}$ is the width of the depletion region across the collector–base junction and v_{Sat} is electron saturation velocity. Figure 5.5(a) shows the HBT equivalent circuit taken out of the cross-sectional diagram for clarity and in figure 5.5(b), the same is drawn in a commonly used form.

At high frequencies, when $h_{fe}(f)=1$, $f=f_T$, the second term is dominant. Then equation (5.14) enables us to write

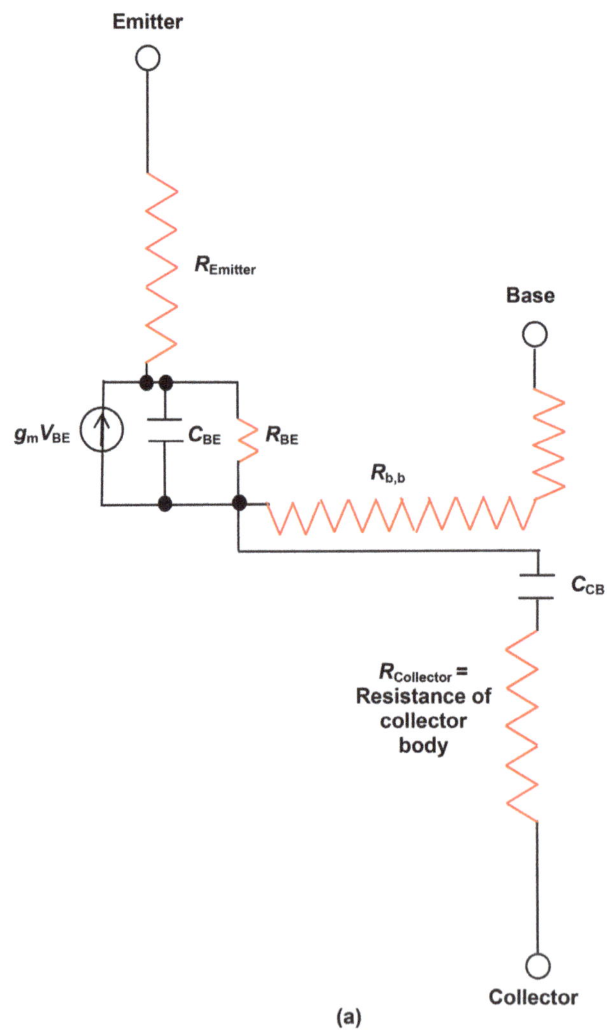

Figure 5.5. HBT equivalent circuits: (a) circuit re-sketched from figure 5.4 for clarity of presentation, (b) circuit of figure 5.5 (a) drawn in more familiar form.

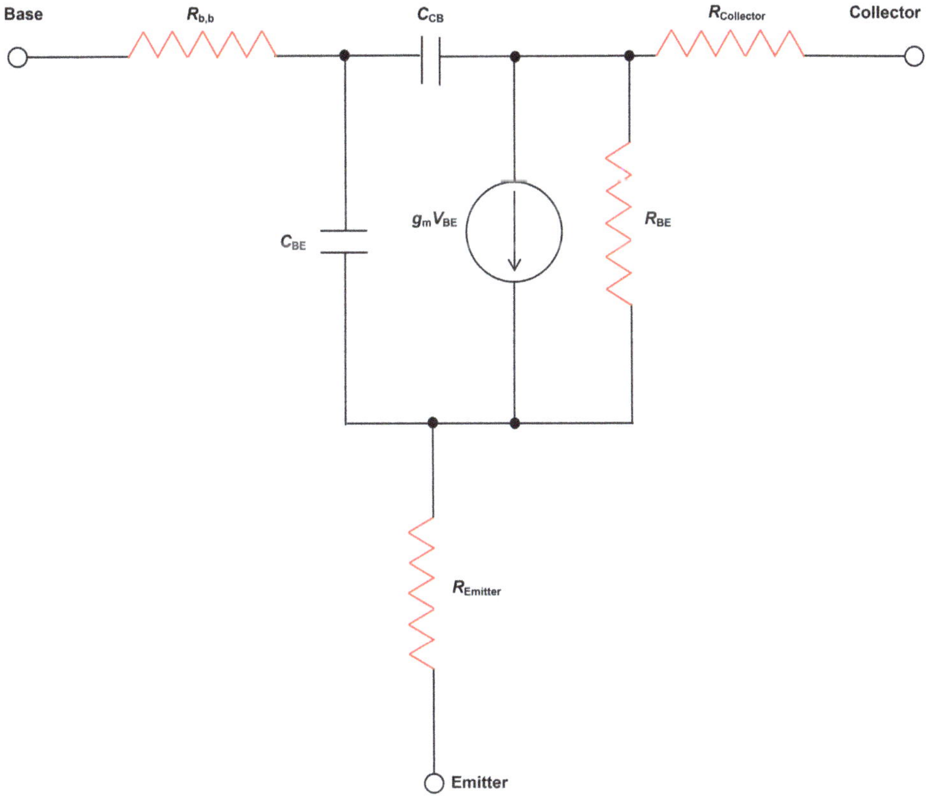

(b)

Figure 5.5. (Continued.)

$$2\pi f_T \left\{ \frac{C_{\mathrm{Diff(BE)}} + C_{\mathrm{Depletion(BE)}} + C_{\mathrm{Depletion(BC)}}}{g_m} + C_{\mathrm{Depletion(BC)}}(R_{\mathrm{Emitter}} + R_{\mathrm{Collector}}) + \frac{W_{\mathrm{Depletion(BC)}}}{2v_{\mathrm{Sat}}} \right\} = 1 \tag{5.16}$$

or,

$$f_T = \frac{1}{2\pi \left\{ \frac{C_{\mathrm{Diff(BE)}}}{g_m} + \frac{C_{\mathrm{Depletion(BE)}} + C_{\mathrm{Depletion(BC)}}}{g_m} + C_{\mathrm{Depletion(BC)}}(R_{\mathrm{Emitter}} + R_{\mathrm{Collector}}) + \frac{W_{\mathrm{Depletion(BC)}}}{2v_{\mathrm{Sat}}} \right\}} \tag{5.17}$$

As expected, all the terms in the denominator of f_T equation representing (capacitance × resistance) products have the dimensions of time.

The maximum oscillation frequency f_{Max} is

$$f_{\mathrm{Max}} \approx \sqrt{\frac{f_{\mathrm{T}}}{8\pi R_{\mathrm{bb}} C_{\mathrm{Depletion(BC)}}}} \tag{5.18}$$

where R_{bb} is the base resistance.

5.5.1 Meaning of the first term in the denominator of f_{T} equation

This term represents the transit time of carriers through the base region of the transistor. It is assigned the symbol τ_{Btt}.

Considering an N–P–N transistor operating in the active region, the emitter–base junction is forward biased (figure 5.6). The electron concentration on the emitter–base boundary, i.e., $x = 0$ is highest. It is denoted by $n_{\mathrm{p}}(0)$. The electron concentration is lowest at the base–collector boundary, i.e., at $x = W_{\mathrm{B}}$ where W_{B} is the base width. The variation of electron concentration with base–emitter voltage v_{BE} is expressed aa

$$n_{\mathrm{p}}(0) = n_{\mathrm{p0}} \exp\left(\frac{e v_{\mathrm{BE}}}{k_{\mathrm{B}} T}\right) \tag{5.19}$$

where n_{p0} is the electron concentration in thermal equilibrium in the P-base, e is the electronic charge, k_{B} is the Boltzmann constant and T is the temperature in Kelvin scale.

The tapered electron concentration profile causes electron diffusion across the base towards the collector. The diffusion current is proportional to the electron concentration gradient $dn_{\mathrm{p}}(x)/dx$. If A is the cross-sectional area of the emitter–base junction and D_{n} is the diffusion coefficient of electrons, the collector current i_{C} is

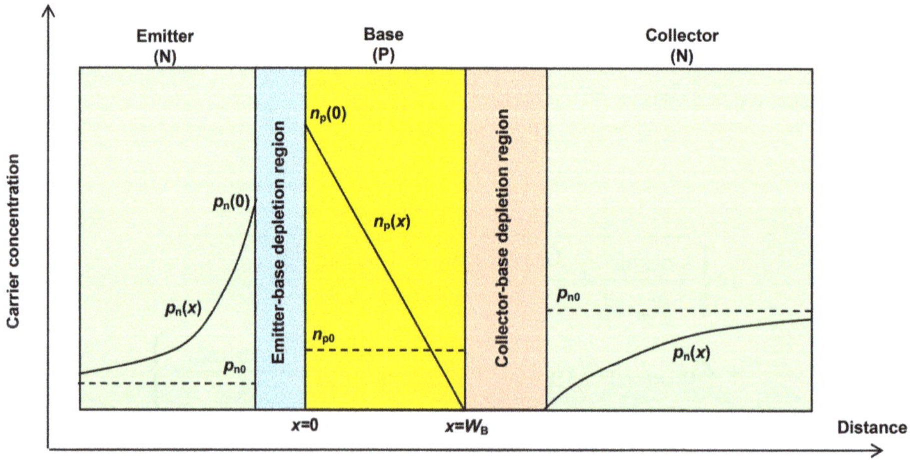

Figure 5.6. Minority-carrier distribution in an NPN bipolar junction transistor in forward biased condition.

$$i_C = AeD_n \frac{dn_p(x)}{dx} = AeD_n \frac{n_p(0)}{W_B} \tag{5.20}$$

Here the assumption is made that

$$\text{Collector current } (i_C) = \text{Emitter current } (i_E) \tag{5.21}$$

or,

$$i_C \approx i_E \tag{5.22}$$

with base transport factor

$$\alpha_T \approx 1 \tag{5.23}$$

Applying equations (5.19) and (5.20), the collector current is written as

$$i_C = AeD_n \frac{n_{p0} \exp\left(\frac{ev_{BE}}{k_B T}\right)}{W_B} = \frac{AeD_n n_{p0}}{W_B} \exp\left(\frac{ev_{BE}}{k_B T}\right) \tag{5.24}$$

Since

$$n_{p0} = \frac{\{\text{Intrinsic carrier concentration}(n_i)\}^2}{\text{P} - \text{base doping concentration}(N_A)} \tag{5.25}$$

or,

$$n_{p0} = \frac{n_i^2}{N_A} \tag{5.26}$$

Equation (5.24) becomes

$$i_C = \frac{AeD_n\left(\frac{n_i^2}{N_A}\right) \exp\left(\frac{ev_{BE}}{k_B T}\right)}{W_B} = \frac{AeD_n n_i^2}{N_A W_B} \exp\left(\frac{ev_{BE}}{k_B T}\right) \tag{5.27}$$

The minority-carrier injection from the emitter to the base creates a diffusion capacitance $C_{Diff(BE)}$ due to stored charges. Assuming a linear distribution of minority carriers in the base, the charge Q_B represents the area of the triangle under the straight line of the charge distribution. It is

$$\begin{aligned} Q_B &= \text{Cross-sectional area of the} \\ &\quad \text{emitter-base junction} \times \text{Electronic charge} \\ &\quad \times \text{Area of the triangle under the straight line distribution} \\ &= Ae \times \frac{1}{2} n_p(0) W_B = \frac{1}{2} Aen_p(0) W_B \end{aligned} \tag{5.28}$$

Substituting for $n_p(0)$ from equation (5.19) in equation (5.28) we get

$$Q_B = \frac{1}{2}Ae\left\{n_{p0}\exp\left(\frac{ev_{BE}}{k_BT}\right)\right\}W_B = \frac{1}{2}Aen_{p0}W_B\exp\left(\frac{ev_{BE}}{k_BT}\right) \tag{5.29}$$

From equations (5.26), (5.27) and (5.29),

$$\begin{aligned} Q_B &= \frac{1}{2}Ae\left(\frac{n_i^2}{N_A}\right)W_B\exp\left(\frac{ev_{BE}}{k_BT}\right) = \frac{Aen_i^2W_B}{2N_A}\exp\left(\frac{ev_{BE}}{k_BT}\right) \\ &= \frac{AeD_nn_i^2}{N_AW_B}\exp\left(\frac{ev_{BE}}{k_BT}\right)\times\frac{1}{2}\frac{W_B^2}{D_n} = i_C\times\frac{1}{2}\frac{W_B^2}{D_n} = \frac{1}{2}\frac{W_B^2}{D_n}i_C \end{aligned} \tag{5.30}$$

The capacitance

$$C_{\text{Diff(BE)}} = \frac{dQ_B}{dv_{BE}} = \frac{d}{dV_{BE}}\left(\frac{1}{2}\frac{W_B^2}{D_n}i_C\right) = \frac{1}{2}\frac{W_B^2}{D_n}\frac{di_C}{dv_{BE}} \tag{5.31}$$

Applying equation (5.15) we get

$$C_{\text{Diff(BE)}} = \frac{1}{2}\frac{W_B^2}{D_n}g_m \tag{5.32}$$

or,

$$\begin{aligned} \frac{C_{\text{Diff(BE)}}}{g_m} &= \frac{1}{2}\frac{W_B^2}{D_n} = \frac{Q_B}{i_C} = \frac{Q_B}{\frac{Q_B}{\tau_{Btt}}} = Q_B\times\frac{\tau_{Btt}}{Q_B} = \tau_{Btt} = \text{Base-transit time } (\tau_{Btt}) \\ &= \text{Time required for injected electrons to diffuse through the base} \end{aligned} \tag{5.33}$$

where equation (5.30) has been applied and i_C is put as the base charge (Q_B)/base-transit time (τ_{Btt}).

5.5.2 Meaning of the second term in the denominator of f_T equation

This represents the emitter charging time denoted by τ_E. It is the time consumed in changing the base potential by charging the capacitances due to depletion regions at the base–emitter and collector–base junctions. These capacitances are charged through the base–emitter differential resistance. The term can be put in the form

$$\tau_E = \frac{C_{\text{Depletion(BE)}} + C_{\text{Depletion(BC)}}}{g_m} = \frac{k_BT}{eI_C}\{C_{\text{Depletion(BE)}} + C_{\text{Depletion(BC)}}\} \tag{5.34}$$

because

$$g_m = \frac{I_C}{\frac{k_BT}{e}} \tag{5.35}$$

as shown below:

Equation (5.27) may be written as

$$i_C = I_S \exp\left(\frac{e v_{BE}}{k_B T}\right) \tag{5.36}$$

where

$$I_S = \frac{A e D_n n_i^2}{N_A W_B} \tag{5.37}$$

The relation between the DC current I_C and DC voltage V_{BE} is stated as

$$I_C = I_S \exp\left(\frac{e V_{BE}}{k_B T}\right) \tag{5.38}$$

On applying an instantaneous voltage v_{br}, the total voltage is

$$v_{BE} = \text{DC voltage} + v_{\text{Instantaneous}} = V_{BE} + v_{be} \tag{5.39}$$

The reader must differentiate between the lowercase and upper case letters/suffixes. The symbol v_{BE} means instantaneous total value, V_{BE} is DC value and v_{be} is value of varying quantity alone.

From equations (5.36), (5.38) and (5.39), we have

$$\begin{aligned} i_C &= I_S \exp\left\{\frac{e(V_{BE} + v_{be})}{k_B T}\right\} = I_S \exp\left(\frac{e V_{BE}}{k_B T}\right) \times \exp\left(\frac{e v_{be}}{k_B T}\right) \\ &= I_C \exp\left(\frac{e v_{be}}{k_B T}\right) = I_C\left(1 + \frac{e v_{be}}{k_B T} + \dots\right) = I_C\left(1 + \frac{v_{be}}{\frac{k_B T}{e}} + \dots\right) \\ &\approx I_C\left(1 + \frac{v_{be}}{\frac{k_B T}{e}}\right) = I_C + \frac{I_C v_{be}}{\frac{k_B T}{e}} \end{aligned} \tag{5.40}$$

where we have applied the expansion of exponential function

$$\exp x = 1 + x + \frac{x^2}{2!} + \frac{x^3}{3!} + \dots \tag{5.41}$$

and retained terms only upto x term. The equation

$$i_C = I_C + \frac{I_C v_{be}}{\frac{k_B T}{e}} \tag{5.42}$$

shows that the collector current comprises the DC component I_C and the signal component i_c, i.e.,

$$i_C = I_C + i_c \tag{5.43}$$

Comparing equations (5.42) and (5.43), we find that

$$i_c = \frac{I_C v_{be}}{\frac{k_B T}{e}} \tag{5.44}$$

from which the ratio

$$\frac{i_c}{v_{be}} = \frac{\text{Instantaneous (signal) current in the collector}}{\text{Instantaneous (signal) voltage across the base-emitter junction}}$$
$$= \text{Transconductance}(g_m) = \frac{I_C}{\frac{k_B T}{e}} = \frac{eI_C}{k_B T} \tag{5.45}$$

or,

$$g_m = \frac{eI_C}{k_B T} \tag{5.46}$$

from which equation (5.35) immediately follows.

5.5.3 Meaning of the third term in the denominator of f_T equation

This represents the time spent in charging the depletion region of the collector. This charging takes place through the combined resistance ($R_{\text{Emitter}} + R_{\text{Collector}}$). The term is denoted by τ_C.

5.5.4 Meaning of the fourth term in the denominator of f_T equation

The fourth term represents the transit time of charge carriers through the space charge or depletion region at the collector–base junction. It is symbolized as $\tau_{\text{Depletion(CB)tt}}$. The mechanism of electron transport through the collector–base depletion region is drifting of charge carriers under the electric field applied across the junction. The current density of electrons is formulated as

$$J_n = \text{Electrical conductivity } (\sigma) \times \text{Electric field } (E)$$
$$= \text{Electron concentration } (n) \times \text{Electronic charge}(e)$$
$$\times \text{Electron mobility } (\mu_n) \times \text{Electric field } (E) = ne\mu_n E \tag{5.47}$$

Since

$$J_n = \frac{\text{Electron current } (I_n)}{\text{Area } (A)} \tag{5.48}$$

the current

$$I_n = J_n A = ne\mu_n EA \tag{5.49}$$

In this equation, the electron mobility can be expressed as

$$\mu_n = \frac{\text{Drift velocity } (v_d)}{\text{Electric field } (E)} \tag{5.50}$$

So, equation (5.49) is written as

$$I_n = ne\left(\frac{v_d}{A}\right)EA = nev_d A \tag{5.51}$$

In the high electric field region prevalent across the collector–base depletion region, the electrons undergo acceleration till they acquire the maximum or saturation velocity v_{Sat}. Hence, equation (5.51) is recast as

$$I_n = nev_{Sat}A \tag{5.52}$$

For a single electron, this electronic current I_n flowing at saturated electron velocity v_{Sat} will take a time $t_{Depletion\ (BC)tt}$ to cross the depletion region

$$t_{Depletion(BC)tt} = \frac{\text{Width of the collector-base depletion region}}{\text{Saturation velocity of electrons}} = \frac{W_{Depletion(BC)}}{v_{Sat}} \tag{5.53}$$

However, the depletion layer transit time is the average delay incurred by all the electrons with respect to their point of entry into the depletion region. It is derived by a rigorous analysis involving the integration of the conduction current inclusive of a frequency-dependent phase across the depletion region divided by $W_{Depletion(BC)}$. Under the assumption that the electron velocity is constant across the collector–base depletion region, a factor of ½ must precede the expression for $t_{Depletion(BC)tt}$ so that the corrected transit time through the depletion region is

$$\tau_{Depletion(BC)tt} = \frac{1}{2}t_{Depletion(BC)tt} = \frac{1}{2}\frac{W_{Depletion(BC)}}{v_{Sat}} \tag{5.54}$$

5.5.5 Overall meaning of the f_T equation

From the above discussions, we can write

$$f_{\mathrm{T}} = \frac{1}{2\pi(\tau_{\mathrm{Btt}} + \tau_{E} + \tau_{\mathrm{C}} + \tau_{\mathrm{Depletion(BC)tt}})} \tag{5.55}$$

Two terms with subscripts E and C correspond to emitter and collector charging times through resistances. The remaining two terms pertain to transit times, one through the base region and another through the collector–base depletion region. These terms have 'tt' in their subscripts.

Example 5.1 (a) In an N-InP emitter/P$^+$-InGaAs base/N$^-$-InGaAs collector HBT, the doping concentrations in the emitter, base and collector are 3×10^{17} cm^{-3}, 8×10^{19} cm^{-3}, and 2×10^{16} cm^{-3}, respectively. The conduction band offset at the InP/InGaAs interface is 0.275 eV (Leibovitch *et al* 1996):

$$\Delta E_{\mathrm{C}} = 0.275\ \mathrm{eV} \tag{5.56}$$

and the valence band offset is 0.325 eV:

$$\Delta E_{\mathrm{V}} = 0.325\ \mathrm{eV} \tag{5.57}$$

Calculate the built-in potentials of emitter–base and collector–base junctions given that:

Effective conduction band density of states in InP

$$\begin{aligned} |N_{\mathrm{C}}|_{\mathrm{InP}} &= 1.1 \times 10^{14} T^{1.5}\ \mathrm{cm}^{-3} = 1.1 \times 10^{14}(300)^{1.5}\ \mathrm{cm}^{-3} \\ &= 1.1 \times 5196.15 \times 10^{14}\ \mathrm{cm}^{-3} = 5.716 \times 10^{17}\ \mathrm{cm}^{-3} \end{aligned} \tag{5.58}$$

Effective valence band density of states in InP

$$\begin{aligned} |N_{\mathrm{V}}|_{\mathrm{InP}} &= 2.2 \times 10^{15} T^{1.5}\ \mathrm{cm}^{-3} = 2.2 \times 10^{15}(300)^{1.5}\ \mathrm{cm}^{-3} \\ &= 2.2 \times 10^{15} \times 5196.15\ \mathrm{cm}^{-3} = 11\,431.53 \times 10^{15}\ \mathrm{cm}^{-3} \\ &= 1.14 \times 10^{19}\ \mathrm{cm}^{-3} \end{aligned} \tag{5.59}$$

Intrinsic carrier concentration in InP

$$|n_i|_{\mathrm{InP}} = 1.3 \times 10^{7}\ \mathrm{cm}^{-3} \tag{5.60}$$

Effective conduction band density of states in InGaAs

$$|N_{\mathrm{C}}|_{\mathrm{InGaAs}} = 2.1 \times 10^{17}\ \mathrm{cm}^{-3} \tag{5.61}$$

Effective valence band density of states in InGaAs

$$|N_{\mathrm{V}}|_{\mathrm{InGaAs}} = 7.7 \times 10^{18}\ \mathrm{cm}^{-3} \tag{5.62}$$

Intrinsic carrier concentration in InGaAs

$$|n_i|_{\text{InGaAs}} = 6.3 \times 10^{11}\ \text{cm}^{-3} \tag{5.63}$$

Take electronic charge (q or e) as 1.602×10^{-19} C.

(b) Find the collector resistance if the mobility of electrons in InGaAs collector is 12 000 $\text{cm}^2\ (\text{V-s})^{-1}$, the collector area is 0.3×0.2 μm and collector thickness is 50 nm.

Solution: (a) The emitter–base junction is a heterojunction. Built-in potential of a heterojunction is

$$\begin{aligned}
V_{bi}|_{\text{BE}} &= \frac{\Delta E_{\text{C}} - \Delta E_{\text{V}}}{2q} + \frac{k_{\text{B}}T}{q}\ln\left(\frac{N_D N_A}{n_{i,N} n_{i,P}}\right) + \frac{k_{\text{B}}T}{2q}\ln\left(\frac{N_{\text{V},N} N_{\text{C},P}}{N_{\text{C},N} N_{\text{V},P}}\right) \\
&= \frac{(0.275 - 0.325) \times 1.602 \times 10^{-19}}{2 \times 1.602 \times 10^{-19}} \\
&\quad + \frac{1.3806 \times 10^{-23} \times 300}{1.602 \times 10^{-19}}\ln\left(\frac{3 \times 10^{17} \times 8 \times 10^{19}}{1.3 \times 10^{7} \times 6.3 \times 10^{11}}\right) \\
&\quad + \frac{1.3806 \times 10^{-23} \times 300}{2 \times 1.602 \times 10^{-19}}\ln\left(\frac{1.14 \times 10^{19} \times 2.1 \times 10^{17}}{5.716 \times 10^{17} \times 7.7 \times 10^{18}}\right) \\
&= -0.025 + \frac{1.3806 \times 300 \times 10^{-23+19}}{1.602}\ln\left(\frac{3 \times 8 \times 10^{17+19-7-11}}{1.3 \times 6.3}\right) \\
&\quad + \frac{1.3806 \times 300 \times 10^{-23+19}}{2 \times 1.602}\ln\left(\frac{1.14 \times 2.1 \times 10^{19+17-17-18}}{5.716 \times 7.7}\right) \\
&= -0.025 + 258.539 \times 10^{-4}\ln(2.93 \times 10^{18}) \\
&\quad + 129.27 \times 10^{-4}\ln(0.0544 \times 10) \\
&= -0.025 + 0.025\,85 \times 42.522 - 0.012\,93 \times 0.6088 \\
&= -0.025 + 1.0992 - 0.007\,87 = 1.066\,33 \approx 1.07\ \text{V}
\end{aligned} \tag{5.64}$$

The collector–base junction is a homojunction. Its built-in potential is

$$\begin{aligned}
V_{bi}|_{\text{BC}} &= \frac{k_{\text{B}}T}{q}\ln\left(\frac{N_D N_A}{n_i^2}\right) = \frac{1.3806 \times 10^{-23} \times 300}{1.602 \times 10^{-19}}\ln\left\{\frac{2 \times 10^{16} \times 8 \times 10^{19}}{(6.3 \times 10^{11})^2}\right\} \\
&= \frac{1.3806 \times 300 \times 10^{-23+19}}{1.602}\ln\left\{\frac{2 \times 8 \times 10^{16+19-22}}{(6.3)^2}\right\} \\
&= 258.539 \times 10^{-4}\ln(0.403\,12 \times 10^{13}) = 0.025\,85 \times 29.0250\,85 \\
&= 0.7503 \approx 0.75\ \text{V}
\end{aligned} \tag{5.65}$$

(b) The collector resistance is

$$R_{\text{Collector}} = \frac{\text{Resistivity of collector material} \times \text{Collector thickness}}{\text{Collector length} \times \text{Collector width}}$$

$$= \frac{\rho_{\text{Collector}} T_{\text{Collector}}}{L_{\text{Collector}} W_{\text{Collector}}}$$

$$= \frac{1}{\text{Collector doping concentration} \times \text{Electronic charge} \times \text{Electron mobility}}$$

$$\times \frac{T_{\text{Collector}}}{L_{\text{Collector}} W_{\text{Collector}}} = \frac{1}{N_{DCollector} e \mu_{\text{n}}} \times \frac{T_{\text{Collector}}}{L_{\text{Collector}} W_{\text{Collector}}} \tag{5.66}$$

$$= \frac{1}{2 \times 10^{16} \times 10^{6} \times 1.602 \times 10^{-19} \times 12000 \times 10^{-4}}$$

$$\times \frac{50 \times 10^{-9}}{0.2 \times 10^{-6} \times 0.3 \times 10^{-6}} = \frac{10^{-16} \times 10^{-6} \times 10^{19} \times 10^{4}}{2 \times 1.602 \times 12000}$$

$$\times \frac{50 \times 10^{-9} \times 10^{6} \times 10^{6}}{0.2 \times 0.3} = \frac{50}{2 \times 1.602 \times 12000 \times 0.2 \times 0.3}$$

$$\times 10^{-16-6+19+4-9+6+6} = 0.021674 \times 10^{4} = 216.74\ \Omega$$

Example 5.2 The emitter area of an HBT (example 5.1) is 0.1 μm × 0.2 μm and the collector area is 0.3 μm × 0.2 μm.

(a) Find the emitter–base and collector–base depletion capacitances at base–emitter voltage $V_{\text{BE}} = 1$ V, and collector–emitter voltage $V_{\text{CE}} = 2$ V. Relative permittivity of InP is 12.5 and that of InGaAs is 13.9.

(b) Also determine the emitter–base diffusion capacitance for 20 nm thick base if the transconductance is 0.8 S and diffusion coefficient of electrons in heavily doped InGaAs is 70 cm^2 s^{-1}.

Solution: We know that

$$V_{\text{CE}} = V_{\text{CB}} + V_{\text{BE}} = -V_{\text{BC}} + V_{\text{BE}} \tag{5.67}$$

or,

$$V_{\text{BC}} = V_{\text{BE}} - V_{\text{CE}} = 1 - 2 = -1V \tag{5.68}$$

At the base–emitter junction, the depletion layer thickness is

$$x_{\text{Depletion}}|_{\text{BE}} = \sqrt{\frac{2\varepsilon_0 \varepsilon_{\text{InP}}\left(V_{bi}|_{\text{BE}} - V_{\text{Applied}}\right)}{qN_D}}$$

$$= \sqrt{\frac{2 \times 8.854 \times 10^{-12} \times 12.5}{1.602 \times 10^{-19} \times 3 \times 10^{17} \times 10^{6}} \times (1.07 - 1)} \tag{5.69}$$

$$= \sqrt{\frac{2 \times 8.854 \times 12.5 \times 10^{-12+19-17-6}}{1.602 \times 3} \times 0.07}$$

$$= \sqrt{3.223\,99 \times 10^{-16}} = 1.7955 \times 10^{-8}\ \text{m}$$

inputting values given and determined in Example 5.1 along with those of Example 5.2.

Base–emitter junction capacitance is

$$
\begin{aligned}
C_{j\mathrm{BE}} &= C_{\mathrm{Depletion(BE)}} = \frac{\varepsilon_0\varepsilon_S A}{x_{\mathrm{Depletion}}|_{\mathrm{BE}}} = \frac{\varepsilon_0\varepsilon_{\mathrm{InP}} A}{x_{\mathrm{Depletion}}|_{\mathrm{BE}}} \\
&= \frac{8.854\times10^{-12}\times12.5\times0.1\times10^{-6}\times0.2\times10^{-6}}{1.7955\times10^{-8}} \\
&= \frac{8.854\times12.5\times0.1\times0.2\times10^{-12-6-6+8}}{1.7955} = 1.2328\times10^{-16}\ \mathrm{F} \\
&\approx 1.23\times10^{-16}\ \mathrm{F}
\end{aligned}
\tag{5.70}
$$

At the collector–base junction, the depletion layer thickness is

$$
\begin{aligned}
W_{\mathrm{Depletion}}|_{\mathrm{BC}} &= \sqrt{\frac{2\varepsilon_0\varepsilon_{\mathrm{InGaAs}}\left(V_{bi}|_{\mathrm{BC}} - V_{\mathrm{Applied}}\right)}{qN_D}} \\
&= \sqrt{\frac{2\times8.854\times10^{-12}\times13.9}{1.602\times10^{-19}\times2\times10^{16}\times10^{6}}\times\{0.75-(-1)\}} \\
&= \sqrt{\frac{2\times8.854\times13.9\times1.75\times10^{-12+19-16-6}}{1.602\times2}} \\
&= \sqrt{134.4404\times10^{-15}} = 3.6666\times10^{-7}\ \mathrm{m}
\end{aligned}
\tag{5.71}
$$

Base–collector junction capacitance is

$$
\begin{aligned}
C_{j\mathrm{BC}} &= C_{\mathrm{Depletion(BC)}} = \frac{\varepsilon_0\varepsilon_S A}{W_{\mathrm{Depletion}}|_{\mathrm{BC}}} = \frac{\varepsilon_0\varepsilon_{\mathrm{InGaAs}} A}{W_{\mathrm{Depletion}}|_{\mathrm{BC}}} \\
&= \frac{8.854\times10^{-12}\times13.9\times0.3\times10^{-6}\times0.2\times10^{-6}}{3.666\times10^{-7}} \\
&= \frac{8.854\times13.9\times0.3\times0.2\times10^{-12-6-6+7}}{3.666} = 2.014\,25\times10^{-17}\ \mathrm{F} \\
&\approx 2\times10^{-17}\ \mathrm{F}
\end{aligned}
\tag{5.72}
$$

(b) Emitter–base diffusion capacitance is

$$
\begin{aligned}
C_{\mathrm{Diff(BE)}} &= \frac{1}{2}\frac{W_{\mathrm{B}}^2}{D_{\mathrm{n}}}g_{\mathrm{m}} = \frac{1}{2}\frac{(20\times10^{-9})^2}{70\times10^{-4}}\times0.8 = \frac{1}{2}\times\frac{400\times10^{-18+4}}{70}\times0.8 \\
&= 2.2857\times10^{-14}\ \mathrm{F}
\end{aligned}
\tag{5.73}
$$

Example 5.3 Calculate the emitter resistance R_{Emitter} from the following values for an HBT: emitter length $L_{\mathrm{Emitter}} = 0.2\ \mu\mathrm{m}$, emitter width $W_{\mathrm{Emitter}} = W_{\mathrm{Emitter\ contact}} = 100$ nm, $W_{\mathrm{Emitter\ junction}} = 95$ nm, and $W_{\mathrm{Electrically\ active\ emitter}} = 90$ nm, specific contact resistivity $\rho_{\mathrm{Contact(Specific)}} = 1.5\ \Omega\text{-}\mu\mathrm{m}^2$; doping concentrations of N^{+}InGaAs, N^{+}InP

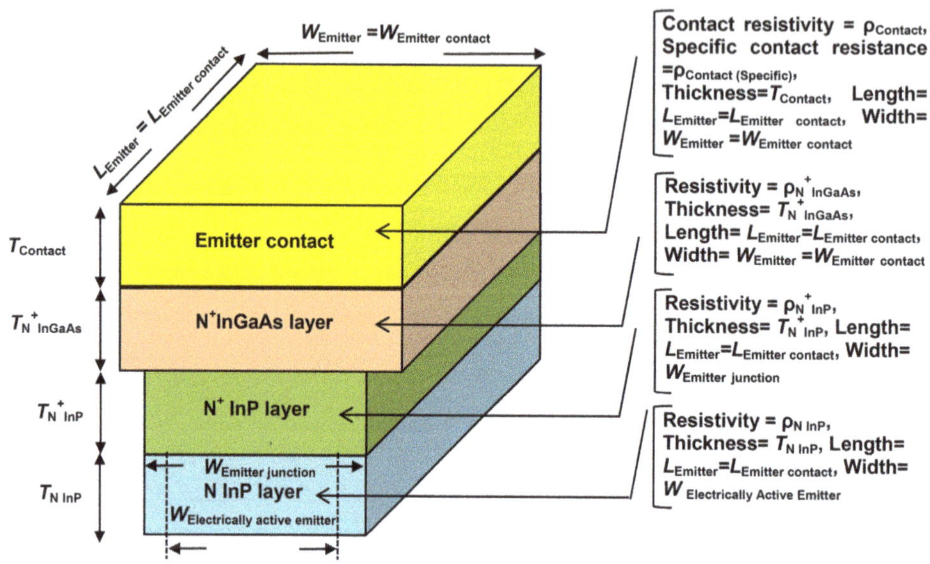

Figure 5.7. Sketch showing the various dimensional and other parameters of the emitter of HBT used in the calculations.

and N InP sublayers: $N_{D\text{N}^+\text{InGaAs}} = 1 \times 10^{19}\,\text{cm}^{-3}$, $N_{D\text{N}^+\text{InP}} = 5 \times 10^{18}\,\text{cm}^{-3}$, and $N_{D\text{NInP}} = 3 \times 10^{17}\,\text{cm}^{-3}$; thicknesses of these layers: $T_{\text{N}^+\text{InGaAs}} = 50\,\text{nm}$, $T_{\text{N}^+\text{InP}} = 20\,\text{nm}$, and $T_{\text{NInP}} = 30\,\text{nm}$. Electron mobility in InGaAs is 12 000 $\text{cm}^2\,(\text{V-s})^{-1}$ while that in InP is 5400 $\text{cm}^2\,(\text{V-s})^{-1}$.

The emitter consists of four sublayers, namely, contact, N$^+$InGaAs and N$^+$InP emitter cap layers, and N InP active emitter layer (figure 5.7).

Let the resistivities, thicknesses and areas of these sublayers be denoted by ρ, T and A with subscripts affixed for the sublayers.

$$
\begin{aligned}
R_{\text{Emitter}} = R_{\text{E}} &= \frac{\rho_{\text{Contact}} T_{\text{Contact}}}{A_{\text{Contact}}} + \frac{\rho_{\text{N}^+\text{InGaAs}} T_{\text{N}^+\text{InGaAs}}}{A_{\text{N}^+\text{InGaAs}}} + \frac{\rho_{\text{N}^+\text{InP}} T_{\text{N}^+\text{InP}}}{A_{\text{N}^+\text{InP}}} \\
&+ \frac{\rho_{\text{NInP}} T_{\text{NInP}}}{A_{\text{NInP}}} = \frac{\rho_{\text{Contact}} T_{\text{Contact}}}{L_{\text{Emitter}} W_{\text{Emitter contact}}} + \frac{\rho_{\text{N}^+\text{InGaAs}} T_{\text{N}^+\text{InGaAs}}}{L_{\text{Emitter}} W_{\text{Emitter contact}}} \\
&+ \frac{\rho_{\text{N}^+\text{InP}} T_{\text{N}^+\text{InP}}}{L_{\text{Emitter}} W_{\text{Emitter junction}}} + \frac{\rho_{\text{NInP}} T_{\text{NInP}}}{L_{\text{Emitter}} W_{\text{Electrically active emitter}}} \\
&= \frac{\rho_{\text{Contact (Specific)}}}{L_{\text{Emitter}} W_{\text{Emitter contact}}} + \frac{T_{\text{N}^+\text{InGaAs}}}{N_{D\text{N}^+\text{InGaAs}} q \mu_{\text{N}^+\text{InGaAs}} L_{\text{Emitter}} W_{\text{Emitter contact}}} \\
&+ \frac{T_{N^+\text{InP}}}{N_{D\text{N}^+\text{InP}} q \mu_{\text{N}^+\text{InP}} L_{\text{Emitter}} W_{\text{Emitter junction}}} \\
&+ \frac{T_{\text{NInP}}}{N_{D\text{NInP}} q \mu_{\text{NInP}} L_{\text{Emitter}} W_{\text{Electrically active emitter}}}
\end{aligned}
\tag{5.74}
$$

because

$$\rho_{\text{Contact(Specific)}} = \text{Contact resistance} \times \text{Area} = \frac{\rho_{\text{Contact}} T_{\text{Contact}}}{L_{\text{Emitter}} W_{\text{Emitter}}} \times L_{\text{Emitter}} W_{\text{Emitter}} = \rho_{\text{Contact}} T_{\text{Contact}} \tag{5.75}$$

$$\rho_{\text{N}^+\text{InGaAs}} = \frac{1}{N_{D\text{N}^+\text{InGaAs}} q \mu_{\text{N}^+\text{InGaAs}}} \tag{5.76}$$

$$\rho_{\text{N}^+\text{InP}} = \frac{1}{N_{D\text{N}^+\text{InP}} q \mu_{\text{N}^+\text{InP}}} \tag{5.77}$$

and

$$\rho_{\text{NInP}} = \frac{1}{N_{D\text{NInP}} q \mu_{\text{NInP}}} \tag{5.78}$$

Note that

$$W_{\text{Emitter junction}} < W_{\text{Emitter contact}} \tag{5.79}$$

This arises from lateral undercutting of the emitter, and occurs during fabrication when etching of emitter–base junction is carried out. Moreover,

$$W_{\text{Electrically active emitter}} < W_{\text{Emitter junction}} \tag{5.80}$$

Surface depletion regions are responsible for this difference.

Equation (5.74) is rewritten as

$$\begin{aligned} R_{\text{Emitter}} = R_{\text{E}} &= \frac{\rho_{\text{Contact(Specific)}}}{L_{\text{Emitter}} W_{\text{Emitter contact}}} \\ &+ \frac{T_{\text{N}^+\text{InGaAs}}}{N_{D\text{N}^+\text{InGaAs}} q \mu_{\text{N}^+\text{InGaAs}} L_{\text{Emitter}} W_{\text{Emitter contact}}} \\ &+ \frac{T_{\text{N}^+\text{InP}}}{N_{D\text{N}^+\text{InP}} q \text{N} \mu_{\text{N}^+\text{InP}} L_{\text{Emitter}} W_{\text{Emitter Junction}}} \\ &+ \frac{T_{\text{NInP}}}{N_{\text{DNInP}} q \mu_{\text{NInP}} L_{\text{Emitter}} W_{\text{Electrically active emitter}}} \end{aligned} \tag{5.81}$$

Putting the given values in equation (5.81),

$$\begin{aligned} R_{\text{Emitter}} = R_{\text{E}} &= \frac{1.5\,\Omega - \mu\text{m}^2}{0.2\,\mu\text{m} \times 100\,\text{nm}} \\ &+ \frac{50\,\text{nm}}{1 \times 10^{19}\,\text{cm}^{-3} \times 1.602 \times 10^{-19}\,\text{C} \times 12000\,\text{cm}^2\,\text{V}^{-1}\,\text{s}^{-1} \times 0.2\,\mu\text{m} \times 100\,\text{nm}} \\ &+ \frac{20\,\text{nm}}{5 \times 10^{18}\,\text{cm}^{-3} \times 1.602 \times 10^{-19}\,\text{C} \times 5400\,\text{cm}^2\,\text{V}^{-1}\,\text{s}^{-1} \times 0.2\,\mu\text{m} \times 95\,\text{nm}} \\ &+ \frac{30\,\text{nm}}{3 \times 10^{17}\,\text{cm}^{-3} \times 1.602 \times 10^{-19}\text{C} \times 5400\,\text{cm}^2\,\text{V}^{-1}\,\text{s}^{-1} \times 0.2\,\mu\text{m} \times 90\,\text{nm}} \end{aligned} \tag{5.82}$$

The unit of first term of equation (5.82) is

$$\frac{\Omega - \mu m^2}{\mu m \times nm} = \frac{\Omega - \mu m^2}{\mu m \times 10^{-3}\mu m} = 10^3\ \Omega \tag{5.83}$$

The second, third and fourth terms of equation (5.82) have the same unit:

$$\begin{aligned}\frac{nm}{cm^{-3} \times C \times cm^2\, V^{-1} s^{-1} \times \mu m \times nm} &= \frac{1}{cm^{-1} \times C \times V^{-1} s^{-1} \times \mu m}\\ &= \frac{1}{cm^{-1} \times C \times V^{-1} s^{-1} \times 10^{-4} cm}\\ &= 10^4 \frac{V}{C \times s^{-1}} = 10^4 \frac{V}{A} = 10^4\ \Omega\end{aligned} \tag{5.84}$$

Changing units in equation (5.82) according to equations (5.83) and (5.84), we get

$$\begin{aligned}R_{\text{Emitter}} = R_E = &\frac{1.5}{0.2 \times 100} \times 10^3\ \Omega\\ &+ \frac{50}{1 \times 10^{19} \times 1.602 \times 10^{-19} \times 12000 \times 0.2 \times 100} \times 10^4\ \Omega\\ &+ \frac{20}{5 \times 10^{18} \times 1.602 \times 10^{-19} \times 5400 \times 0.2 \times 95} \times 10^4\ \Omega\\ &+ \frac{30}{3 \times 10^{17} \times 1.602 \times 10^{-19} \times 5400 \times 0.2 \times 90} \times 10^4\ \Omega = 75\\ &+ 1.3 + 2.4336 + 64.22 = 142.9536 \approx 143\ \Omega\end{aligned} \tag{5.85}$$

Example 5.4 Calculate the base resistance if emitter length $L_{\text{Emitter}} = 0.2\ \mu m$, emitter width $W_{\text{Emitter}} = 100$ nm, base contact width $W_{\text{b, Contact}}$ is 70 nm, emitter–base spacing $W_{\text{Emitter–base spacing}}$ is 30 nm, base contact resistivity $\rho_{\text{b, Contact}}$ is 2 Ω-μm^2, and base sheet resistance $R_{\text{b, Sheet}}$ is 800 $\Omega\ sq^{-1}$.

Base resistance R_{bb} is the sum of three terms (figure 5.8): contact resistance $R_{\text{b, Contact}}$, emitter–base gap resistance $R_{\text{b, gap}}$ and spreading resistance under the emitter $R_{\text{b, Spreading}}$.

$$R_{bb} = R_{b,\text{Contact}} + R_{b,\text{gap}} + R_{b,\text{Spreading}} \tag{5.86}$$

where

$$R_{b,\text{Contact}} = \frac{\sqrt{R_{b,\text{Sheet}}\rho_{b,\text{Contact}}}}{2L_{\text{Emitter}}} \coth\left(W_{b,\text{Contact}}\sqrt{\frac{R_{b,\text{Sheet}}}{\rho_{b,\text{Contact}}}}\right) \tag{5.87}$$

Since

$$\coth(x) = x^{-1} + \frac{1}{3}x - \frac{1}{45}x^3 + \dots \tag{5.88}$$

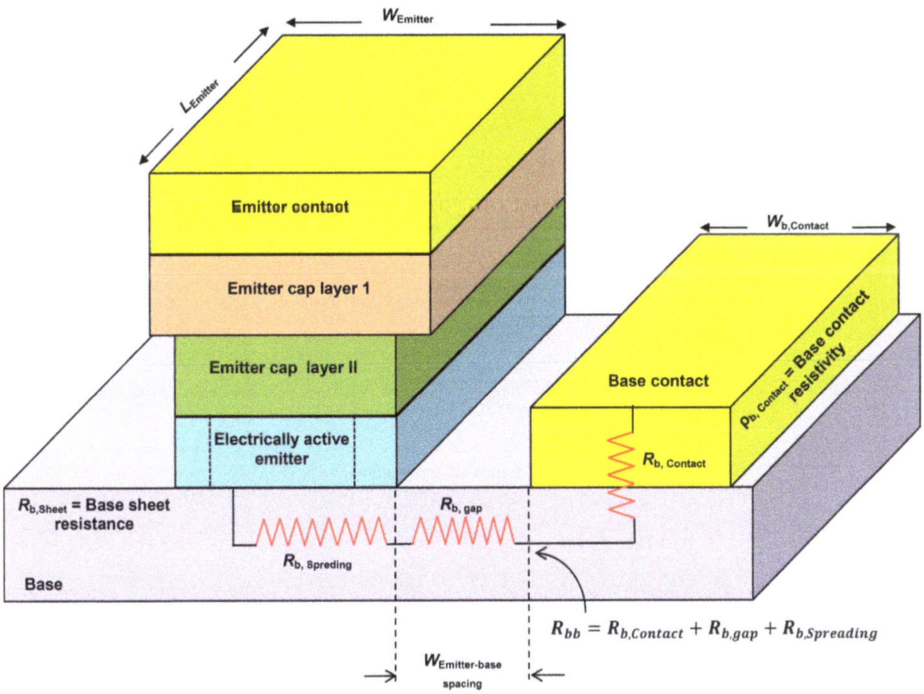

Figure 5.8. Emitter, sub-emitter and base parameters used in calculation of base resistance of an HBT.

expanding up to first two terms

$$\coth\left(W_{b,\text{ Contact}}\sqrt{\frac{R_{b,\text{ Sheet}}}{\rho_{b,\text{Contact}}}}\right) \approx \frac{1}{W_{b,\text{ Contact}}\sqrt{\frac{R_{b,\text{ Sheet}}}{\rho_{b,\text{Contact}}}}} + \frac{1}{3}W_{b,\text{ Contact}}\sqrt{\frac{R_{b,\text{ Sheet}}}{\rho_{b,\text{Contact}}}} \tag{5.89}$$

$$\begin{aligned}
\therefore R_{b,\text{ Contact}} &= \frac{\sqrt{R_{b,\text{ Sheet}}\rho_{b,\text{Contact}}}}{2L_{\text{Emitter}}} \times \frac{1}{W_{b,\text{ Contact}}\sqrt{\frac{R_{b,\text{ Sheet}}}{\rho_{b,\text{Contact}}}}} \\
&\quad + \frac{\sqrt{R_{b,\text{ Sheet}}\rho_{b,\text{Contact}}}}{2L_{\text{Emitter}}} \times \frac{1}{3}W_{b,\text{ Contact}}\sqrt{\frac{R_{b,\text{ Sheet}}}{\rho_{b,\text{Contact}}}} \\
&= \frac{\sqrt{R_{b,\text{ Sheet}}\rho_{b,\text{Contact}}}}{2L_{\text{Emitter}}} \times \frac{1}{W_{b,\text{ Contact}}}\sqrt{\frac{\rho_{b,\text{Contact}}}{R_{b,\text{ Sheet}}}} + \frac{W_{b,\text{ Contact}}R_{b,\text{ Sheet}}}{6L_{\text{Emitter}}} \\
&= \frac{\rho_{b,\text{Contact}}}{2L_{\text{Emitter}}W_{b,\text{ Contact}}} + \frac{W_{b,\text{ Contact}}R_{b,\text{ Sheet}}}{6L_{\text{Emitter}}}
\end{aligned} \tag{5.90}$$

Further,

$$R_{b,\text{ gap}} = \frac{R_{b,\text{ Sheet}} W_{\text{Emitter-Base spacing}}}{2L_{\text{Emitter}}} \tag{5.91}$$

$$R_{b,\text{ Spreading}} = \frac{R_{b,\text{ Sheet}} W_{\text{Emitter}}}{12L_{\text{Emitter}}} \tag{5.92}$$

Combining together equations (5.90), (5.91) and (5.92), we get

$$\begin{aligned} R_{\text{bb}} &= \frac{\rho_{b,\text{Contact}}}{2L_{\text{Emitter}} W_{b,\text{ Contact}}} + \frac{W_{b,\text{ Contact}} R_{b,\text{ Sheet}}}{6L_{\text{Emitter}}} + \frac{R_{b,\text{ Sheet}} W_{\text{Emitter-Base spacing}}}{2L_{\text{Emitter}}} \\ &+ \frac{R_{b,\text{ Sheet}} W_{\text{Emitter}}}{12L_{\text{Emitter}}} = \frac{\rho_{b,\text{Contact}}}{2L_{\text{Emitter}} W_{b,\text{ Contact}}} \\ &+ \frac{R_{b,\text{ Sheet}}}{2L_{\text{Emitter}}}\left(\frac{W_{b,\text{ Contact}}}{3} + W_{\text{Emitter-Base spacing}} + \frac{W_{\text{Emitter}}}{6}\right) \end{aligned} \tag{5.93}$$

Equation (5.93) is rewritten as

$$\begin{aligned} R_{\text{bb}} = &\frac{\rho_{b,\text{Contact}}}{2L_{\text{Emitter}} W_{b,\text{ Contact}}} \\ &+ \frac{R_{b,\text{ Sheet}}}{2L_{\text{Emitter}}}\left(\frac{W_{b,\text{ Contact}}}{3} + W_{\text{Emitter-Base spacing}} + \frac{W_{\text{Emitter}}}{6}\right) \end{aligned} \tag{5.94}$$

Inputting the given values in equation (5.94) we get

$$\begin{aligned} R_{\text{bb}} &= \frac{2\,\Omega - \mu\text{m}^2}{2 \times 0.2\,\mu\text{m} \times 70\text{ nm}} + \frac{800\,\Omega\text{ sq}^{-1}}{2 \times 0.2\,\mu\text{m}}\left(\frac{70\text{ nm}}{3} + 30\text{ nm} + \frac{100\text{ nm}}{6}\right) \\ &= \frac{2\,\Omega - \mu\text{m}^2}{2 \times 0.2\,\mu\text{m} \times 70 \times 10^{-3}\mu\text{m}} \\ &+ \frac{800\,\Omega\text{ sq}^{-1}}{2 \times 0.2\,\mu\text{m}}\left(\frac{70}{3} + 30 + \frac{100}{6}\right) \times 10^{-3}\mu\text{m} \\ &= 71.428\,57 + 2000(23.33 + 30 + 16.67) \times 10^{-3}\,\Omega \\ &= 71.428\,57 + 2 \times 70 = 211.428\,57\,\Omega = 211.43\,\Omega \end{aligned} \tag{5.95}$$

Example 5.5 Collect the relevant parameters and terms in f_{T} and f_{Max} equations for the HBT structure considered in examples 5.1 to 5.4, and apply them to calculate both the frequencies for this HBT device. Which terms in f_{T} equation are most crucial in enhancing the HBT performance at high frequencies? Hence justify the preference of using HBT over BJT for this regime of frequencies.

Solution: Values of the following parameters/terms are of interest in this calculation:

R_{C}~216.74 Ω (from equation 5.66), $g_{\text{m}} = 0.8$ (example 5.2b), $C_{\text{jBE}} \sim 1.23 \times 10^{-16}$ F (from equation 5.70), $W_{\text{Depletion}}|_{\text{BC}} = 3.6666 \times 10^{-7}$ m (from equation 5.71),

$C_{jBC} \sim 2\times10^{-17}$ F (from equation 5.72), $C_{Diff(BE)} = 2.2857 \times 10^{-14}$ F (from equation 5.73), $R_{Emitter} \sim 143\ \Omega$ (from equation 5.85), $R_{bb} \sim 211.43\ \Omega$ (from equation 5.95), $L_{Emitter} = 0.2$ μm (example 5.4), $W_{Emitter} = 0.1$ μm (example 5.4), $v_{Sat}|_{InGaAs} = 3 \times 10^7$ ms^{-1} (given).

Inputting the values in the denominator terms of equation (5.17),

$$\frac{C_{Diff(BE)}}{g_m} = \frac{2.2857 \times 10^{-14}\ \text{F}}{0.8\ \text{S}} = 2.857 \times 10^{-14}\frac{\text{F}}{\text{S}} = 2.857 \times 10^{-14}\frac{\text{s}^4\ \text{A}^2\ \text{m}^{-2}\ \text{kg}^{-1}}{\text{kg}^{-1}\ \text{m}^{-2}\ \text{s}^3\ \text{A}^2} = 2.857 \times 10^{-14}\ \text{s} \tag{5.96}$$

$$\frac{C_{Depletion(BE)} + C_{Depletion(BC)}}{g_m} = \frac{C_{jBE} + C_{jBC}}{g_m} = \frac{1.23 \times 10^{-16}\ \text{F} + 2 \times 10^{-17}\ \text{F}}{0.8\ \text{S}} = \frac{1.43 \times 10^{-16}}{0.8}\frac{\text{F}}{\text{S}} = 1.7875 \times 10^{-16}\ \text{s} \tag{5.97}$$

$$C_{Depletion(BC)}(R_{Emitter} + R_{Collector}) = C_{jBC}(R_{Emitter} + R_{Collector}) = 2 \times 10^{-17}\ \text{F} \times (143\ \Omega + 216.74\ \Omega) = 2 \times 10^{-17} \times 359.74\ \text{F} \times \Omega = 719.48 \times 10^{-17}\ \text{s}^4\ \text{A}^2\ \text{m}^{-2}\ \text{kg}^{-1} \times \text{kg}\ \text{m}^2\ \text{s}^{-3}\ \text{A}^{-2} = 7.2 \times 10^{-15}\ \text{s} \tag{5.98}$$

$$\frac{W_{Depletion(BC)}}{2v_{Sat}} = \frac{W_{Depletion}|_{BC}}{2v_{Sat}} = \frac{3.6666 \times 10^{-7}\text{m}}{2 \times 3 \times 10^7\text{ms}^{-1}} = 0.6111 \times 10^{-7-7} = 6.111 \times 10^{-15}\text{s} \tag{5.99}$$

Substituting the values from equations (5.96–5.99) into equation (5.17)

$$\begin{aligned} f_T &= \frac{1}{2 \times 3.141\,59\{2.857 \times 10^{-14}\ \text{s} + 1.7875 \times 10^{-16}\ \text{s} + 7.2 \times 10^{-15}\ \text{s} + 6.111 \times 10^{-15}\ \text{s}\}} \\ &= \frac{1}{2 \times 3.141\,59\{2.857 \times 10^{-14}\ \text{s} + 0.0178\,75 \times 10^{-14}\ \text{s} + 0.72 \times 10^{-14}\ \text{s} + 0.6111 \times 10^{-14}\ \text{s}\}} \\ &= \frac{10^{14}}{2 \times 3.141\,59 \times 4.205975\ \text{s}} = 0.03784 \times 10^{14}\ \text{Hz} = 3.784 \times 10^{12}\ \text{Hz} \\ &\approx 3.8\ \text{THz} \end{aligned} \tag{5.100}$$

Putting the values from equations (5.95) and (5.100) in equation (5.18),

$$
\begin{aligned}
f_{\mathrm{Max}} &\approx \sqrt{\frac{3.8\times 10^{12}\ \mathrm{Hz}}{8\times 3.141\,59\times 211.43\ \Omega\times 2\times 10^{-17}\ \mathrm{F}}} \\
&= \sqrt{\frac{3.8\ \times 10^{12+17}\ \mathrm{Hz}}{8\times 3.141\,59\times 211.43\times 2\times \Omega\times \mathrm{F}}} \\
&= \sqrt{\frac{3.8\times 10^{29}\ \mathrm{Hz}}{8\times 3.141\,59\times 211.43\times 2\times \mathrm{s}}} = \sqrt{0.0003576\times 10^{29}\ \mathrm{Hz}^2} \\
&= \sqrt{35.76\ \times 10^{24}\ \mathrm{Hz}^2} = 5.9799\times 10^{12}\ \mathrm{Hz}\approx 5.98\times 10^{12}\ \mathrm{Hz} \\
&= 5.98\ \mathrm{THz}
\end{aligned}
\tag{5.101}
$$

As we can see, the first term in the denominator of cut-off frequency equation (5.100) is

$$= \frac{2.857\times 10^{-14}\ \mathrm{s}}{0.017875\times 10^{-14}\ \mathrm{s}} = 159.83\ \text{times the second term}$$

$$= \frac{2.857\times 10^{-14}\ \mathrm{s}}{0.72\times 10^{-14}\ \mathrm{s}} = 3.968\ \text{times the third term}$$

and

$$= \frac{2.857\times 10^{-14}\ \mathrm{s}}{0.6111\times 10^{-14}\ \mathrm{s}} = 4.675\ \text{times the fourth term} \tag{5.102}$$

So the first term plays a decisive role in controlling the cut-off frequency followed by the third term. A closer inspection of this term shows that the first term represents the transit time of carriers through the base and contains the square of base width in the numerator. Hence the base width critically determines the cut-off frequency. The permission to choose a high base doping concentration in an HBT enables us to thin down the base appreciably and hence decrease the transit time of carriers through the base considerably with respect to a homojunction transistor.

Arguing on similar lines we note that the equation for f_{Max} contains the base resistance R_{bb} in the denominator. A scrutiny of the base resistance expression shows that all the three contributory terms, viz., the contact resistance, the spreading resistance under the emitter and the resistance of the gap between emitter and base must be kept as low as possible to get high maximum oscillation frequency. Both the spreading and gap resistances being dependent on sheet resistance of the base, its value needs careful choice and needs to be closely monitored during the process for design optimization. These remarks are in harmony with the comments regarding HBT in section 5.1.2.

5.6 DHBT IC technologies

All the required transmitter and receiver functionalities can be realized in an HBT IC platform because:

(i) HBTs fabricated with InP collector offer a higher breakdown voltage at a given current gain cut-off frequency than equivalent HEMT devices; and
(ii) the noise figures of HBTs are on a par with those of similar bandwidth HEMTs for HBTs working at 10%–20% of their maximum current density (Urteaga *et al* 2010).

5.6.1 250 nm InP DHBT IC technology

5.6.1.1 HBTs

The InP/InGaAs/InP HBT contains a wide bandgap InP emitter/collector and narrow bandgap InGaAs base (Seo *et al* 2010, 2011). The DC beta was 25 and common-emitter breakdown voltage was >4 V. Epitaxial layers were grown on InP substrates by molecular beam epitaxy. The HBT base layer was 30 nm thick. It was doped with carbon. Compositional grading was used to reduce transit time of carriers through the base. Thickness of N-collector was 150 nm. The transistor operates at current density >10 mA μm^{-2} and power density >20 mW μm^{-2}. The extrapolated f_T of the HBTs was 0.392 THz and f_{max} was 0.859 THz.

5.6.1.2 HBT oscillators

The HBT oscillators were designed in a differential topology with integrated output buffer. They were series-tuned because series tuning offers greater flexibility of layout in transmission line-based design than parallel tuning. It can therefore be easily scaled to various frequencies. Two schemes were adopted, viz., fixed-frequency and varactor-tuned. The differential implementation is advantageous over single-ended design because a virtual ground is produced during differential operation along the plane of symmetry. Hence the operation is not affected by common-mode impedance, e.g., the emitter-base bias circuit. Further, it provides a lossless AC ground avoiding ohmic and radiation losses.

At DC power $\leqslant$ 115 mW, the output powers were −5.1, −6.9, − 9.2, and −11.0 dBm, respectively, for 0.2674, 0.2868, 0.3102, and 0.3462 THz designs (Seo *et al* 2010). The output power was −6.2 dBm for 0.3102 THz oscillator. It was −5.6 dBm for the 0.4129 THz oscillator and −19.2 dBm for the 0.5731 THz oscillator. Up to 0.3 THz, the tuning bandwidths for varactor-tuned oscillators were 0.0106–0.0123 THz. Above this frequency, they were smaller (Seo *et al* 2011).

5.6.2 130 nm InP DHBT IC technology

5.6.2.1 HBTs

The base of the HBT was 25 nm thick carbon-doped InGaAs layer with compositional and doping grading (Urteaga *et al* 2016). Conduction band discontinuities at emitter–base and collector–base junctions were smoothened by using chirped superlattice. A superlattice is a structure comprising alternating layers of two different materials; here the materials are InGaAs and InAlAs. 'CHIRP' stands for 'coherent heterointerfaces for reflection and penetration'. In a chirped superlattice, the periodicity of alternating layers is gradually varied. The emitter–base grading improves the turn-on voltage. A 33 nm-thick portion of the collector consists of an InGaAs layer followed by a superlattice grading layer. The remaining 67 nm thick portion is an N^- InP layer. The total collector layer thickness is 33 + 67 = 100 nm.

The peak current gain was 15–20 and common emitter breakdown voltage was 3.5 V. The HBT operates at current densities >30 mA μm^{-2} and power densities >50 mW μm^{-2}. The f_T of the HBTs was 0.521 THz and f_{max} was 1.15 THz.

5.6.2.2 HBT amplifiers

A 9-stage common-base amplifier was fabricated. At 0.67 THz, the gain was 22 dB. The saturated output power was 0.4 mW at 0.585 THz. It decreased to 0.18 mW at 0.67 THz. Fundamental oscillators were also demonstrated (Urteaga *et al* 2016).

5.7 Discussion and conclusions

The HBT structure is delved into and inspected in detail. Different components in the transistor equivalent circuit are explained and their roles in influencing the high frequency operation of devices are elucidated. The knowledge gained is helpful both in understanding the design and process requirements for building high frequency transistors, and devising suitable techniques to fabricate these devices. HBT IC technologies have reached a gratifying level of advancement to fulfill the expectations from these transistors at these frequencies.

References

Ashburn P 2003 *SiGe Heterojunction Bipolar Transistors* (Chichester: Wiley) pp 71–92

Leibovitch M, Kronik L, Mishori B, Shapira Y, Hanson C M, Clawson A R and Ram P 1996 Determining band offsets using surface photovoltage spectroscopy: the InP/$In_{0.53}Ga_{0.47}As$ heterojunction *Appl. Phys. Lett.* **69** 2587–89

Pulfrey D 2010 HBT basics *Understanding Modern Transistors and Diodes* (Cambridge: Cambridge University Press) pp 153–68

Rücker H and Heinemann B 2018 SiGe HBT technology *Silicon-Germanium Heterojunction Bipolar Transistors for mm-Wave Systems: Technology, Modeling and Circuit Applications* eds N Rinaldi and M Schröter (Gistrup, Denmark: River Publishers) pp 11–54

Schwierz F and Liou J J 2001 Semiconductor devices for RF applications: evolution and current status *Microelectron. Reliab.* **4** 145–68

Seo M, Urteaga M, Young A, Jain V, Griffith Z, Hacker J, Rowell P, Pierson R and Rodwell M 2010 >300 GHz fixed-frequency and voltage-controlled fundamental oscillators in an InP DHBT process *2010 IEEE MTT-S Int. Microwave Symp. (23–28 May, 2010 Anaheim, CA)* 272–5

Seo M, Urteaga M, Hacker J, Young A, Griffith Z, Jain V, Pierson R, Rowell P, Skalare A, Peralta A, Lin R, Pukala D and Rodwell M 2011 InP HBT IC technology for terahertz frequencies: fundamental oscillators up to 0.57 THz *IEEE J. Solid-State Circ* **46** 2203–14

Urteaga M, Hacker J, Griffith Z, Young A, Pierson R, Rowell P, Seo M and Rodwell M J W 2016 A 130 nm InP HBT integrated circuit technology for THz electronics *2016 IEEE Int. Electron Devices Meeting (IEDM) (3–7 December 2016 San Francisco, CA)* 29.2.1–4

Urteaga M, Seo M, Hacker J, Griffith Z, Young A, Pierson R, Rowell P, Skalare A and Rodwell M J W 2010 InP HBT Integrated Circuit Technology for Terahertz Frequencies *2010 IEEE Compound Semiconductor Integrated Circuit Symp. (CSICS) (3–6 Oct., 2010 Monterey, CA)* 1–4

IOP Publishing

Practical Terahertz Electronics: Devices and Applications, Volume 1

Solid-state devices and vacuum tubes

Vinod Kumar Khanna

Chapter 6

Metal-oxide semiconductor field-effect transistors

At high frequencies, the operation of silicon metal-oxide semiconductor field-effect transistors (MOSFETs) deteriorates because the time taken by carriers in journey through the channel is so long that they cannot follow the fast cyclic changes. A different regime of MOSFET operation exploits the propagation of plasma waves in the channel. Dyakonov and Shur showed that the steady electron motion during DC current flow in a MOSFET can become unstable with respect to production of plasma waves. This instability called the Dyakonov and Shur instability, is the basis of a new mechanism of production of electromagnetic waves at terahertz frequencies using nanoscale MOSFETs. Besides terahertz emission, plasma waves in FET channel also open the possibility of use of MOSFETs in terahertz detection. An incoming electromagnetic radiation induces an AC voltage on the source side launching a plasma wave. The alternating radiation field in the channel of an antenna-coupled MOSFET is rectified due to its non-linear properties paving the way to detection of terahertz waves. Applications of MOSFET devices as terahertz sources and detectors are described, and the underlying physical phenomena are explained.

We shall briefly describe the working of the MOSFET. Since we are primarily interested in its high-frequency operation, we shall draw the equivalent circuit of a MOSFET, formulate the MOSFET model and apply it to assess the MOSFET performance at these frequencies.

6.1 MOSFET construction and operation

The MOSFET is a four-terminal device (Ytterdal *et al* 2003). The four terminals are called source, drain, gate and body or substrate. Consider an N-channel

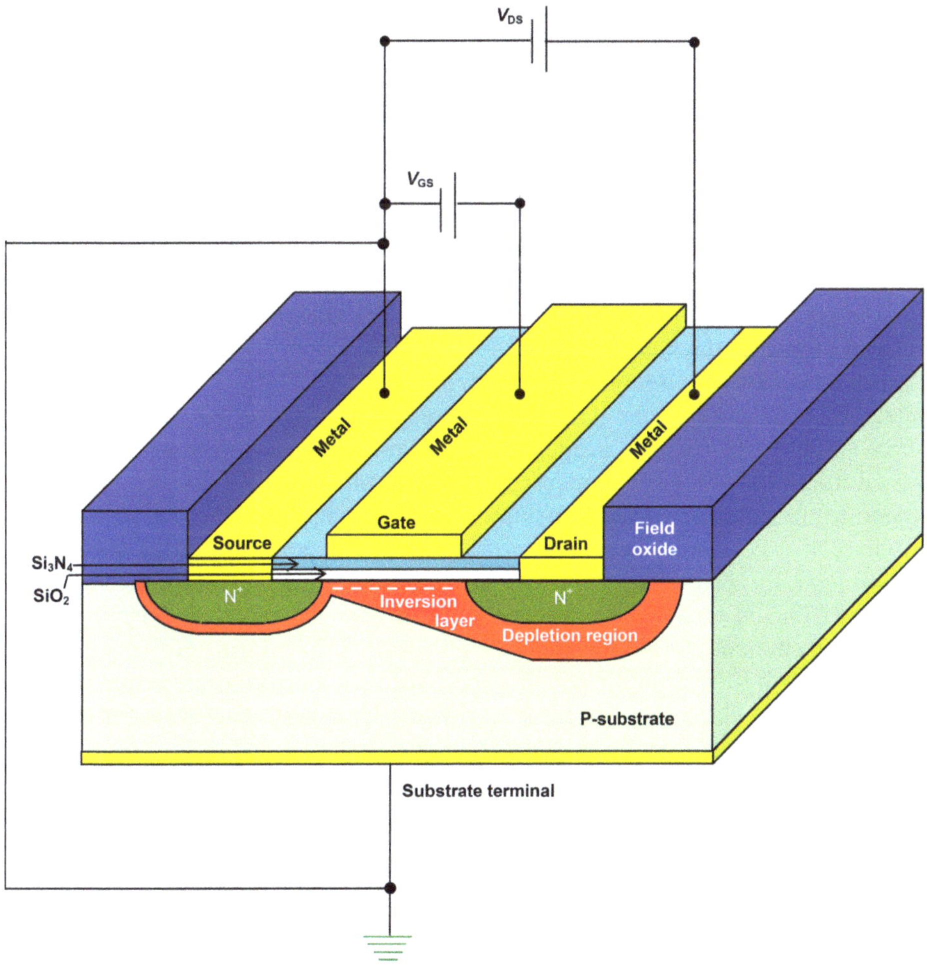

Figure 6.1. An N-channel MOSFET fabricated in a P-type silicon substrate. Construction of the MOSFET and its electrical connections are shown.

enhancement mode device (figure 6.1). The starting material is a P-type substrate. In this P-type substrate, N^+ doped source and drain regions are formed. A thin layer of silicon dioxide followed by silicon nitride covers the gate region. So we have a silicon dioxide–silicon nitride dual dielectric gate. Silicon nitride is a passivation film, which is necessary because silicon dioxide is moisture sensitive. But silicon nitride cannot be directly used as the gate insulator because of the poor quality of silicon/silicon nitride interface. Thick silicon dioxide called field oxide is grown in border regions separating individual devices. Metal film is deposited over the N^+ doped source and drain regions as well as the gate dielectric. Source, drain and drain contacts are formed by photolithography and chemical etching of this film to delineate the pattern.

Let us now describe the operation of an N-channel enhancement mode MOSFET in active region. The source and body terminals of the device are shorted together and connected to ground. A positive voltage is applied on the drain terminal with respect to the source. This is the drain–source voltage V_{DS}. A small positive voltage is applied on the gate electrode. This is the gate–source voltage V_{GS}.

The positive gate–source voltage repels the positively charged majority carrier holes in the P-type substrate from the region below the gate downwards in the substrate. Owing to this repulsion, a depletion region is formed underneath the gate terminal. The depletion region is deeper near the drain end and becomes shallower near the source end due to voltage dropped in the substrate as we move from the drain end towards the source end.

Simultaneously the minority carrier free electrons in the P type substrate are pulled upwards towards the gate insulator due to attraction by the positive gate–source voltage. The electron accumulation builds up and a thin sheet of electrons is formed adjoining the gate insulator/P-substrate interface. As the positive voltage applied to the gate is increased, a stage is reached when the electron population in the electron sheet layer increases so much that the polarity of this sheet region changes from P-type to N-type. As the polarity is reversed, an N-type inversion layer is formed. This N-type inversion layer containing a high electron density acts as a conducting channel connecting the N^+ source with N^+ drain. No sooner than a continuous path for current flow is set up between source and drain terminals via the channel, does an electronic current start flowing from source to drain resulting in a conventional current flow from drain to source, hence called drain–source current I_{DS}. The positive gate–source voltage necessary to create an inversion layer and establish a conducting channel between source and drain terminals is called the threshold voltage of MOSFET.

6.2 Short-circuit current gain

Figure 6.2 shows the equivalent circuit of the MOSFET at high frequencies. In figure 6.2(a), various components of the equivalent circuit are shown on the cross-sectional diagram of the MOSFET to bring out their association with the relevant portion of the device structure. These include the gate-to-source capacitance C_{GS}, the gate-to-drain capacitance C_{GD}, the series resistances R_S and R_D of the source and drain, respectively, the small-signal drain current $g_m\, v_{GS}$ due to gate–source voltage v_{GS}, (g_m is the transconductance) and the drain-source conductance g_{DS}. In figure 6.2(b), the equivalent circuit is drawn out clearly, as used for analysis. In figure 6.2(c), a simplified version of the circuit of figure 6.2(b) is shown in which series resistances R_S, R_D are assumed to be zero; hence deleted. In figure 6.2(d), the drain-source conductance g_{DS} is replaced by a short-circuit for short-circuit gain calculation. Applying Kirchoff's current law, and summing currents at input node 1,

$$i_{IN} - v_{GS} j\omega C_{GS} - v_{GS} j\omega C_{GD} = 0 \tag{6.1}$$

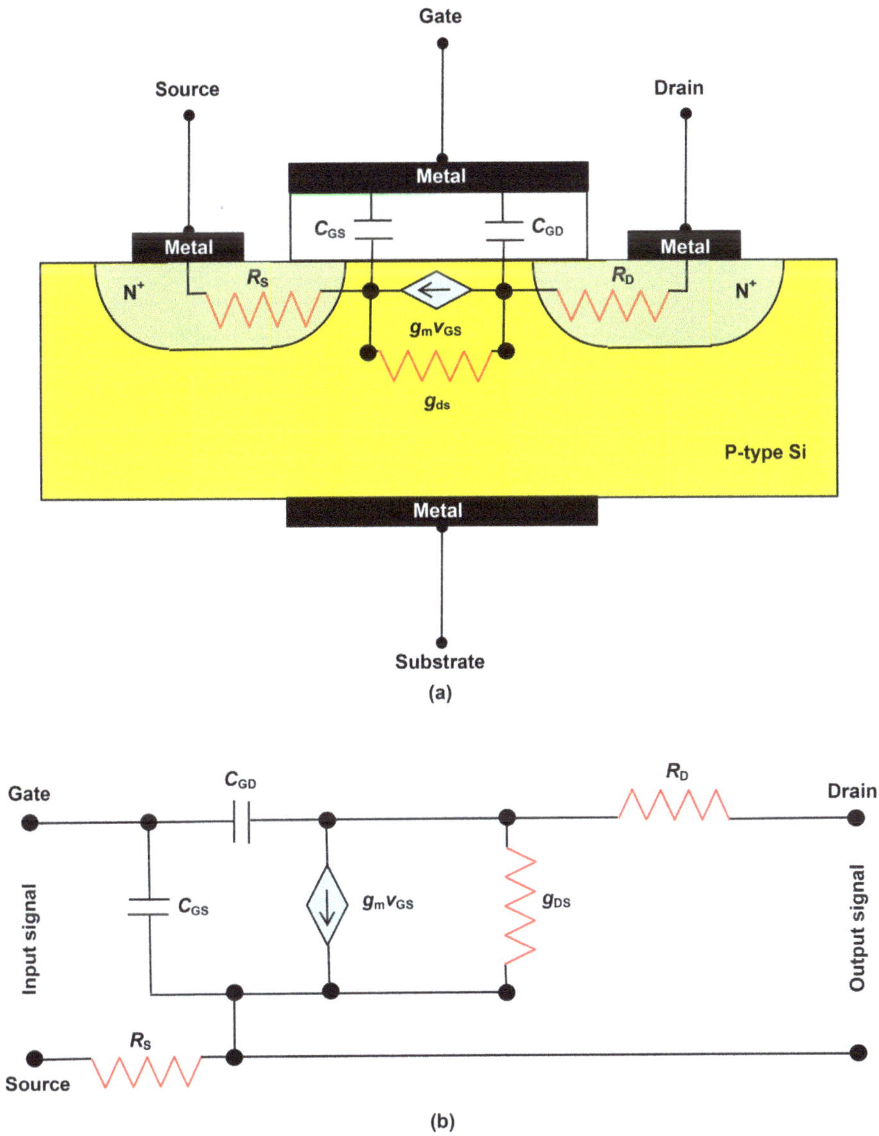

Figure 6.2. MOSFET equivalent circuit at high operating frequencies: (a) cross-sectional layout of the MOSFET showing how the various components are related to the device structure, (b) the circuit diagram, (c) simplified circuit assuming that source and drain series resistances are zero, and (d) diagram for calculation of short-circuit current gain.

where i_{IN} is the input current, ω is the angular frequency of AC source; also C_{GS}, C_{GD} are the gate–source and gate–drain capacitances, and v_{GS} is the gate-to-source voltage, as mentioned earlier. Equation (6.1) is rearranged to write

$$i_{\mathrm{IN}} = v_{GS} j\omega C_{GS} + v_{\mathrm{GS}} j\omega C_{\mathrm{GD}} = v_{\mathrm{GS}} j\omega (C_{\mathrm{GS}} + C_{\mathrm{GD}}) \tag{6.2}$$

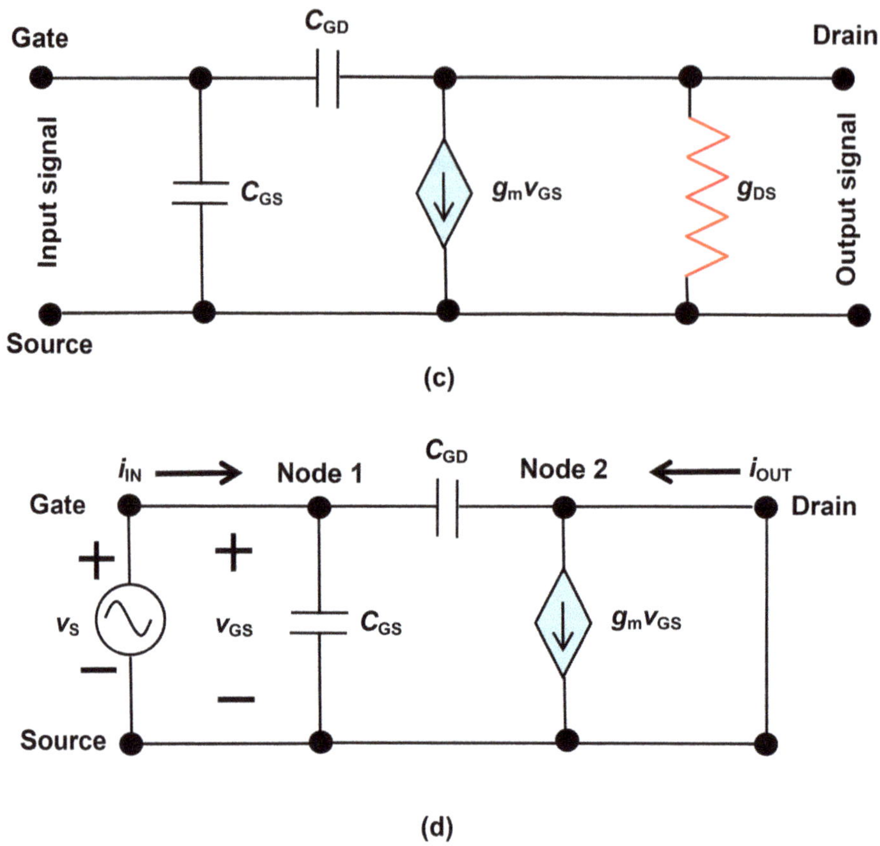

Figure 6.2. (Continued.)

Similarly summing currents at output node 2,

$$i_{OUT} - g_m v_{GS} + v_{GS} j\omega C_{GD} = 0 \tag{6.3}$$

where i_{OUT} is the output current. Rearranging equation (6.3) we get

$$i_{OUT} = g_m v_{GS} - v_{GS} j\omega C_{GD} = v_{GS}(g_m - j\omega C_{GD}) \tag{6.4}$$

Dividing equation (6.4) by equation (6.2), we get the short-circuit current gain

$$\begin{aligned} A_i = \frac{i_{OUT}}{i_{IN}} = \frac{v_{GS}(g_m - j\omega C_{GD})}{v_{GS} j\omega(C_{GS} + C_{GD})} = \frac{g_m - j\omega C_{GD}}{j\omega(C_{GS} + C_{GD})} = \frac{g_m}{j\omega(C_{GS} + C_{GD})} \\ - \frac{j\omega C_{GD}}{j\omega(C_{GS} + C_{GD})} = \frac{g_m}{j\omega(C_{GS} + C_{GD})} - \frac{C_{GD}}{(C_{GS} + C_{GD})} \end{aligned} \tag{6.5}$$

This equation shows that:

$$\text{As } \omega \to 0, \; A_i \to \infty \tag{6.6}$$

As zero frequency corresponds to DC, this result is as per expectation because the DC current flowing through the gate is zero except for a very small leakage current.

6.3 MOSFET capacitances

6.3.1 Gate-to-channel capacitance

The gate-to-channel capacitor is a parallel-plate capacitor of area

$$A = \text{Channel width} \times \text{Channel length} = WL \tag{6.7}$$

with the SiO_2 film of relative permittivity ε_{ox} and thickness d as the dielectric. It is expressed as

$$C_{\text{Gate-to-channel}} = \frac{\varepsilon_0 \varepsilon_{ox} A}{d} = \frac{\varepsilon_0 \varepsilon_{ox} WL}{d} \tag{6.8}$$

where ε_0 is the permittivity of free space and ε_{ox} is the dielectric constant of oxide film.

6.3.2 Gate–source and gate–drain capacitances

We shall write down the equations for C_{GS} and C_{GD} by considering the various factors contributing to these capacitances. For a MOSFET operating in the linear region of its current–voltage characteristics, the gate-to-channel capacitance is divided equally between the source and drain regions. By this subdivision one half of the capacitance in equation (6.8) can be ascribed to source and the other half to drain. When writing the equation for C_{GS}, another contribution to C_{GS} coming from overlap capacitance must be included; hence

$$C_{GS} = \frac{1}{2} \frac{\varepsilon_0 \varepsilon_{ox} WL}{d} + C_{\text{Overlap}} \tag{6.9}$$

where C_{overlap} is the overlap capacitance arising from the overlapping of the gate electrode over the heavily-doped source region. For short-geometry MOSFETs, the overlap capacitance can be as high as 50% of the total gate-to-source capacitance. A similar capacitance term is added to the gate–drain capacitance. Therefore

$$C_{GD} = C_{GS} = \frac{1}{2} \frac{\varepsilon_0 \varepsilon_{ox} WL}{d} + C_{\text{Overlap}} \tag{6.10}$$

But during forward active mode of operation of the MOSFET, the channel pinches off at the drain end. So the gate-to-channel capacitance contains only the gate–source capacitance and that also of smaller value because the charge in the channel decreases linearly from the source end to the point near the drain where the channel pinches off. A rigorous analysis shows that when the MOSFET is in strong inversion and under saturation,

$$C_{GS} = \frac{2}{3} \frac{\varepsilon_0 \varepsilon_{ox} WL}{d} + C_{\text{Overlap}} \tag{6.11}$$

and

$$C_{\mathrm{GD}} = \frac{1}{3}\frac{\varepsilon_0\varepsilon_{\mathrm{ox}} WL}{d} + C_{\mathrm{Overlap}} \approx C_{\mathrm{Overlap}} \tag{6.12}$$

6.4 Cut-off frequency

6.4.1 The cut-off frequency formula

f_{T} is the frequency at which

$$|A_i| = 1 \tag{6.13}$$

In equation (6.5) for A_{i}, the second term arises from the current flowing through C_{GD}. This term is generally very small and hence negligible. Then

$$A_i = \frac{g_m}{j\omega(C_{\mathrm{GS}} + C_{\mathrm{GD}})} \tag{6.14}$$

At cut-off frequency

$$f = f_T \tag{6.15}$$

or,

$$2\pi f = 2\pi f_T \tag{6.16}$$

or,

$$\omega = \omega_T \tag{6.17}$$

So, equations (6.13), (6.14) and (6.17) give

$$\frac{g_m}{\omega_T(C_{\mathrm{GS}} + C_{\mathrm{GD}})} = 1 \tag{6.18}$$

or,

$$\frac{g_m}{2\pi f_T(C_{\mathrm{GS}} + C_{\mathrm{GD}})} = 1 \tag{6.19}$$

or,

$$f_T = \frac{g_m}{2\pi(C_{\mathrm{GS}} + C_{\mathrm{GD}})} \tag{6.20}$$

where equations (6.15)–(6.16) are applied.

6.4.2 Physical meaning of cut-off frequency formula

Equation (6.20) can be approximated as

$$f_T \cong \frac{g_m}{2\pi C_{\text{GS}}} \tag{6.21}$$

Substituting for C_{GS} from equation (6.11) in equation (6.21)

$$f_T = \frac{g_m}{2\pi\left(\frac{2}{3}\frac{\varepsilon_0\varepsilon_{\text{ox}}WL}{d} + C_{\text{Overlap}}\right)} \cong \frac{g_m}{2\pi\left(\frac{2}{3}\frac{\varepsilon_0\varepsilon_{\text{ox}}WL}{d}\right)} = \frac{g_m}{2\pi\left(\frac{2}{3}C_{\text{ox}}WL\right)} \tag{6.22}$$

where

$$C_{\text{ox}} = \text{Oxide capacitance per unit area} = \frac{\varepsilon_0\varepsilon_{\text{ox}}}{d} \tag{6.23}$$

Further, the transconductance g_{m} of MOSFET of channel length L and channel width W is obtained from the equation for its drain–source current

$$I_{\text{DS}} = \mu C_{\text{ox}}\frac{W}{L}\left(V_{\text{GS}} - V_{\text{Th}} - \frac{V_{\text{DS}}}{2}\right)V_{\text{DS}} \tag{6.24}$$

where μ is the mobility of charge carriers in the channel; V_{GS}, V_{DS}, V_{Th} are the gate-to-source voltage, drain–source voltage and threshold voltage of the MOSFET, respectively. The drain current saturates when

$$V_{\text{DS}} = V_{\text{GS}} - V_{\text{Th}} \tag{6.25}$$

Then equation (6.24) becomes

$$\begin{aligned} I_{\text{DS}} &= \mu C_{\text{ox}}\frac{W}{L}\left(V_{\text{GS}} - V_{\text{Th}} - \frac{V_{\text{GS}} - V_{\text{Th}}}{2}\right)(V_{\text{GS}} - V_{\text{Th}}) \\ &= \mu C_{\text{ox}}\frac{W}{L}\left\{(V_{\text{GS}} - V_{\text{Th}})^2 - \frac{(V_{\text{GS}} - V_{\text{Th}})^2}{2}\right\} \\ &= \mu C_{\text{ox}}\frac{W}{L}\frac{(V_{\text{GS}} - V_{\text{Th}})^2}{2} \end{aligned} \tag{6.26}$$

The transconductance is the variation of drain current with gate-to-source voltage keeping the drain–source voltage constant

$$\begin{aligned} g_m &= \left.\frac{\partial I_{\text{DS}}}{\partial V_{\text{GS}}}\right]_{V_{\text{DS}} = \text{Constant}} = \frac{\partial}{\partial V_{\text{GS}}}\left\{\mu C_{\text{ox}}\frac{W}{L}\frac{(V_{\text{GS}} - V_{\text{Th}})^2}{2}\right\} \\ &= \mu C_{\text{ox}}\frac{W}{L} \times \frac{\partial}{\partial V_{\text{GS}}}\left\{\frac{(V_{\text{GS}} - V_{\text{Th}})^2}{2}\right\} = \mu C_{\text{ox}}\frac{W}{L} \times 2\frac{V_{\text{GS}} - V_{\text{Th}}}{2}(1 - 0) \\ &= \mu C_{\text{ox}}\frac{W}{L}(V_{\text{GS}} - V_{\text{Th}}) \end{aligned} \tag{6.27}$$

Substituting for g_m from equation (6.27) in equation (6.22) we get

$$f_T = \frac{g_m}{2\pi\left(\frac{2}{3}C_{ox}WL\right)} = \frac{\mu C_{ox}\frac{W}{L}(V_{GS} - V_{Th})}{2\pi\left(\frac{2}{3}C_{ox}WL\right)}$$

$$= \mu C_{ox}\frac{W}{L}(V_{GS} - V_{Th}) \times \frac{1}{2\pi\left(\frac{2}{3}C_{ox}WL\right)} = \frac{\mu\frac{3}{2}\frac{V_{GS} - V_{Th}}{L}}{2\pi L} \quad (6.28)$$

$$= \frac{\mu < \text{Electric field} >}{2\pi L} = \frac{\text{Carrier velocity}}{2\pi L} = \frac{1}{2\pi}\left(\frac{v_d}{L}\right)$$

because electric field is the gradient of potential and carrier mobility is drift velocity per unit electric field.

Expressing the drift velocity of carriers in terms of their source-to-drain transit time

$$f_T = \frac{1}{2\pi}\left(\frac{\dfrac{L}{\text{Transit time of carriers from source to drain}}}{L}\right)$$

$$= \frac{1}{2\pi}\frac{L}{\text{Transit time of carriers from source to drain}} \times \frac{1}{L} = \frac{1}{2\pi\tau_t} \quad (6.29)$$

Thus the cut-off frequency is inversely proportional to the transit time of carriers under the gate, and can be reduced by decreasing the gate length L.

Example 6.1 (a) Calculate the gate–source and drain–source capacitances of an N-channel MOSFET operating under saturation assuming that the overlap capacitance is equal to the gate-to-channel capacitance. The MOSFET has gate width $W = 1$ μm, gate length $L = 50$ nm, gate oxide thickness $d = 3$ nm, and the dielectric constant of SiO_2 is 3.9. (b) Find the transconductance for $V_{GS} = 1.0$ V, threshold voltage = 0.5 V, and electron mobility in the channel = 675 cm^2 (V-s)$^{-1}$. (c) What is the cut-off frequency of the MOSFET?

Solution: (a) From equation (6.11) and replacing overlap capacitance by gate–source capacitance, we get

$$C_{GS} = \frac{2}{3}\frac{\varepsilon_0\varepsilon_{ox}WL}{d} + C_{Overlap} = 2 \times \frac{2}{3}\frac{\varepsilon_0\varepsilon_{ox}WL}{d}$$

$$= 2 \times \frac{2}{3} \times \frac{8.854 \times 10^{-12} \times 3.9 \times 1 \times 10^{-6} \times 50 \times 10^{-9}}{3 \times 10^{-9}} \quad (6.30)$$

$$= \frac{2 \times 2 \times 8.854 \times 3.9 \times 5}{3} \times 10^{-12-6+1} = 230.204 \times 10^{-17}$$

$$= 2.3 \times 10^{-15}\ \text{F}$$

Taking 2/3 contribution of oxide capacitance towards gate–source capacitance and 1/3 towards gate–drain capacitance, i.e., taking gate–drain capacitance as half the gate–source capacitance, we have

$$C_{GD} = \frac{2.3 \times 10^{-15}}{2} = 1.15 \times 10^{-15}\,\text{F} \tag{6.31}$$

(b) Applying equation (6.27)

$$\begin{aligned} g_m &= \mu C_{ox}\frac{W}{L}(V_{GS} - V_{Th}) = \mu\frac{\varepsilon_0\varepsilon_{ox}}{d}\frac{W}{L}(V_{GS} - V_{Th}) \\ &= 675 \times (10^{-2})^2\frac{8.854 \times 10^{-12} \times 3.9}{3 \times 10^{-9}}\frac{10^{-6}}{50 \times 10^{-9}}(1.0 - 0.5) \\ &= \frac{675 \times 8.854 \times 3.9 \times 0.5}{3 \times 5} \times 10^{-4-12-6+9-1+9} = 776.9385 \times 10^{-23+18} \\ &= 776.9385 \times 10^{-5} = 0.00777\,\text{S} \end{aligned} \tag{6.32}$$

(c) By equation (6.20)

$$\begin{aligned} f_T &= \frac{g_m}{2\pi(C_{GS} + C_{GD})} = \frac{0.00777}{2 \times 3.14159(2.3 \times 10^{-15} + 1.15 \times 10^{-15})} \\ &= \frac{0.00777 \times 10^{15}}{2 \times 3.14159(3.45)} = 0.000\,358\,4 \times 10^{15} = 3.584 \times 10^{11}\,\text{Hz} \\ &= \frac{3.584 \times 10^{11}}{10^9}\text{GHz} = 3.584 \times 10^2\,\text{GHz} = 358.4\,\text{GHz} \end{aligned} \tag{6.33}$$

Example 6.2 Calculate the cut-off frequency of the MOSFET of example 6.1 from the formula of equation (6.28) given that the electron saturation velocity in silicon is 1×10^5 m s^{-1}.

Solution: Applying equation (6.28)

$$\begin{aligned} f_T &= \frac{1}{2\pi}\left(\frac{v_d}{L}\right) = \frac{1}{2 \times 3.14159}\left(\frac{1 \times 10^5}{50 \times 10^{-9}}\right) = \frac{10^{5+9-1}}{2 \times 3.14159 \times 5} \\ &= 0.03183 \times 10^{13}\,\text{Hz} = 3.183 \times 10^{11}\,\text{Hz} = \frac{3.183 \times 10^{11}}{10^9} \\ &= 3.183 \times 10^2 = 318.3\,\text{GHz} \end{aligned} \tag{6.34}$$

Note: RF performance of practical MOSFETs is influenced by parasitic elements in the form of fringing capacitance which lowers the cut-off frequency because it is added to the capacitances in the denominator of equation (6.20) as

$$f_T = \frac{g_m}{2\pi(C_{GS} + C_{GD} + C_{Parasitic})} \tag{6.35}$$

so that cut-off frequency of transit-time MOSFETs typically lies in the GHz range, and may reach up to 0.5–0.8 THz with nanoscale devices. Equation (6.20) represents a gross over-simplification. Nonetheless, it gives an idea about the frequency response of the device and gives a first-cut estimate of the switching speed attainable by downscaling.

6.5 Circumventing the MOSFET speed limitations due to long electron transit time

Although it may seem difficult to use MOSFETs in the terahertz frequency range in the carrier transit mode despite aggressive reduction of dimensions, there is a silver lining in the cloud. The channel of a MOSFET serves as a resonant cavity in which plasma waves propagate with a velocity which is up to an order-of-magnitude higher than the electron drift velocity. For terahertz operation, FET devices must utilize plasma wave propagation instead of the customary trend of depending on electron drift in the electric field.

To understand these ideas, we proceed step-by-step by first learning about a resonant cavity and knowing how standing waves are formed in such a cavity? Then we ask what is meant by plasma waves? And more related questions.

6.5.1 Resonant cavity

A resonant cavity or cavity resonator is a hollow enclosure, e.g., a space bound by metallic boundaries and filled with dielectric material to confine electromagnetic waves. The dimensions of the enclosure are comparable to or greater than the wavelength of the electromagnetic radiation. The waves bounce backwards and forwards undergoing reflections from the metallic walls of the cavity. The waves propagating in opposite directions combine to form standing or stationary waves.

6.5.2 Standing waves

A standing wave, as its name implies, is a wave in which oscillations take place around a fixed location in space and the disturbance as a whole does not move from its position. The standing wave formed in a fully enclosed or partially bounded mass of water can be seen as water level bouncing upwards and downwards (figure 6.3). In the graphs displaying the waves, position x of a particle on the water wave along the horizontal direction is plotted on the X-axis. The vertical displacement y, i.e., the elevation/depression of the particle on the surface of the water wave is plotted on the Y-axis. The still water line is placed at zero point of the Y-axis, i.e., along the X-axis.

A standing wave is formed by the superimposition of two waves of the same amplitude and frequency travelling in opposite directions, such as in a resonant cavity. Interference takes place between these waves. During constructive interference, their energies are added together while in destructive interference, they become zero. The resulting wave contains points of no motion termed nodes and points of maximum amplitude called antinodes.

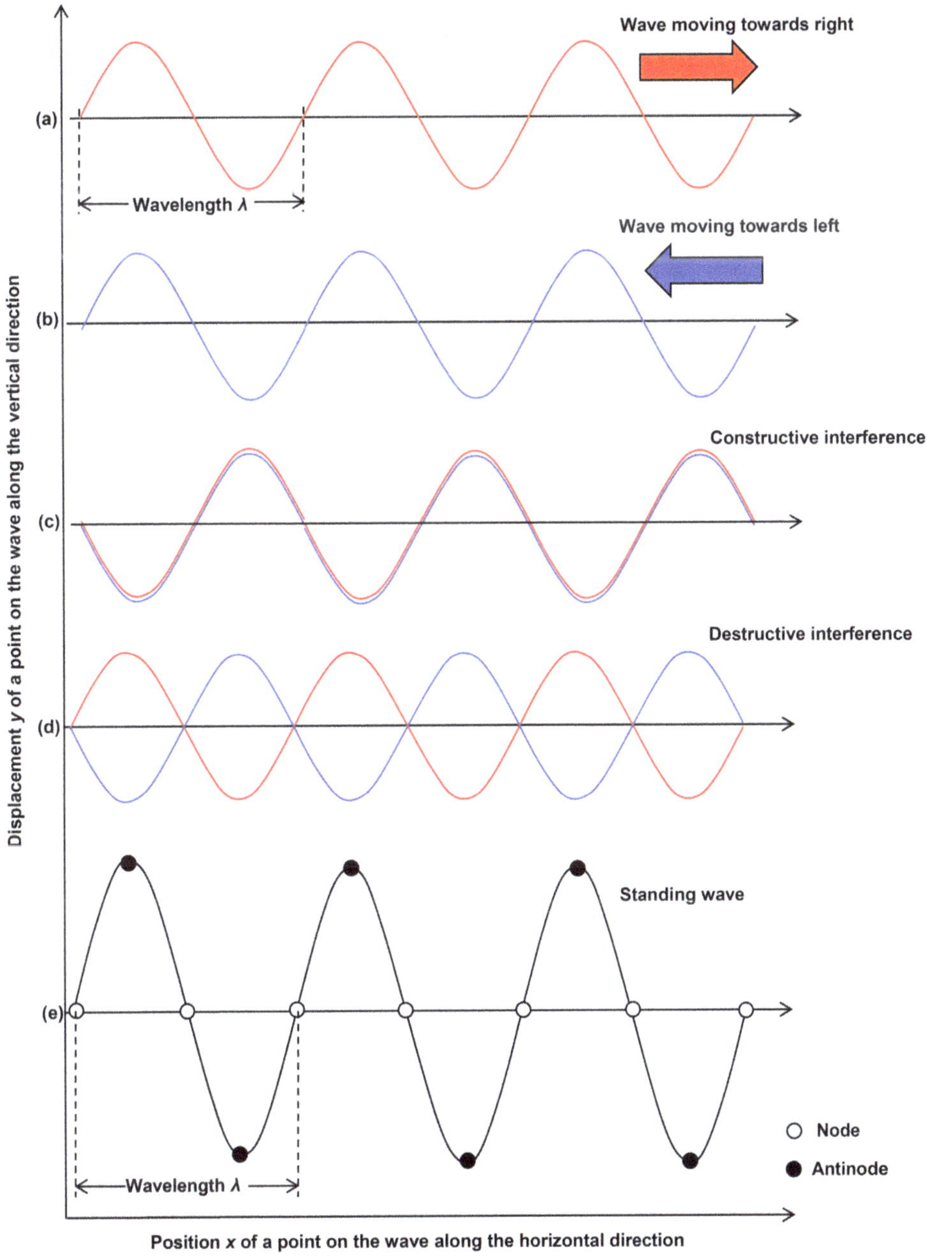

Figure 6.3. Formation of standing waves by superimposition of two travelling waves of the same amplitude and frequency moving in opposite directions: (a) a wave moving rightward, (b) a wave moving leftward, (c) interaction of two waves by constructive interference, (d) destructive interference and (e) typical resulting vibration pattern of the standing wave having twice the amplitude of any one of the participating waves.

Formation of standing waves requires that certain conditions must be satisfied. It happens when the energy is fed into the system at a suitable frequency. This frequency is the natural frequency of the system. Thus a resonance state is essential. So standing wave formation is a resonance phenomenon. The lowest frequency of production of standing waves is called fundamental resonance frequency and the higher frequencies are known as harmonics or overtones.

6.5.3 Stationary wave formation in a medium under different boundary conditions

(a) Medium bounded at both ends: Water filled in a tank or a rope tied to fixed supports at its ends may be imagined. Refer to figure 6.4.

The simplest standing wave has two nodes (minimum amplitude) at the two ends and one antinode (maximum amplitude) in the center. The length L of the system equals half a wavelength $\lambda/2$, i.e.,

$$L = \frac{\lambda}{2} \tag{6.36}$$

This is the fundamental mode or first harmonic. Equation (6.36) means that half the wavelength fits in the length L of the system. The next possibility is a standing wave with three nodes, two at the ends and one in the middle along with two antinodes. The length L is

$$L = \frac{2\lambda}{2} \tag{6.37}$$

This is the second harmonic. The third possible standing wave has four nodes and three antinodes, and the length

$$L = \frac{3\lambda}{2} \tag{6.38}$$

This is the third harmonic. Thus the lengths of the system for the fundamental mode and harmonics are integral multiples of half wavelengths as

$$\text{Lengths} = \frac{\lambda}{2}, \frac{2\lambda}{2}, \frac{3\lambda}{2}, \frac{4\lambda}{2}, \frac{5\lambda}{2}, \ldots\ldots\ldots \frac{n\lambda}{2} \tag{6.39}$$

The equation must be carefully understood. The length L of the system remains the same but increasing number of half wavelengths fit inside the same length L for these cases, e.g., for the nth harmonic n number of half wavelengths comprise the system length L.

(b) Medium free at both ends: Arguing on the same lines as in the preceding case, one arrives at the same result except that the nodes are interchanged with antinodes and vice versa (figure 6.5).

(c) Medium bound at one end but fixed at the other end: The previous cases (a) and (b) represented symmetrical boundary conditions because both ends of the system were similar, either free or fixed. But here in case (c) we come across asymmetrical

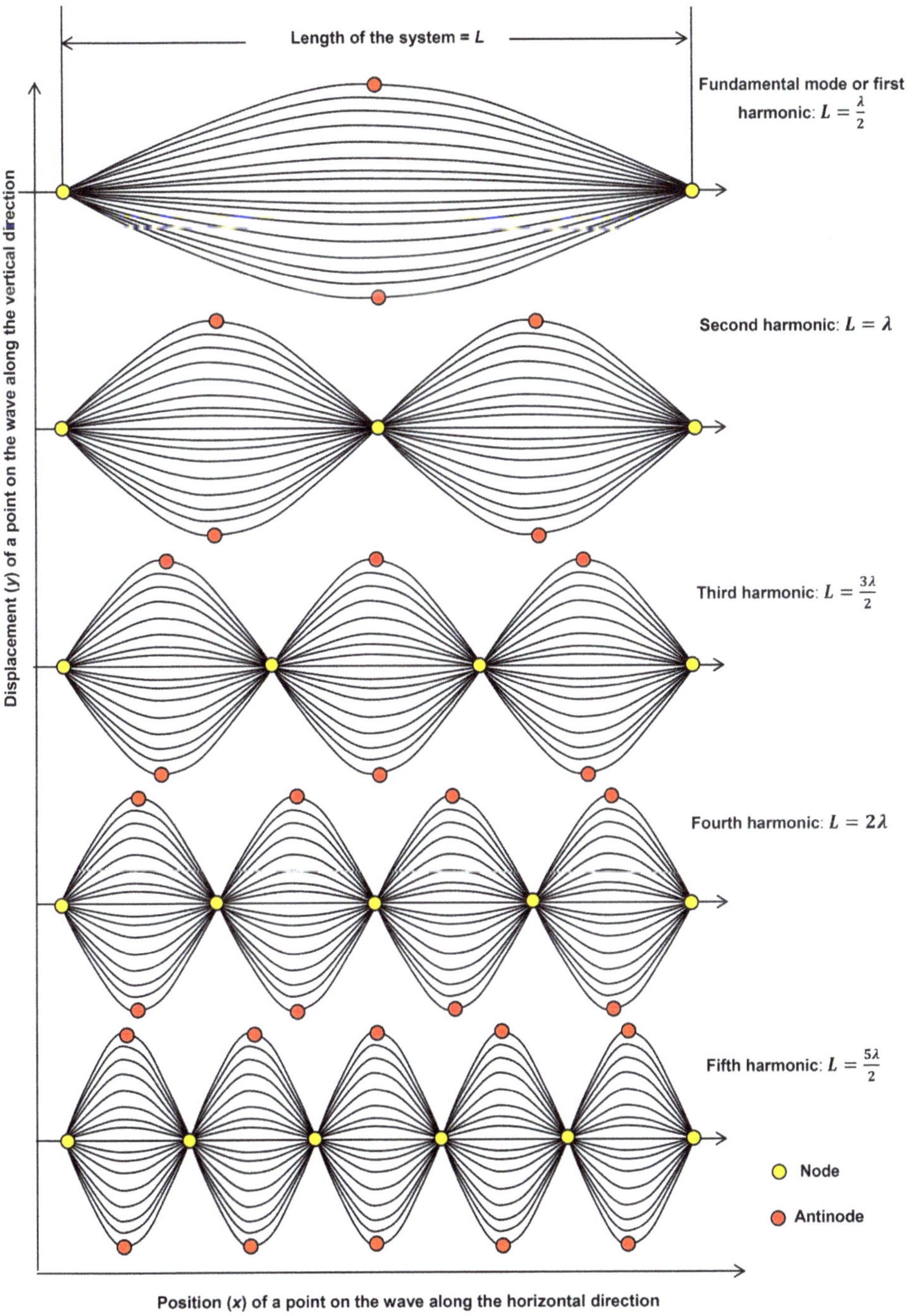

Figure 6.4. Standing wave patterns of the fundamental mode and harmonics for a system bounded at both ends showing the relationships between the length L of the system and wavelength λ for these modes; various snapshots of the waves are shown.

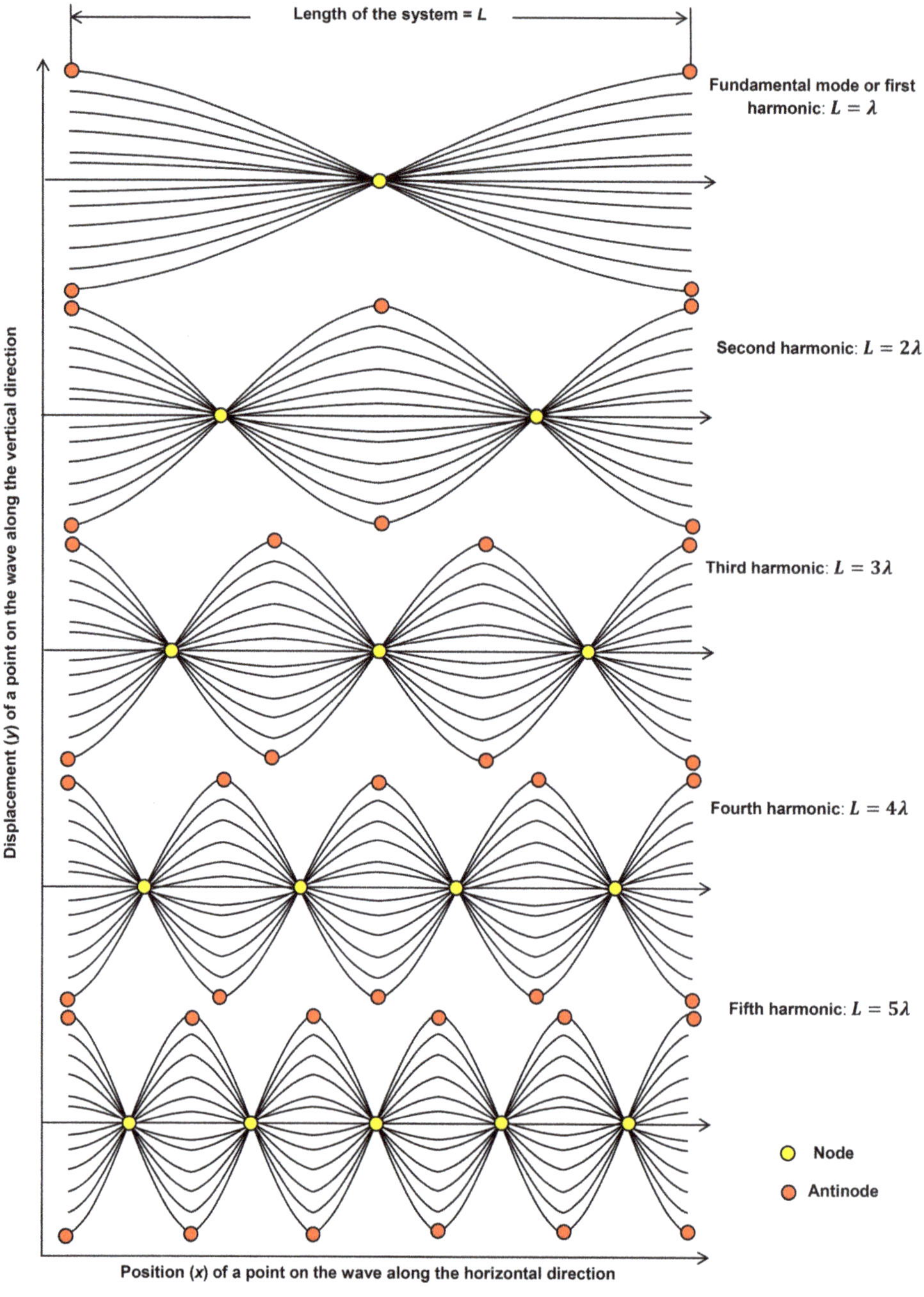

Figure 6.5. Standing wave patterns of the fundamental mode and harmonics for a system free at both ends showing the relationships between the length L of the system and wavelength λ for these modes. The diagram shows various snapshots of the waves.

boundary conditions in which the two ends of the system differ. The outcome is depicted in figure 6.6.

The simplest situation for case (c) has a node at the fixed end and an antinode at the free end. The length L is easily seen to be a quarter wavelength. So, for the fundamental mode or first harmonic, we can write

$$L = \frac{\lambda}{4} \tag{6.40}$$

The next possible wave for the second harmonic has two nodes and two antinodes, and the length of the system accommodates three quarters of a wavelength so that

$$L = \frac{3\lambda}{4} \tag{6.41}$$

Following the same procedure, the third harmonic contains five quarters of a wavelength as

$$L = \frac{5\lambda}{4} \tag{6.42}$$

For the fourth harmonic, we have

$$L = \frac{7\lambda}{4} \tag{6.43}$$

and so on. Hence the lengths are

$$\text{Lengths} = \frac{\lambda}{4}, \frac{3\lambda}{4}, \frac{5\lambda}{4}, \frac{7\lambda}{4}, \frac{9\lambda}{4}, \ldots\ldots.. \tag{6.44}$$

or,

$$\text{Length} = \frac{n\lambda}{4} \text{ where } n = 1,3,5,7, \ \ldots.. \tag{6.45}$$

It is evident that the wavelengths of the harmonics are fractional multiples of the fundamental mode. These wavelengths have odd numbers in the numerator.

6.5.4 Excitation of plasma waves in an FET channel

After learning about resonant cavity and standing waves, let us now turn our attention to bring out the relevance of the discussions in the perspective of plasma waves in a MOSFET channel. As we know, the plasma is a collective set of ions and free electrons. Since a large population of free electrons exists in a MOSFET channel, we can say that a plasma comprising the two-dimensional electron gas exists in the channel of a field-effect transistor. The electrons in the plasma behave as a fluid in exactly the same manner as shallow water (Dyakonov and Shur 1993). By shallow water is meant water in which the wavelength or the spatial range of variation in the level of water is much larger than the depth of water. The equations of motion for the 2D electron fluid are coincident with the hydrodynamic equations for shallow water.

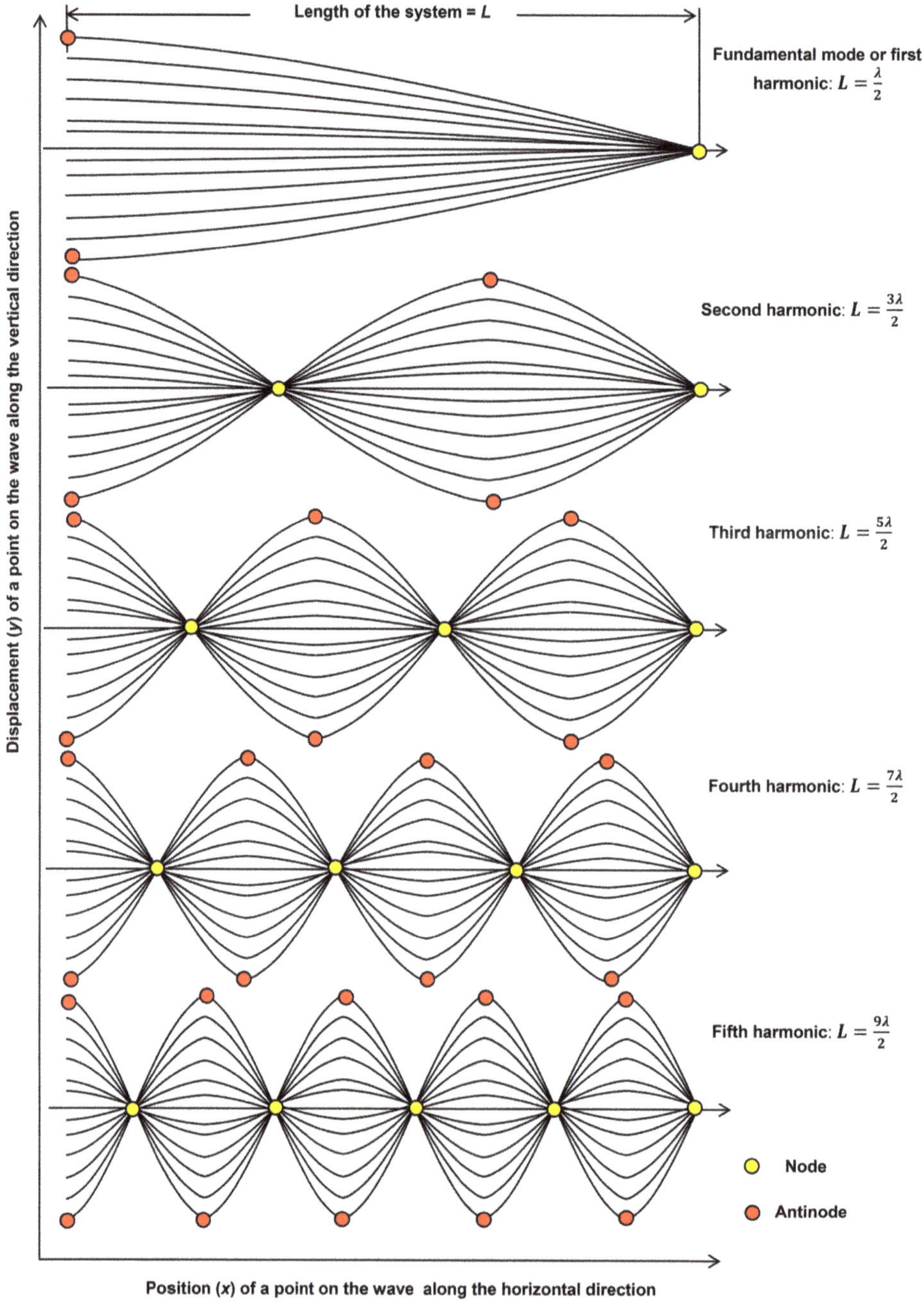

Figure 6.6. Standing wave patterns of the fundamental mode and harmonics for a system free at one end and bound at the other end showing the relationships between the length L of the system and wavelength λ for these modes; various snapshots of the waves are displayed.

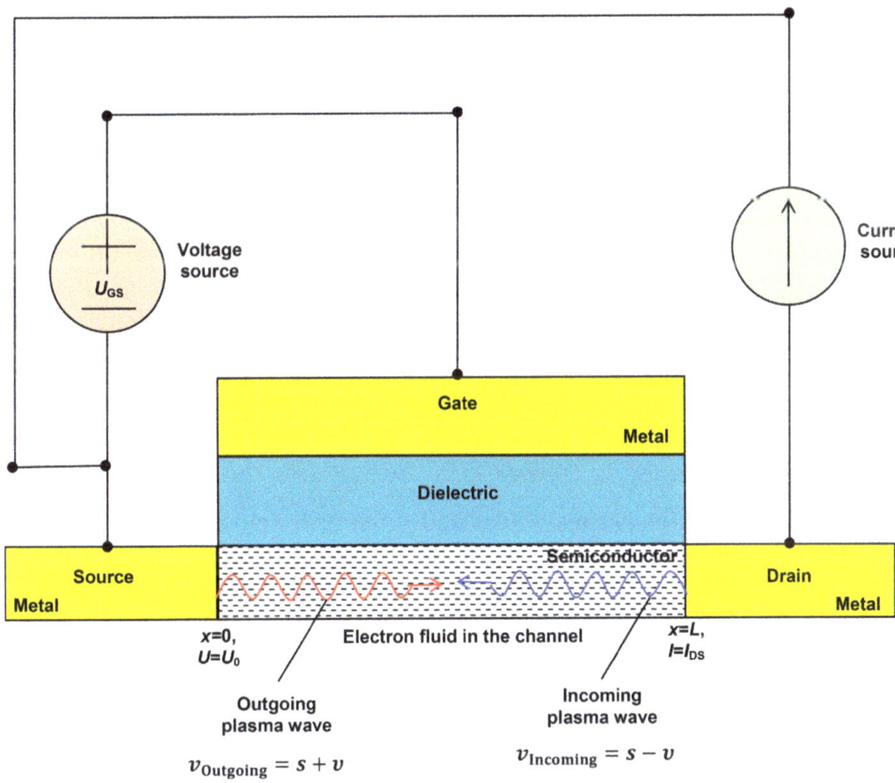

Figure 6.7. Plasma waves in a MOSFET channel.

Waves move in the electron fluid in the same way as they move in water. But here the waves consist of variations of electron density in space and time. These electron density waves are known as plasma waves. A plasma wave contains an interconnected set of charged particles and electric and magnetic fields. These particles and fields propagate in a periodic repeated fashion constituting wave motion.

The question arises: How are plasma waves excited in a FET? We will explain this by referring to an N-channel MOSFET shown in figure 6.7. The gate–source junction of this device is fed by a voltage source while the drain–source junction ($x = L$) is fed by a current source. An ideal voltage source is a two-terminal device having a special property. This distinctive property is that it maintains a constant voltage across its terminals regardless of the current flowing through it. The ideal current source is a properly designed circuit element with a defined goal. Its specialty is that it maintains a constant current flow irrespective of the voltage across its terminals. Consequently, there exists a constant voltage at the source $x = 0$. At the drain $x = L$, a constant current I_{DS} is obtained. For this arrangement, the gate signal U_{GS} contains the positive DC gate–source voltage U_0 of sufficient magnitude to induce strong inversion and hence formation of a channel containing a sheet of large density of electrons. The AC variation at $x = 0$ is possible even for a constant current in the external circuit. This happens because the AC current at $x = 0$ is short-circuited to the gate by the DC voltage source.

When the electrons are moving from source to drain, they experience collisions with each other *en route*. Dyakonov and Shur (1993) showed that the steady flow of DC current in a ballistic MOSFET is likely to be unstable. As a consequence, a low drain current may induce oscillations of electron plasma causing a periodic variation of channel charge. Therefore, the advancing electron fluid in the channel is accompanied by plasma wave generation. The plasma waves start from the source end, and travel towards the drain end. From the drain end, the plasma waves are reflected back.

In the hydrodynamic analogy, the water level is equivalent to the gate-to-channel voltage. The still water level of a water wave pertains to the undisturbed electron plasma level. The elevations and depressions of water surface with reference to still water level refer to the upper and lower levels of electron plasma which in turn correspond to placid gate-to-channel voltage and upper/lower levels of this voltage.

The outgoing waves from the source end travel with a velocity

$$v_{\text{Outgoing}} = s + v \tag{6.46}$$

where s is the velocity of plasma waves and v the drift velocity of electrons. The incoming waves from the drain end move with a velocity

$$v_{\text{Incoming}} = s - v \tag{6.47}$$

Due to the difference in upstream and downstream velocities of plasma wave propagation, the reflection coefficients at the boundaries may exceed unity. In the seminal paper by Dyakonov and Shur (1993), it was shown that the amplitude of the plasma wave remains unaltered at the fixed voltage, zero impedance, short-circuit point $x = 0$ but changes at the fixed current, infinite impedance, open-circuit point $x = L$ in accordance with the ratio of velocities of outgoing and incoming waves. At $x = L$, the

$$\text{Ratio of amplitudes} = \frac{s + v}{s - v} = \frac{1 + \frac{v}{s}}{1 - \frac{v}{s}} \tag{6.48}$$

When $v < s$, v/s is a fraction and the ratio >1, suggesting that the plasma waves undergo amplification.

The time taken by plasma wave to travel from source to drain is

$$\tau_{\text{Forward}} = \frac{L}{s + v} \tag{6.49}$$

while the taken by the wave to return from the drain to source is

$$\tau_{\text{Backward}} = \frac{L}{s - v} \tag{6.50}$$

The total time spent in the round trip of the channel is obtained from equations (6.49) and (6.50) as

$$\tau = \tau_{\text{Forward}} + \tau_{\text{Backward}} = \frac{L}{s + v} + \frac{L}{s - v} \tag{6.51}$$

In time τ, the wave completes one round trip. Therefore in unit time, the number of round trips completed is $1/\tau$, and during an interval of time t, the number of round trips is t/τ. Since the amplification achieved in one round trip is given by equation (6.48), the amplification obtained in t/τ trips is (Dyakonov 2010)

$$\frac{1+\frac{v}{s}}{1-\frac{v}{s}} \times \frac{1+\frac{v}{s}}{1-\frac{v}{s}} \times \frac{1+\frac{v}{s}}{1-\frac{v}{s}} \ldots\ldots\ldots\ldots \frac{t}{\tau}\text{times} = \left(\frac{1+\frac{v}{s}}{1-\frac{v}{s}}\right)^{\frac{t}{\tau}} \tag{6.52}$$

Working against the amplification mechanism of waves, there is a decay mechanism also. It has two components: an external component arising from the scattering of electrons by phonons or impurities in the semiconductor, and an internal component originating from the viscosity of the electron fluid. The growth of the plasma wave is impeded by its decay due to these two components. Accordingly, the electron velocity is incremented or decremented. When the electron velocity crosses a threshold limit, an instability of current-carrying steady state develops against the spontaneous production of plasma waves. The conditions for this instability are:

(i) the existence of asymmetric boundary conditions at the source and drain, and

(ii) the damping of plasma waves must be small, i.e.,

$$\text{Frequency of plasma waves } (\omega) > \frac{1}{\text{Momentum relaxation time } (\tau_{\text{Relaxation}})} \tag{6.53}$$

or,

$$\omega > \frac{1}{\tau_{\text{Relaxation}}} \tag{6.54}$$

or,

$$\omega\tau_{\text{Relaxation}} > 1 \tag{6.55}$$

(ii) The velocity of electrons in the steady state is greater than a threshold value. This value depends on the damping time $\tau_{\text{Relaxation}}$.

The instability provides a new mechanism of generation of terahertz waves. The plasma oscillations induce a periodic variation of the charge in the channel. As a consequence, a mirror image charge is created in the gate contact. The opposite charges in the channel and the gate contact constitute an electric dipole. The moment of this dipole changes in a periodic manner. These dipole moment changes are accompanied by the emission of electromagnetic waves at the frequency of plasma waves. The method offers an advantage of being voltage tunable because the voltage applied to the gate of the MOSFET can be varied to change the frequency of radiated terahertz waves. Note that the length of the MOSFET is smaller than the wavelength of the electromagnetic waves emitted at the frequency of plasma waves.

Hence, the device can be considered as a point or linear source of terahertz waves. Several such devices must be combined in the form of an array for raising the power level of terahertz emission to the desired magnitude. Thus the frequency limit imposed on the MOSFET by the transit time of electrons is overcome to reach terahertz frequencies.

The plasma waves are characterized by three important parameters: wavelength, velocity and frequency. Let us see how we can calculate these parameters.

6.5.5 Determination of wavelength of plasma waves

The length of the FET channel L governs the wavelength of plasma waves. A standing plasma wave is generated in the same fashion as outlined for standing waves in section 6.5.2. The channel region supports this standing wave by acting as a resonant cavity. It may be imagined as a localized time-varying perturbation of electron concentration. This is a resonance phenomenon. It has a fundamental frequency and harmonics. But to know these frequencies we must note the boundary conditions prevailing in the FET channel (Dyakonov and Shur 2001). The boundary conditions here are antisymmetric because the source end is a short circuit and with the transistor driven into saturation, the pinched off drain region is an open circuit. The former represents zero impedance and the latter infinite impedance. These bear analogies to the free end and tied end of case (c) in section 6.5.3. In view of this resemblance to the situation in case (c), we can straightway say that the fundamental mode occurs at a channel length of a quarter wavelength and the harmonics are produced at odd number multiples of the quarter wavelength. Therefore, the wavelength of the different modes is obtained from equation (6.45) as

$$L = \frac{n\lambda}{4} \tag{6.56}$$

or

$$\lambda = \frac{4L}{n} \tag{6.57}$$

6.5.6 Determination of velocity of plasma waves

The velocity of these plasma waves is determined by the effective DC part of the gate-to-channel voltage swing U_0, i.e., the difference between the applied gate-source voltage (V_{GS}) and the threshold voltage of the device (V_{Th}), i.e.,

$$U_0 = V_{GS} - V_{Th} \tag{6.58}$$

The kinetic energy E of an electron of effective mass m_e^* moving with a velocity s is given by

$$E = \frac{1}{2} m_e^* s^2 \tag{6.59}$$

The work done on the electron of charge q under an applied potential U_0 is

$$W = qU_0 \tag{6.60}$$

Since

$$E = W \tag{6.61}$$

equations (6.59) and (6.60) enable us to write

$$\frac{1}{2}m_e^* s^2 = qU_0 \tag{6.62}$$

or,

$$s^2 = \frac{2qU_0}{m_e^*} \tag{6.63}$$

$$\therefore s = \sqrt{\frac{2qU_0}{m_e^*}} \tag{6.64}$$

Applying equation (6.58) we get

$$s = \sqrt{\frac{2q(V_{\mathrm{GS}} - V_{\mathrm{Th}})}{m_e^*}} \tag{6.65}$$

6.5.7 Determination of frequency of plasma waves

The foregoing derivations in section 6.5.5 and 6.5.6 of wavelength and velocity of plasma waves help us to find the frequency of the plasma waves in a MOSFET of given channel length L because plasma waves obey a linear dispersion law, which means that the angular frequency ω, velocity s and wavevector k are related as

$$\omega = sk \tag{6.66}$$

The magnitude of the wavevector k is given by

$$k = \frac{2\pi}{\lambda} \tag{6.67}$$

From equations (6.66) and (6.67)

$$\omega = s\left(\frac{2\pi}{\lambda}\right) = \frac{2\pi s}{\lambda} \tag{6.68}$$

To find the linear frequency ν, we write

$$\omega = 2\,\pi\nu \tag{6.69}$$

which combined with equation (6.68) gives

$$2\pi\nu = \frac{2\pi s}{\lambda} \tag{6.70}$$

or,

$$\nu = \frac{s}{\lambda} \tag{6.71}$$

Equations (6.57) and (6.71) help us to express the frequency of plasma waves in terms of the channel length L of the MOSFET to ascertain its dependency on L

$$\nu = \frac{s}{\frac{4L}{n}} = \frac{ns}{4L} \tag{6.72}$$

Example 6.3 (a) The channel length of an N-MOSFET is $L = 80$ nm. Plasma waves are excited in this device. Find the wavelengths of the first three modes of plasma waves. (b) If the threshold voltage of the MOSFET is 0.3 V, calculate the velocity of plasma waves for an applied gate-source voltage of 0.8 V. For silicon, the effective mass m_e* for conductivity calculations = 0.26 times electron rest mass m_0. (c) What are the frequencies of plasma waves for the three modes? (d) If the gate–source voltage is raised to 1.5 V, what is the velocity of plasma waves? (e) Find the frequencies of the three modes of plasma waves for the gate-source voltage of 1.5 V.
Solution: (a) Applying equation (6.57)

$$\lambda = \frac{4L}{n} = \frac{4 \times 80}{n}\text{nm} = \frac{320}{n}\text{nm} \tag{6.73}$$

For $n = 1, 3, 5$

$$\lambda = \frac{320}{1}, \frac{320}{3}, \frac{320}{5} = 320, 106.67, 64 \text{ nm} \tag{6.74}$$

(b) Equation (6.65) gives

$$\begin{aligned} s &= \sqrt{\frac{2q(V_{\text{GS}} - V_{\text{Th}})}{m_e^*}} = \sqrt{\frac{2 \times 1.602 \times 10^{-19}(0.8 - 0.3)}{0.26 \times 9.109 \times 10^{-31}}} \\ &= \sqrt{\frac{2 \times 1.602 \times 10^{-19+31} \times 0.5}{0.26 \times 9.109}} = \sqrt{0.6764 \times 10^{12}} = 822435.408 \\ &= 8.22 \times 10^5 \text{ ms}^{-1} \end{aligned} \tag{6.75}$$

(c) By equation (6.72),

$$\nu = \frac{ns}{4L} = \frac{n \times 8.22 \times 10^{5}}{4 \times 80 \times 10^{-9}} = \frac{n \times 8.22 \times 10^{5+9}}{4 \times 80} = n \times 0.025\,687 \times 10^{14}\ \text{Hz} \tag{6.76a}$$
$$= n \times 2.5687 \times 10^{12}\ \text{Hz}$$

For $n = 1, 3, 5$

$$\nu = 1 \times 2.5687 \times 10^{12}\ \text{Hz},\ 3 \times 2.5687 \times 10^{12}\ \text{Hz},\ 5 \times 2.5687 \times 10^{12}\ \text{Hz}$$
$$= 2.57\ \text{THz},\ 7.71\ \text{THz},\ 12.84\ \text{THz} \tag{6.76b}$$

(d) From equation (6.65)

$$s = \sqrt{\frac{2 \times 1.602 \times 10^{-19}(1.5 - 0.3)}{0.26 \times 9.109 \times 10^{-31}}} = \sqrt{\frac{2 \times 1.602 \times 10^{-19+31} \times 1.2}{0.26 \times 9.109}} \tag{6.77}$$
$$= \sqrt{1.6234 \times 10^{12}} = 1\,274127.15 = 1.274 \times 10^{6}\ \text{ms}^{-1}$$

(e) Using equation (6.72)

$$\nu = \frac{ns}{4L} = \frac{n \times 1.274 \times 10^{6}}{4 \times 80 \times 10^{-9}} = \frac{n \times 1.274 \times 10^{6+9}}{4 \times 80} = n \times 0.003\,981 \times 10^{15}\ \text{Hz} \tag{6.78}$$
$$= n \times 3.981 \times 10^{12}\ \text{Hz}$$

For $n = 1, 3, 5$

$$\nu = 1 \times 3.981 \times 10^{12}\ \text{Hz},\ 3 \times 3.981 \times 10^{12}\ \text{Hz},\ 5 \times 3.981 \times 10^{12}\ \text{Hz}$$
$$= 3.98\ \text{THz},\ 11.94\ \text{THz},\ 19.905\ \text{THz} \tag{6.79}$$

Example 6.4 The channel length L of a MOSFET is 60 nm and channel width W is 50 μm. The gate oxide is 7 nm thick. (a) Find the wavelength of plasma waves for excitation in fundamental mode. (b) If the threshold voltage of the device is 0.2 V and the applied DC gate voltage is 0.5 V, determine the velocity of plasma waves. (c) Calculate the frequency of fundamental mode oscillations. (d) Find the maximum radiated power. Dielectric constant of silicon dioxide is $\varepsilon_{ox} = 3.9$. Electron saturation velocity in silicon is $\upsilon = 1 \times 10^{5}$ m s^{-1}.
Solution: (a) By equation (6.57),

$$\lambda = \frac{4L}{n} = \frac{4 \times 60}{1}\text{nm} = 240\ \text{nm} \tag{6.80}$$

(b) Applying equation (6.65)

$$s = \sqrt{\frac{2q(V_{\rm GS} - V_{\rm Th})}{m_e^*}} = \sqrt{\frac{2 \times 1.602 \times 10^{-19}(0.5-0.2)}{0.26 \times 9.109 \times 10^{-31}}}$$
$$= \sqrt{\frac{2 \times 1.602 \times 10^{-19+31} \times 0.3}{0.26 \times 9.109}} = \sqrt{0.405\,854 \times 10^{12}} = 637066.72 \tag{6.81}$$
$$= 6.37 \times 10^5 \text{ ms}^{-1}$$

(c) Equation (6.72) gives

$$\nu = \frac{ns}{4L} = \frac{n \times 6.37 \times 10^5}{4 \times 60 \times 10^{-9}} = \frac{1 \times 6.37 \times 10^{5+9}}{4 \times 60} = 0.026\,542 \times 10^{14} \text{ Hz}$$
$$= 2.6542 \times 10^{12} \text{ Hz} \approx 2.65 \text{ THz} \tag{6.82}$$

(d) If $C_{\rm ox}$ is the capacitance of the gate oxide per unit area and it acquires a charge q on applying a DC gate voltage U_0, we can write

$$q = C_{\rm ox} U_0 \tag{6.83}$$

Work done in delivering an infinitesimal charge dq to the gate capacitor per unit area is

$$\text{Work} = \text{Charge} \times \text{Potential difference} = dq \times U_0 = U_0 dq = \frac{q}{C_{\rm ox}} dq \tag{6.84}$$

by applying equation (6.83).

Hence, total work done in delivering a charge q to the gate capacitor per unit area is

$$\text{Total work} = \int_0^q \frac{1}{C_{\rm ox}} q dq = \frac{1}{C_{\rm ox}} \int_0^q q dq = \frac{1}{C_{\rm ox}} \left| \frac{q^2}{2} \right|_0^q = \frac{1}{C_{\rm ox}} \left(\frac{q^2}{2} - 0 \right)$$
$$= \frac{q^2}{2C_{\rm ox}} \tag{6.85}$$

If the channel length is L and channel width is W, the electrostatic energy stored in the gate capacitor is obtained by multiplying equation (6.85) by the area WL, i.e.,

$$\text{Energy} = \frac{q^2 WL}{2C_{\rm ox}} \tag{6.86}$$

which can be rewritten using equation (6.83) as

$$\text{Energy} = \frac{(C_{\rm ox} U_0)^2 WL}{2C_{\rm ox}} = \frac{C_{\rm ox} U_0^2 WL}{2} \tag{6.87}$$

This energy is radiated in a time = the electron transit time through the channel of length L, and is written as

$$\text{Radiation emission time} = \frac{\text{Channel length}}{\text{Electron velocity}} = \frac{L}{v} \tag{6.88}$$

Hence, the power radiated is determined with the help of equations (6.87) and (6.88) as

$$\text{Power radiated} = \frac{\text{Energy}}{\text{Radiation emission time}} = \frac{\frac{C_{\mathbf{ox}} U_0^2 WL}{2}}{\frac{L}{v}} = \frac{C_{\mathbf{ox}} U_0^2 WL}{2} \times \frac{v}{L} \tag{6.89}$$

$$= \frac{C_{\mathbf{ox}} U_0^2 Wv}{2}$$

The maximum power radiated is obtained by replacing U_0 by the wave amplitude or peak value U_1 which is related to U_0 as

$$U_1 = \sqrt{2}\, U_0 \tag{6.90}$$

for a sinusoidal waveform. In terms of U_1, we have

$$\text{Maximum power radiated } P_{\text{Maximum}} = \frac{C_{\mathbf{ox}} U_1^2 Wv}{2} = \frac{C_{\mathbf{ox}}(\sqrt{2}\, U_0)^2 Wv}{2} = C_{\mathbf{ox}} U_0^2 Wv \tag{6.91}$$

If the thickness of gate oxide is t_{ox} and the dielectric constant of silicon dioxide is ε_{ox}, we can write

$$C_{ox} = \frac{\varepsilon_0 \varepsilon_{ox}}{t_{ox}} \tag{6.92}$$

Substituting for C_{ox} from equation (6.92) in equation (6.91) we get

$$P_{\text{Maximum}} = C_{\mathbf{ox}} U_0^2 Wv = \frac{\varepsilon_0 \varepsilon_{ox}}{t_{ox}} U_0^2 Wv \tag{6.93}$$

Inputting the given values in equation (6.93)

$$\begin{aligned} P_{\text{Maximum}} &= \frac{8.854 \times 10^{-12} \times 3.9}{7 \times 10^{-9}} (0.5)^2 \times 50 \times 10^{-6} \times 1 \times 10^5 \\ &= \frac{8.854 \times 3.9 \times 0.25 \times 50 \times 10^{-12+9-6+5}}{7} = 61.66 \times 10^{-4}\ \text{W} \\ &= 61.66 \times 10^{-4} \times 10^3\ \text{mW} = 6.166\ \text{mW} \approx 6.2\ \text{mW} \end{aligned} \tag{6.94}$$

Note: The estimated power represents only a crude guess because the radiation efficiency is a function of antenna design and other parameters, and may be very small. Albeit, the feasibility of the method is obvious.

6.6 Terahertz MOSFET detectors

6.6.1 TeraFET

Terahertz MOSFET called TeraFET has been fabricated using complementary metal-oxide-semiconductor (CMOS) technology (Ikamas *et al* 2018). The device chosen is N-channel MOS (NMOS). Its channel length is 100 nm and the channel width is 500 nm. The terahertz radiation is coupled to the device through hemispherical silicon lens. A bowtie antenna with an opening angle of 90° receives the radiation. This antenna is integrated with the top metal/insulator stack. The rectified voltage is produced between the source and drain terminals.

The threshold voltage of the MOSFET is 0.445 V. In a broad span of gate voltages above the threshold voltage, the relation between the detected signal and the power of terahertz radiation is linear. For gate bias less than threshold voltage, a superlinear increase in the detected signal is recorded.

The technique of illumination of a photoconductive GaAs emitter with 30 fs infrared laser pulses was applied to generate the terahertz radiation. The average power of radiation is 1.25 μW when the antenna is biased at 50 V. The optical responsivity of the TeraFET is 100 V W^{-1} and the noise equivalent power is 67 pW $Hz^{-1/2}$ at 0.6 THz (Ikamas *et al* 2018).

6.6.2 Dual-frequency CMOS terahertz detector

A dual-frequency detector is fabricated in a 180 nm standard CMOS process with on-chip integrated antennas, a 0.65 THz metal antenna with bow-tie structure and a 0.22 THz heavily-doped polysilicon antenna having fan–rod structure (Huang *et al* 2019). The process involves the steps of shallow trench isolation, formation of deep N-well and P-well, gate oxidation, polysilicon deposition and etching, source/drain implantation, silicide contacts, and metal antenna deposition. The maximum responsivity achieved with the metal antenna is 400 V W^{-1} at 0.65 THz while that with the silicon antenna is 2000 V W^{-1} at 0.22 THz. At a modulation frequency of 1 kHz for the terahertz radiation, the noise equivalent power is 110 pW $Hz^{-1/2}$ for the metal antenna; the same is 23 pW $Hz^{-1/2}$ for the silicon antenna (Huang *et al* 2019).

6.7 Discussion and conclusions

The important role played by plasma waves in semiconductor materials and devices becomes more relevant when thinking about applying them for extending the range of operational frequencies of devices. Instead of treading the conventional path in applying MOSFET from electron drift under an applied electric field, the perspective has changed to using plasma waves propagating in the channel. The Dyakonov–Shur instability of plasma waves in a MOSFET channel has led to exciting developments in the realm of MOSFETs. The plasma wave-based regime of MOSFET operation opens a new field of plasma wave electronics for terahertz frequencies. In the next chapter, we shall consider the high-electron-mobility transistor (HEMT), which is also a field-effect device. Hence it can also act as a terahertz emitter like the MOSFET utilizing plasma waves, as will be described in section 7.5.3.

References

Dyakonov M I 2010 Generation and detection of terahertz radiation by field effect transistors *C. R. Phys.* **11** 413–20

Dyakonov M and Shur M 1993 Shallow water analogy for a ballistic field effect transistor: new mechanism of plasma wave generation by DC current *Phys. Rev. Lett.* **71** 2465–8

Dyakonov M and Shur M S 2001 Plasma wave electronics for terahertz applications *Terahertz Sources and Systems. NATO Science Series (Series II: Mathematics, Physics and Chemistry)* ed R E Miles, P Harrison and D Lippens vol 27 (Dordrecht: Springer) pp 1–21

Huang R, Ji X, Liao Y, Peng J, Wang K, Xu Y and Yan F 2019 Dual-frequency CMOS terahertz detector with silicon-based plasmonic antenna *Opt. Express* **27** 23250–61

Ikamas K, Nevinskas I, Krotkus A and Lisauskas A 2018 Silicon field effect transistor as the nonlinear detector for terahertz autocorellators *Sensors* **18** 11

Ytterdal T, Cheng Y and Fjeldly T A 2003 *Device Modeling for Analog and RF CMOS Circuit Design* (Chichester: Wiley) pp 1–117

IOP Publishing

Practical Terahertz Electronics: Devices and Applications, Volume 1

Solid-state devices and vacuum tubes

Vinod Kumar Khanna

Chapter 7

High-electron-mobility transistors

The working of the high-electron-mobility transistor (HEMT) at high frequencies is pondered from its equivalent circuit representation. The equivalent circuit consists of intrinsic and extrinsic capacitances. Other components are the series resistances and the output conductance. An expression for cut-off frequency of HEMT is derived. The total time delay of the device is partitioned into intrinsic, extrinsic and parasitic terms. The approach to minimize the time delay is outlined. Transistor gate length must be reduced within technological constraints while ensuring the avoidance of short-channel effects. Formulae for calculation of built-in potential of the Schottky contact and junction capacitances of HEMT are presented. A detailed analysis of a HEMT structure is performed from typical dimensional parameters and doping profiles to illustrate the calculation of capacitances, cut-off frequency and delays. Technological advancements in the development of HEMTs on InP/InAlAs/InGaAs system are surveyed. HEMT performance has evolved over the years through several milestones to reach the 1 THz mark. The present status of InP HEMTs is described.

Let us run through the elementary principles of operation of the metal–semiconductor-field-effect transistor (MESFET) and the HEMT to gear up for discussions on HEMT working at high frequencies.

7.1 MESFET and HEMT basics

7.1.1 The MESFET

Figure 7.1 shows the MESFET structure. The substrate used is semi-insulating gallium arsenide. Ion implantation is used for formation of the N^+ source and N^+ drain along with the channel. The metallization scheme decided for the device takes care of the difference between gate and source/drain contact metals. The gate contact is a Schottky metal while the source and drain terminals have ohmic contacts.

doi:10.1088/978-0-7503-3171-5ch7

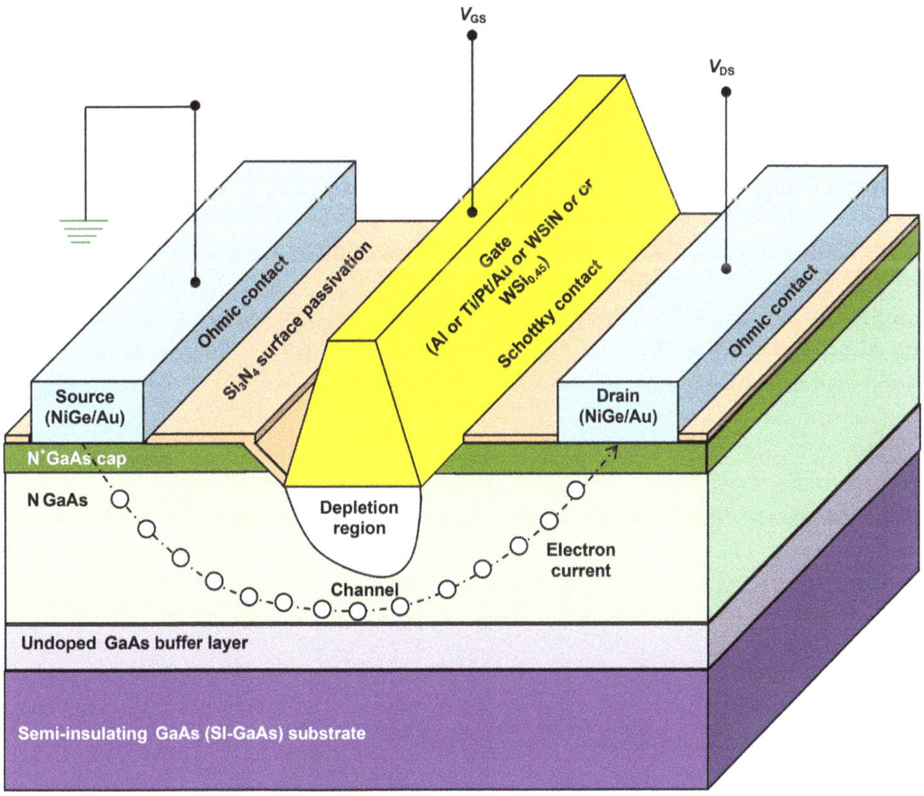

Figure 7.1. A GaAs metal–semiconductor field-effect transistor (MESFET).

During operation of the MESFET, the metal–semiconductor Schottky diode junction at the gate terminal is reverse biased. The thickness of the depletion region formed in the channel region underneath the gate depends on this reverse bias. Therefore, by varying the gate–source bias, the thickness of this depletion region can be changed as desired and thereby the drain–source current can be altered, either decreased or increased. Let us see how this happens? On increasing the reverse bias, the depletion region is enlarged. As the depletion region stretches deeper into the channel, the channel shrinks and hence the drain–source current decreases. Conversely, on decreasing the reverse bias, the depletion region constricts whereby the channel broadens and drain–source current increases. Thus the drain–source current is modulated and controlled by the applied gate bias.

As seen from the MESFET structure, the channel is separated from the gate terminal by the depletion region. This separation of channel from the gate terminal prevents scattering of electrons from defects at the metal–semiconductor interface. The benefit derived is that the mobility of electrons in the channel is higher than in a MOSFET.

7.1.2 The HEMT

This device is a MESFET combined with a heterojunction to provide the free electrons for electrical conduction, i.e., for creating the channel (Nirmal and Ajayan 2019).

Band offset between a wide bandgap semiconductor and a narrow bandgap semiconductor together with doping of the wide bandgap semiconductor is utilized to create a conducting channel.

To know about its construction, let us look at the various layers from bottom to the top of the HEMT structure (figure 7.2). The bottom substrate is made of semi-insulating GaAs. On the SI-GaAs layer, the P-doped GaAs buffer layer is laid down. The GaAs buffer layer is covered with the undoped GaAs channel layer. Over the GaAs channel layer is the undoped AlGaAs spacer layer. Then the next layer is the AlGaAs barrier layer with silicon delta doping. This is the donor layer, which supplies free electrons. On the N^+ doped AlGaAs layer, the Schottky contact for the gate electrode is formed so that the metal–semiconductor junction comprises the layers: metal/AlGaAs. On N^+ GaAs layers, the ohmic contacts of the source and drain terminals are formed.

Heterojunction is formed between AlGaAs and GaAs layers. Some of the free electrons in the doped wide bandgap semiconductor AlGaAs diffuse to the undoped narrow bandgap semiconductor GaAs. As a result, the region of GaAs adjoining the AlGaAs/GaAs interface becomes negatively charged by the gain of electrons. Similarly, the region of AlGaAs close to the AlGaAs/GaAs becomes positively charged due to loss of electrons. Thus a potential difference is created at the

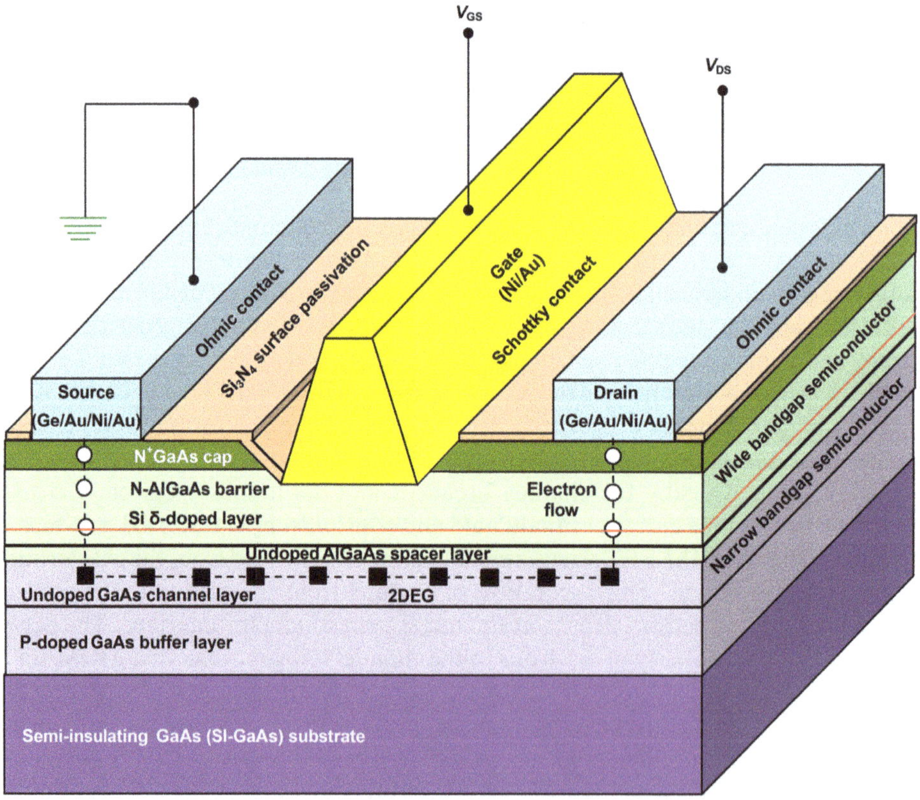

Figure 7.2. AlGaAs/GaAs high-electron-mobility transistor (HEMT).

AlGaAs/GaAs interface inhibiting further transfer of electrons. The overall consequence of this electron transferring process is that a two-dimensional sheet of charge called the two-dimensional electron gas (2DEG) is formed near the AlGaAs/GaAs interface on the GaAs side. This 2DEG makes the GaAs layer conducting. It is therefore the channel of the HEMT device.

The reverse-biased Schottky diode serves as the gate terminal of the HEMT. It is used to modulate the current flowing through the channel. By varying the voltage applied on the gate electrode, the channel conductivity is controlled. In turn the drain–source current is controlled. This is similar to the MESFET action.

Note that here also the channel is located away from the metal–semiconductor contact at the gate terminal. This remoteness of channel location protects the electrons in the channel from suffering any degradation of mobility through scattering from defects at the metal–semiconductor interface. In this respect, the HEMT is similar to the MESFET and better than MOSFET. But the HEMT wins over both MESFET and MOSFET for high-speed switching, as we shall see in section 7.1.3 below.

More HEMTs are examined in figure 7.3 which shows the stacks of layers comprising AlGaN/GaN and InAlAs/InGaAs HEMTs. Look at the channel

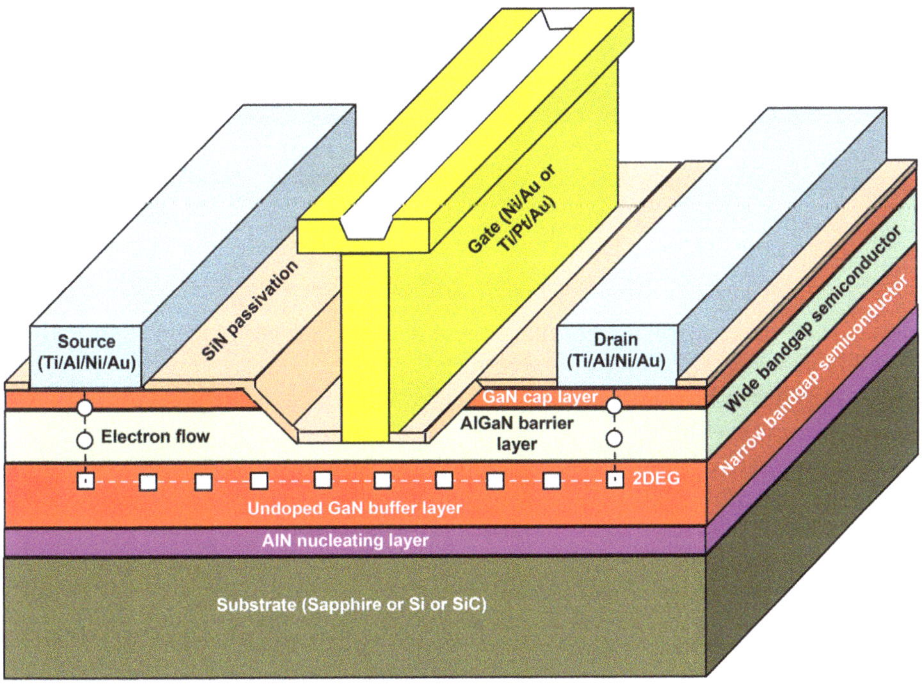

(a)

Figure 7.3. Two types of popular HEMTs with T-gates: (a) AlGaN/GaN HEMT and (b) InAlAs/InGaAs HEMT.

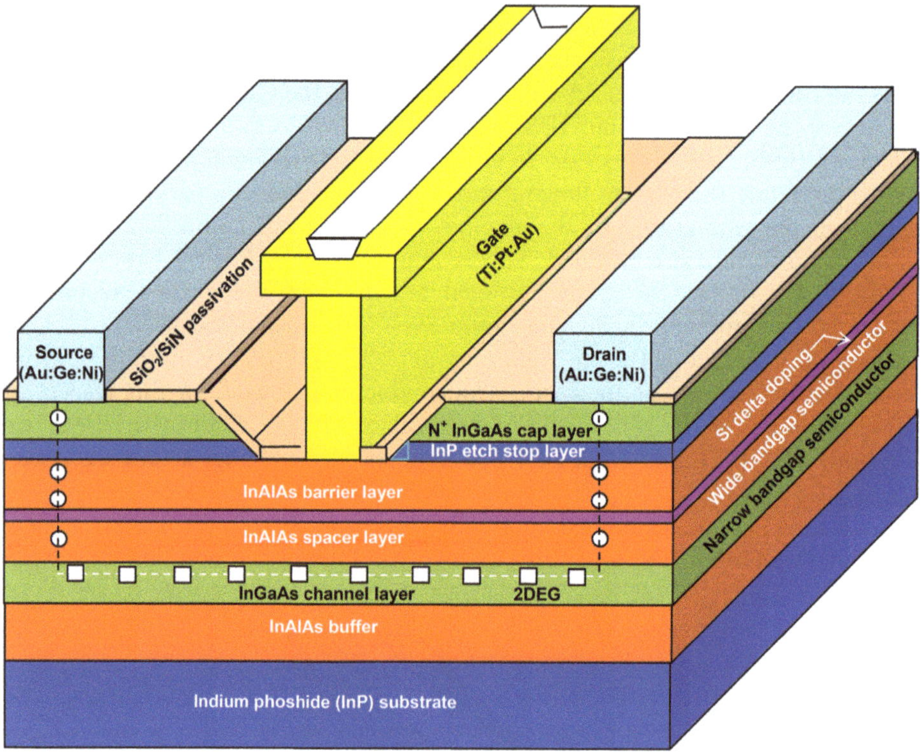

(b)

Figure 7.3. (Continued.)

locations and the electron flow paths in these devices. Reckoned from bottom to top, the layers in these HEMTs are, respectively:

Substrate/AlN nucleating layer/GaN buffer layer (channel layer)/AlGaN barrier layer/GaN cap layer, and

Substrate/InAlAs buffer layer/InGaAs channel layer/InAlAs spacer/delta doped layer/InAlAs barrier layer/InGaAs cap layer

7.1.3 Superiority of HEMT over MESFET

As mentioned in its name itself, the high-electron-mobility transistor (HEMT) offers the advantage of colossal electron mobility. This is a distinct advantage for working on this device for terahertz wave amplification and generation. The HEMT device achieves the enormous electron mobility by a unique feature of modulation doping in which the free electrons are spatially separated from their parent donor impurity ions. The spatial separation is provided by the confinement of the two-dimensional electron gas in the quantum well created at the interface between two semiconductors of different bandgaps. Ionized impurity scattering is therefore avoided and the mobility is considerably improved over the value for a doped semiconductor. Like the metal–semiconductor field-effect transistor (MESFET), the HEMT is also a field-effect transistor but it gives the additional advantage of fast speed of electrons and high switching speed favoring its utilization in the terahertz range over MESFET. Mobility as high as 3.5×10^7 $cm^2 (V\text{-}s)^{-1}$ has been measured in AlGaAs/GaAs heterostructures (Umansky *et al* 2009, del Alamo 2011).

7.2 HEMT operation at high frequencies

7.2.1 High-frequency model of HEMT

Figure 7.4 presents a simplified analytical model for HEMT for analyzing its high-frequency behavior (Kim *et al* 2011). The model is based on its small-signal equivalent circuit. The model provides a clear physical understanding of the principal factors that determine how fast the transistor works. Utilizing this understanding, necessary design and process improvements can be made.

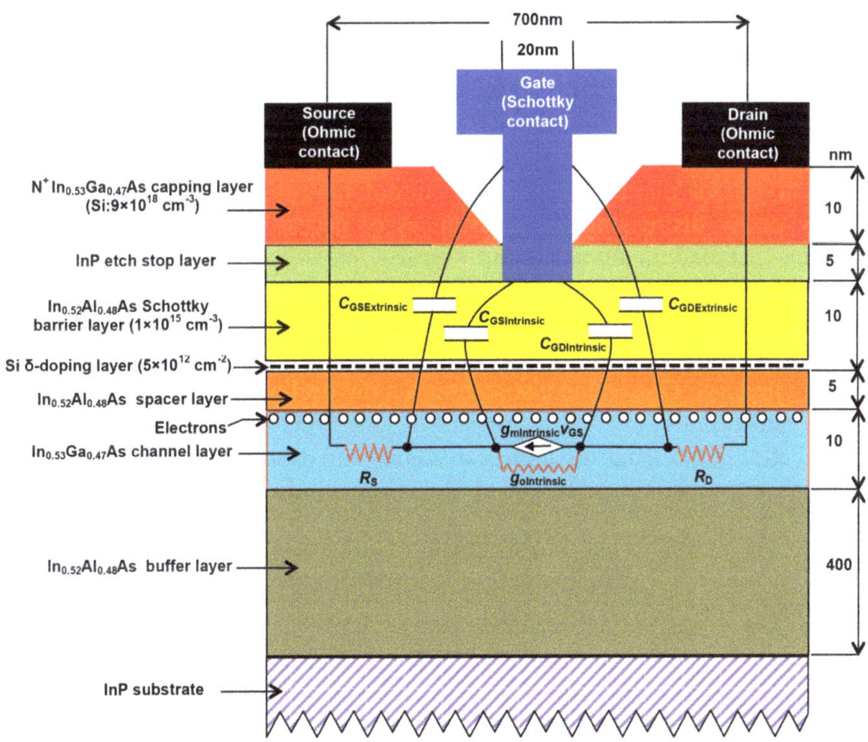

Figure 7.4. High-frequency model of HEMT drawn on its cross-sectional diagram to assist in establishing equivalence between the structural elements of HEMT and the circuit components. Physical parameters for an InAlAs/InGaAs HEMT are given for mathematical analysis of the HEMT (not to scale).

Figure 7.5(a) shows the same model drawn out clearly without the device cross-section in background and figure 7.5(b) presents it in the commonly used form.

In the models shown in figures 7.4–7.5, the gate–source and gate–drain capacitances C_{GS}, C_{GD} are subdivided into intrinsic and extrinsic components as

$$C_{GS} = C_{GS(Intrinsic)} + C_{GS(Extrinsic)} \tag{7.1}$$

$$C_{GD} = C_{GD(Intrinsic)} + C_{GD(Extrinsic)} \tag{7.2}$$

The intrinsic components refer to the capacitances of the gate with respect to the intrinsic channel. These are the capacitances through which the gate electrode creates or controls the channel, i.e., the channel charges or discharges.

The extrinsic components pertain to fringing capacitances and those associated with lateral extensions of the gate region, as determined by the gate structure used. For the T-gate structure of planar HEMT, the extrinsic gate capacitance consists of two main components (Song *et al* 2014).The first component is a fringing capacitance between the stem of the T-gate and the access regions of the device. The second component arises from the parallel-plate capacitance between the T-gate and neighboring electrodes. Of these two components, the first one plays a crucial role in high-frequency performance of HEMT. Its effect is minimized by employing a T-gate with stem height >200 nm. Using a dielectric with low relative permittivity for device passivation is another way of reducing the fringing capacitance. This capacitance is always associated with the geometrical layout of planar HEMTs. It cannot be got rid off by optimizing the geometry or by decreasing the resistances of access regions to zero. So, it is sometimes included in intrinsic capacitance. However, we shall treat it as an extrinsic portion.

The capacitance C_{GD} is called the feedback capacitance or reverse transfer capacitance. A large C_{GD} value results in slow rise of drain current after turning on the gate. The large C_{GD} value also causes a slow decline of drain current after the gate is turned off.

R_S and R_D are the series resistances of source and drain terminals, and $g_{oIntrinsic}$ is the intrinsic output conductance. First, let us assume that R_S and R_D are both zero. Then

$$\begin{aligned}\text{Input current} = i_i &= \text{Applied gate} - \text{source voltage/Capacitive reactance}\\ &= v_{GS}/X_C = \frac{v_{GS}}{\frac{1}{\omega(C_{GS}+C_{GD})}} = v_{GS}\omega(C_{GS}+C_{GD})\end{aligned} \tag{7.3}$$

where ω is the angular frequency of V_{GS}. Let us short circuit the conductance $g_{oIntrinsic}$ to obtain the short-circuit current gain A_i. If the output current is i_L, the current gain A_i is

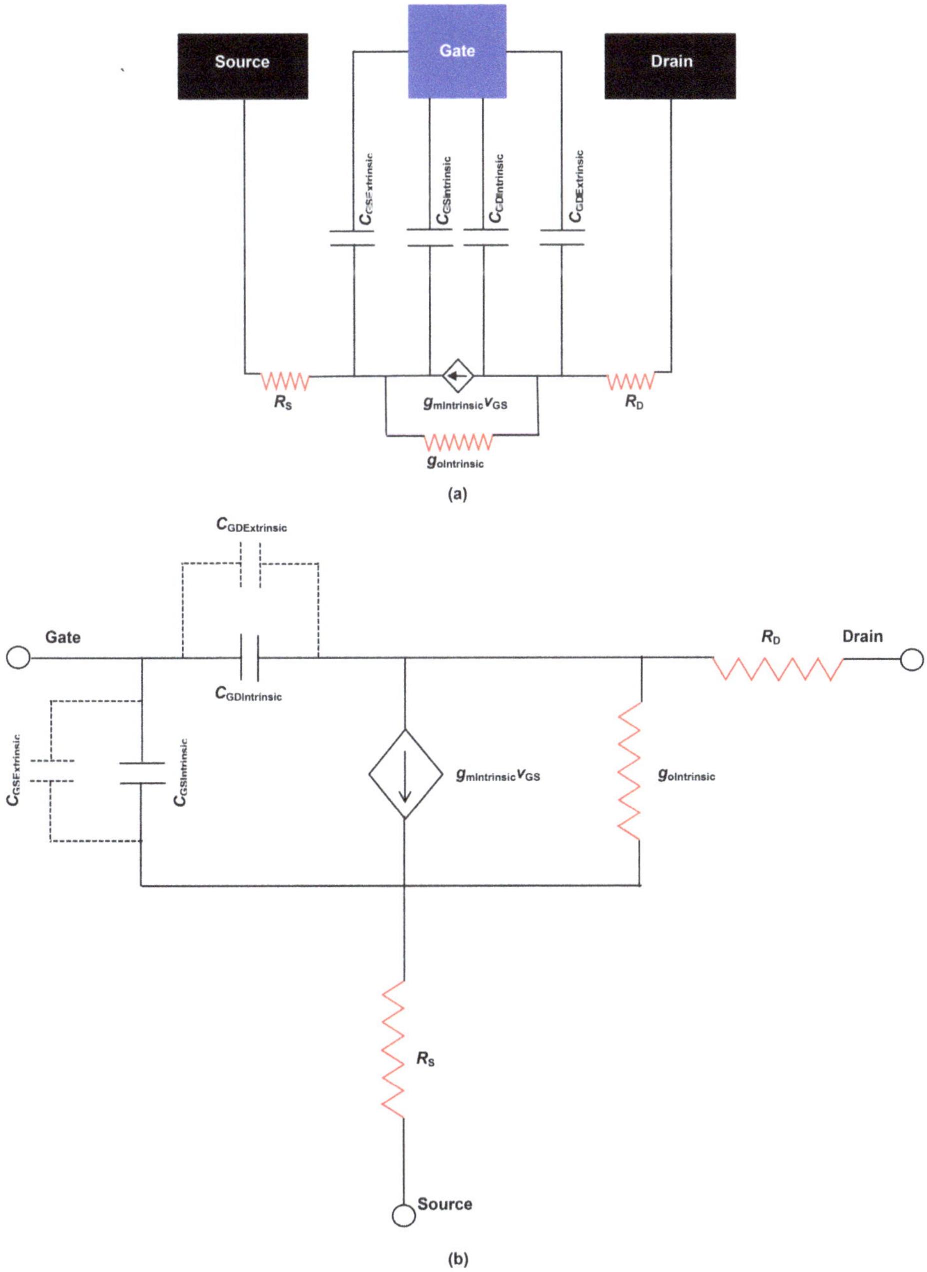

Figure 7.5. Electrical equivalent circuits of HEMT: (a) circuit of figure 7.4 redrawn for clarity and (b) circuit of figure 7.5(a) presented in familiar form.

$$A_i = \frac{i_L}{i_i} = \frac{i_L}{v_{GS}\omega(C_{GS} + C_{GD})} \tag{7.4}$$

But

$$\frac{i_L}{v_{GS}} = g_{m\text{intrinsic}} \tag{7.5}$$

where $g_{m\text{intrinsic}}$ is the intrinsic transconductance. Hence, equation (7.4) reduces to

$$A_i = \frac{i_L}{v_{GS}} \times \frac{1}{\omega(C_{GS} + C_{GD})} = \frac{g_{m\text{intrinsic}}}{\omega(C_{GS} + C_{GD})} = \frac{g_{m\text{intrinsic}}}{2\pi f \times (C_{GS} + C_{GD})} \tag{7.6}$$

where f is the linear frequency of the signal applied at the gate. At the cut-off frequency $f = f_T$, the short-circuit current gain A_i is unity. Therefore, from equation (7.6),

$$1 = \frac{g_{m\text{intrinsic}}}{2\pi f_T \times (C_{GS} + C_{GD})} \tag{7.7}$$

giving

$$f_T = \frac{g_{m\text{intrinsic}}}{2\pi(C_{GS} + C_{GD})} \tag{7.8}$$

In the above analysis, the series resistances R_S, R_D as well as the intrinsic output conductance g_{oi} were not considered. These will lower the cut-off frequency. Their effect will be incorporated in a parasitic capacitance term added to $C_{GS} + C_{GD}$ in equation (7.8) as

$$f_T = \frac{g_{m\text{intrinsic}}}{2\pi(C_{GS} + C_{GD} + C_{\text{Parasitic}})} \tag{7.9}$$

The parasitic capacitance $C_{\text{Parasitic}}$ is expressed as the sum of two capacitances

$$C_{\text{Parasitic}} = C_{\text{Miller}} + C_{g_{SD}} \tag{7.10}$$

where C_{Miller} is the capacitance arising from the amplification of the gate–drain capacitance C_{GD} by multiplication with intrinsic transconductance $g_{m\text{intrinsic}}$ through the Miller effect, and is given by

$$C_{\text{Miller}} = g_{m\text{intrinsic}} C_{GD}(R_S + R_D) \tag{7.11}$$

Further, C_{gSD} is the capacitance originating from the leakage of charging current for the gate capacitance $C_{GS} + C_{GD}$ through the output source-drain conductance g_{SD}, and is written as

$$C_{g_{SD}} = \{C_{GS}(R_S + R_D) + C_{GD}(R_S + R_D)\}g_{0i} = \{(C_{GS} + C_{GD})(R_S + R_D)\}g_{0i} \quad (7.12)$$

where g_{oi} is the intrinsic output conductance.

Combing together equations (7.9–7.12), we get

$$\begin{aligned} f_T &= \frac{1}{2\pi}\frac{g_{m\text{intrinsic}}}{C_{GS} + C_{GD} + g_{m\text{intrinsic}}C_{GD}(R_S + R_D) + \{(C_{GS} + C_{GD})(R_S + R_D)\}g_{0i}} \\ &= \frac{1}{2\pi}\frac{g_{m\text{intrinsic}}}{\begin{array}{c} C_{GS(\text{Intrinsic})} + C_{GD(\text{Intrinsic})} + C_{GS(\text{Extrinsic})} + C_{GD(\text{Extrinsic})} \\ + g_{m\text{intrinsic}}C_{GD}(R_S + R_D) + \{(C_{GS} + C_{GD})(R_S + R_D)\}g_{0i} \end{array}} \\ &= \frac{1}{2\pi}\frac{g_{m\text{intrinsic}}}{\begin{array}{c} C_{GS(\text{Intrinsic})} + C_{GD(\text{Intrinsic})} + C_{GS(\text{Extrinsic})} + C_{GD(\text{Extrinsic})} \\ + g_{m\text{intrinsic}}C_{GD}(R_S + R_D)\left[1 + \{(C_{GS} + C_{GD})\}\frac{g_{0i}}{g_{m\text{intrinsic}}C_{GD}}\right] \end{array}} \end{aligned} \quad (7.13)$$

from which the total time delay τ can be written as

$$\begin{aligned} \tau = \frac{1}{2\pi f_T} &= \frac{\begin{array}{c} C_{GS(\text{Intrinsic})} + C_{GD(\text{Intrinsic})} + C_{GS(\text{Extrinsic})} + C_{GD(\text{Extrinsic})} \\ + g_{m\text{intrinsic}}C_{GD}(R_S + R_D)\left[1 + \{(C_{GS} + C_{GD})\}\frac{g_{0i}}{g_{m\text{intrinsic}}C_{GD}}\right] \end{array}}{g_{m\text{intrinsic}}} \\ &= \tau_{\text{Intrinsic}} + \tau_{\text{Extrinsic}} + \tau_{\text{Parasitic}} \end{aligned} \quad (7.14)$$

The three components of delay can be partitioned into intrinsic, extrinsic and parasitic categories as

$$\tau_{\text{Intrinsic}} = \frac{C_{GS(\text{Intrinsic})} + C_{GD(\text{Intrinsic})}}{g_{m\text{intrinsic}}} \quad (7.15)$$

$$\tau_{\text{Extrinsic}} = \frac{C_{GS(\text{Extrinsic})} + C_{GD(\text{Extrinsic})}}{g_{m\text{intrinsic}}} \quad (7.16)$$

and

$$\begin{aligned} \tau_{\text{Parasitic}} &= \frac{g_{m\text{intrinsic}}C_{GD}(R_S + R_D)\left[1 + \{(C_{GS} + C_{GD})\}\frac{g_{0i}}{g_{m\text{intrinsic}}C_{GD}}\right]}{g_{m\text{intrinsic}}} \\ &= C_{GD}(R_S + R_D)\left[1 + \{(C_{GS} + C_{GD})\}\frac{g_{0i}}{g_{m\text{intrinsic}}C_{GD}}\right] \end{aligned} \quad (7.17)$$

7.2.2 Minimization of delay components in HEMT

Looking at equation (7.15), an obvious strategy to decrease intrinsic delay is to minimize the intrinsic capacitances and maximize transconductance. To achieve this aim, a parameter which has received most attention is the gate length L_G (Lee *et al* 2014). Gate length downscaling diminishes the capacitance. At the same time, it increases transconductance by raising the lateral electric field. The increase in electric field is also beneficial because it accelerates the electrons. The higher electron drift velocity through the channel results in a smaller electron transit time. However, there is a restriction to downscaling L_G for sub-100 nm transistors because of the accompanying worsening of short channel effects (SCEs), notably the increased drain-induced barrier lowering (DIBL) and subthreshold swing (SS). Suppression of short-channel effects is accomplished by maintaining the ratio

$$\begin{aligned}\alpha &= \frac{\text{Gate length}(L_G)}{\text{Gate} - \text{to} - \text{channel distance } (d_{GC}) + \text{Channel layer thickness } (t_C)} \\ &= \frac{L_G}{d_{GC} + t_C} > 1\end{aligned} \tag{7.18}$$

i.e., $\alpha > 1$. Equation (7.18) shows that SCE reduction calls for decreasing both d_{GC} and t_C. But a design trade-off is needed. When a thin channel layer is used, electron scattering increases thereby degrading the carrier mobility μ_n in the quantum well, which in turn lowers transconductance.

7.3 Built-in potential and capacitances

7.3.1 Built-in potential of the gate electrode/semiconductor junction of HEMT

In the ideal case, Schottky barrier height of the metal–semiconductor interface is determined by the Schottky–Mott rule

$$\phi_B = \phi_M - \chi_S \tag{7.19}$$

where ϕ_M is the work function of the metal and χ_S is the electron affinity of the semiconductor.

In practice, the chemical reaction between the metal and the semiconductor, and the presence of a thin layer of native oxide or surface states at the interface, cause the Fermi level of the metal to be aligned near the edge of the semiconductor conduction band. This effect called Fermi level pinning liberates the band bending in the semiconductor from the metal effects for a large change in metal work function. The influence of surface states is quantified through a coefficient γ. In terms of γ, a modified Schottky–Mott equation is written as (Tung 2014)

$$\phi_B = \gamma(\phi_M - \chi_S) + (1 - \gamma)(E_g - \phi_0) \tag{7.20}$$

where ϕ_0 is the energy of charge neutrality level (CNL) reckoned from valence band maximum (VBM); the CNL is the energy level at which the character of the states alters from predominantly donor-like to acceptor-like (Islam *et al* 2014). The value $\gamma = 1$ represents the perfect Schottky contact. When this happens, the second term in equation (7.20) becomes zero. The symbol E_g stands for the energy bandgap of the semiconductor.

The built-in potential at the metal–semiconductor interface is given by

$$V_{\mathrm{bi}} = \phi_{\mathrm{B}} - (E_{\mathrm{C}} - E_{\mathrm{F}}) = \phi_{\mathrm{B}} - \frac{k_{\mathrm{B}}T}{q}\ln\left(\frac{N_{\mathrm{C}}}{N_{\mathrm{D}}}\right) \tag{7.21}$$

where ϕ_{B} = barrier height and $E_{\mathrm{C}} - E_{\mathrm{F}}$ = energy of conduction band edge – energy of Fermi level, N_{C} = effective density of states (DOS) in the conduction band of the semiconductor while N_{D} = concentration of donor impurity in the semiconductor.

7.3.2 Junction capacitances

From the equation for junction capacitance of a Schottky diode (Moloi and McPherson 2013), the gate–source capacitance is modeled as a voltage-dependent capacitor

$$C_{\mathrm{GS}} = A\sqrt{\frac{q\varepsilon_0\varepsilon_{\mathrm{S}}N_{\mathrm{D}}}{2\left(V_{\mathrm{bi}} + V_{\mathrm{Applied\ bias}}\right)}} = A\sqrt{\frac{q\varepsilon_0\varepsilon_{\mathrm{S}}N_{\mathrm{D}}}{2(V_{\mathrm{bi}} + |V_{\mathrm{GS}}|)}} \tag{7.22}$$

where A denotes the junction area, q stands for the electronic charge, ε_0 is permittivity of free space, ε_{S} is relative permittivity of the semiconductor, and V_{GS} is the gate–source voltage. On similar lines and using identical symbols, the gate–drain capacitance is

$$C_{\mathrm{GD}} = A\sqrt{\frac{q\varepsilon_0\varepsilon_{\mathrm{S}}N_{\mathrm{D}}}{2(V_{\mathrm{bi}} + |V_{\mathrm{GD}}|)}} \tag{7.23}$$

where V_{GD} is the gate–drain voltage.

7.4 Analysis of an HEMT structure

Defining the HEMT structural parameters for the problem: A HEMT device is fabricated with a platinum gate electrode on $In_{0.52}Al_{0.48}As$. Because of Fermi level pinning, the Schottky barrier height between platinum and unintentionally-doped $In_{0.52}Al_{0.48}As$ is 0.623 eV (Gao *et al* 1995). Let the un-intentionally doped $In_{0.52}Al_{0.48}As$ barrier layer thickness be 10 nm and its carrier concentration be 1×10^{15} cm^{-3}. The $In_{0.52}Al_{0.48}As$ spacer layer is 5 nm thick. Between the barrier and spacer layers is incorporated a silicon atomic plane with activated sheet doping concentration 5×10^{12} cm^{-2} by delta doping technique (Mei *et al* 2015). Below the spacer layer is the $In_{0.53}Ga_{0.47}As$ channel layer. The effective density of states (DOS) for electrons in $In_{0.52}Al_{0.48}As$ is 5.2×10^{17} cm^{-3} (Movassaghi *et al* 2017), the dielectric constant of $In_{0.52}Al_{0.48}As$ is 12.2 and that of $In_{0.53}Ga_{0.47}As$ is 13.9 (Abdulwahid *et al* 2018). A large conduction band discontinuity ΔE_{C} = 0.48 eV exists between $In_{0.52}Al_{0.48}As$ and $In_{0.53}Ga_{0.47}As$ lattice matched to InP substrate (Datta *et al* 1998). Saturation velocity of electrons in $In_{0.53}Ga_{0.47}As$ is 2×10^7 cm s^{-1}

(Ono *et al* 2004, Yeon *et al* 2007). The gate length is 20 nm and the gate width is 50 μm. The applied gate–source voltage V_{GS} is 0.9 V and the drain–source voltage V_{DS} is 2 V. Electronic charge $q = 1.602 \times 10^{-19}$ C.

Example 7.1 Built-in potential, depletion width, capacitances and transconductance calculation: Let us determine the following:

(i) the built-in potential of the $Pt/In_{0.52}Al_{0.48}As$ Schottky diode and threshold voltage of HEMT;
(ii) zero-bias depletion region width;
(iii) gate–source capacitance in fF and fF mm^{-1};
(iv) gate–drain capacitance in fF and fF mm^{-1};
(v) saturation transconductance in Siemen and S mm^{-1}, supposing that at the given biases, the HEMT is operating in saturation region.

Solution: (i) Note that here

$$\phi_B \neq \phi_{Pt} - \chi_{InAlAs} = 5.3 - 4.25 = 1.015 \text{ eV} \tag{7.24}$$

Instead,

$$\phi_B = 0.623 \text{ eV} \tag{7.25}$$

The built-in potential of the Pt/undoped InAlAs junction is calculated from equation (7.21) as

$$\begin{aligned} V_{bi} &= \frac{\phi_B}{q} - \frac{k_B T}{q}\ln\left(\frac{N_C}{N_D}\right) \\ &= \frac{0.623 \times 1.602 \times 10^{-19}}{1.602 \times 10^{-19}} - \frac{1.3806 \times 10^{-23} \times 300}{1.602 \times 10^{-19}}\ln\left(\frac{5.2 \times 10^{17}\text{ cm}^{-3}}{1 \times 10^{15}\text{ cm}^{-3}}\right) \\ &= 0.623 - \frac{1.3806 \times 10^{-23+19} \times 300}{1.602}\ln(5.2 \times 10^{2}) \\ &= 0.623 - 258.54 \times 10^{-4} \times 6.2538 = 0.623 - 0.0259 \times 6.2538 \\ &= 0.623 - 0.162 = 0.461 \text{ V} \end{aligned} \tag{7.26}$$

Threshold voltage of HEMT is given by (Ajayan *et al* 2018)

$$V_{Th} = \frac{\phi_B - \Delta E_C}{q} - \frac{qN(\delta)d_{Gate-Channel}}{\varepsilon_0 \varepsilon_{Barrier\ layer}} \tag{7.27}$$

where ϕ_B is the Schottky barrier height of the Pt/InAlAs interface, ΔE_C the discontinuity of conduction band edge at the InAlAs/InGaAs heterojunction, $N(\delta)$ the delta doping density, $d_{Gate\text{-}Channel}$ is the separation between Pt gate and InGaAs channel layer, and $\varepsilon_{Barrier\ layer}$ is the relative permittivity of InAlAs barrier layer. From figure 7.4, the separation between the gate and the channel layer is 10 + 5 = 15 nm.

Substituting the values in equation (7.27),

$$V_{\mathrm{Th}} = \frac{(0.623\ -0.48) \times 1.602 \times 10^{-19}}{1.602 \times 10^{-19}}\frac{\mathrm{C} \times \mathrm{V}}{\mathrm{C}} - \frac{1.602 \times 10^{-19} \times 5 \times 10^{12} \times 10^{4} \times 15 \times 10^{-9}}{8.854 \times 10^{-12} \times 12.2}\frac{\mathrm{C\,m^{-2}} \times \mathrm{m}}{\mathrm{Fm^{-1}}}$$

$$= 0.143\ \mathrm{V} - \frac{1.602 \times 5 \times 15 \times 10^{-19+12+4-9+12}}{8.854 \times 12.2}\frac{\mathrm{Cm^{-2}} \times \mathrm{m}}{\mathrm{s^4A^2m^{-2}kg^{-1}m^{-1}}}$$

$$= 0.143\ \mathrm{V} - 1.1123\frac{\mathrm{C} \times \mathrm{m}}{\mathrm{s^4\,A^2\,kg^{-1}\,m^{-1}}} = 0.143\ \mathrm{V} - 1.1123\frac{\mathrm{C} \times \mathrm{m}}{\mathrm{s^4\,C^2\,s^{-2}kg^{-1}\,m^{-1}}} \tag{7.28}$$

$$= 0.143\ \mathrm{V} - 1.1123\frac{\mathrm{kg\,m^2}}{\mathrm{s^2\,C}} = 0.143\ \mathrm{V} - 1.1123\frac{\mathrm{kg\,m^2}}{\mathrm{s^3\frac{C}{s}}}$$

$$= 0.143\ \mathrm{V} - 1.1123\frac{\mathrm{kg\,m^2}}{\mathrm{s^3\,A}} = 0.143\ \mathrm{V} - 1.1123\ \mathrm{kg\,m^2\,s^{-3}A^{-1}}$$

$$= 0.143\ \mathrm{V} - 1.1123\ \mathrm{V} = -0.9693\ \mathrm{V}$$

Negative threshold voltage implies that this HEMT is a normally-on or depletion mode device.

(ii) Diameter of silicon atom = 222 pm = 0.222 nm. Taking the silicon atomic plane as 0.3 nm thick, we find that 0.3 nm × 1 cm^2 area corresponds to a volume of 0.3 × 10^{-9} × 100 cm × 1 cm^2 = 0.3 × 10^{-7} cm^3 = 3 × 10^{-8} cm^3. Since this volume contains 5 × 10^{12} Si atoms, the silicon delta doping concentration = 5 × 10^{12}/(3 × 10^{-8}) = 1.67 × 10^{20} cm^{-3}. When distributed over the 15 nm thick InAlAs (barrier + spacer) layers, the effect of this concentration is the same as that of the concentration 1.67 × 10^{20} cm^{-3}/(15/0.3) = 0.0334 × 10^{20} = 3.34 × 10^{18} cm^{-3}.

Zero-bias depletion region width is obtained as

$$x_d = \sqrt{\frac{2\varepsilon_0\varepsilon_{\mathrm{S}}V_{\mathrm{bi}}}{qN_{\mathrm{D}}}} = \sqrt{\frac{2 \times 8.854 \times 10^{-12} \times 12.2 \times 0.461}{1.602 \times 10^{-19} \times 3.34 \times 10^{18} \times 10^{6}}}$$

$$= \sqrt{\frac{2 \times 8.854 \times 12.2 \times 0.461 \times 10^{-12+19-18-6}}{1.602 \times 3.34}} = \sqrt{18.613 \times 10^{-17}} \tag{7.29}$$

$$= 13.643 \times 10^{-9}\ \mathrm{m} = 13.6\ \mathrm{nm}$$

(iii) When viewed from the gate terminal, the HEMT structure looks like a parallel-plate capacitor. What are the two plates of this parallel plate capacitor? One plate is the gate metal electrode and the other plate is the channel. Suppose an AC signal is applied to the gate electrode kept at a potential of V_{GS} = 0.9 V. Then the gate potential swings between positive and negative voltage values about the mean position V_{GS}. These voltage excursions lead to forward and reverse biasing of the Schottky diode gate and produce charge storage effects which are represented by the gate–source and gate–drain capacitances. Considering maximum voltage swing for V_{GS} as 0.9 V, the gate–source capacitance for V_{GS} = 0.9 V is determined from equation (7.22) as

$$
\begin{aligned}
C_{GS} &= A\sqrt{\frac{q\varepsilon_0\varepsilon_S N_D}{2(V_{bi} + |0.9|)}} \\
&= 50 \times 10^{-6} \times 20 \\
&\quad \times 10^{-9}\sqrt{\frac{1.602 \times 10^{-19} \times 8.854 \times 10^{-12} \times 12.2 \times 3.34 \times 10^{18} \times 10^{6}}{2(0.461 + 0.9)}} \\
&= 1 \times 10^{-12}\sqrt{\frac{1.602 \times 8.854 \times 12.2 \times 3.34 \times 10^{-19-12+18+6}}{2(1.361)}} \\
&= 1 \times 10^{-12}\sqrt{\frac{1.602 \times 8.854 \times 12.2 \times 3.34 \times 10^{-7}}{2.722}} \\
&= 1 \times 10^{-12}\sqrt{212.334 \times 10^{-7}} = 1 \times 10^{-12} \times 4.61 \times 10^{-3} \\
&= 4.61 \times 10^{-15}\,\text{F}
\end{aligned}
\tag{7.30}
$$

Gate–source capacitance for 50 μm gate width is 4.61×10^{-15} F. So, the capacitance for 1 μm gate width is $4.61 \times 10^{-15}/50$ and that for 1 mm = 1000 μm gate width is $4.61 \times 10^{-15} \times 1000/50 = 92.2 \times 10^{-15}$ F = 92.2 fF. Hence, the capacitance per unit gate width is 92.2 fF mm^{-1}.

(iv) Since $V_{GS} = 0.9$ V and $V_{DS} = 2$ V,

$$
\begin{aligned}
V_{GD} &= V_G - V_D = V_{GS} + V_S - V_D = V_{GS} + V_D - V_{DS} - V_D = V_{GS} - V_{DS} \\
&= 0.9 - 2 = -1.1\,\text{V}
\end{aligned}
\tag{7.31}
$$

since

$$V_{GS} = V_G - V_S \tag{7.32}$$

$$V_{GD} = V_G - V_D \tag{7.33}$$

and

$$V_{DS} = V_D - V_S \tag{7.34}$$

For $V_{GD} = -1.1$ V, the gate–drain capacitance is found from equation (7.23) as

$$
\begin{aligned}
C_{GD} &= A\sqrt{\frac{q\varepsilon_0\varepsilon_S N_D}{2(V_{bi} + |-1.1|)}} \\
&= 50 \times 10^{-6} \times 20 \\
&\quad \times 10^{-9}\sqrt{\frac{1.602 \times 10^{-19} \times 8.854 \times 10^{-12} \times 12.2 \times 3.34 \times 10^{18} \times 10^{6}}{2(0.461 + 1.1)}} \\
&= 1 \times 10^{-12}\sqrt{\frac{1.602 \times 8.854 \times 12.2 \times 3.34 \times 10^{-7}}{2(1.561)}} \\
&= 1 \times 10^{-12}\sqrt{185.129 \times 10^{-7}} = 1 \times 10^{-12} \times 4.3 \times 10^{-3} = 4.3 \times 10^{-15}\,\text{F}
\end{aligned}
\tag{7.35}
$$

The gate–drain capacitance for 1 mm gate width is $4.3 \times 10^{-15} \times 1000/50 = 86 \times 10^{-15}$ F = 86 fF, which can be stated as 86 fF mm^{-1}.

(v) In order that the HEMT is operating in saturation mode, V_{GS} must be $> V_{Th}$ and $V_{DS} > V_{GS}\text{-}V_{Th}$. Here V_{GS} (0.9 V) is $> V_{Th}$ (− 0.9693 V) and $V_{DS} = 2$ V is $> V_{GS}\text{-}V_{Th}$ = Saturation voltage $V_{D(Sat)} = 0.9$ V-(−0.9693 V) = 0.9 + 0.9693 V = 1.8693 V. So the device is under saturation.

Under saturation condition, the intrinsic transconductance is approximately given by

$$\begin{aligned} g_{m\text{Intrinsic}} &= \frac{\varepsilon_0 \varepsilon_{\text{InGaAs}} v_{\text{sat}} W}{d_{\text{Gate-Channel}}} = \frac{8.854 \times 10^{-12} \times 13.9 \times 2 \times 10^5 \times 50 \times 10^{-6}}{15 \times 10^{-9}} \\ &= \frac{8.854 \times 13.9 \times 2 \times 50}{15} \times 10^{-12+5-6+9} = 820.471 \times 10^{-4} \\ &= 0.082 \text{ S} \end{aligned} \tag{7.36}$$

where $\varepsilon_{\text{InGaAs}}$ is the relative permittivity of InGaAs, v_{sat} is the saturation velocity of electrons in InGaAs, W is the channel width and $d_{\text{Gate-Channel}}$ is the gate-to-channel distance. Compare equation (7.36) with the MOSFET equation (6.27) for similarities and differences. Seek a correlation between the parameters in the two equations and try to understand the implications of equation (7.36).

Transconductance can be expressed in units of Siemens per channel width which is the same as Siemens per gate width. Here, the transconductance is 0.082 S for 50 μm gate width which means that the transconductance for 1 μm gate width is 0.082 S/50 = 0.001 64 S = 1.64 mS, written as 1.64 mS μm^{-1}. Hence, transconductance for 1 mm = 1000 μm gate width is 1.64 mS × 1000 = 1.64 S; this is stated as 1.64 S m^{-1}m^{-1}.

Example 7.2 Frequencies and delays calculation for HEMT structure of example 7.1: Find: (i) the cut-off frequency assuming that extrinsic capacitance as well as the sum-total source and drain series resistance $R_S + R_D$ is zero, (ii) the cut-off frequency taking the extrinsic capacitance contribution as 40% of the total gate capacitance, (iii) the resistance $R_S + R_D$ considering that a very small value of on-resistance = 140 Ω-μm is achieved and (iv) the cut-off frequency after accounting for $R_S + R_D$, (v) intrinsic delay, (vi) extrinsic delay and (vii) parasitic delay.

Solution: (i) From equations (7.8), (7.30), (7.35) and (7.36)

$$\begin{aligned} f_T &= \frac{g_{m\text{Intrinsic}}}{2\pi(C_{GS} + C_{GD})} = \frac{0.082}{2 \times 3.141\,59 \times (4.61 + 4.3) \times 10^{-15}} \\ &= \frac{0.082}{2 \times 3.141\,59 \times 8.91 \times 10^{-15}} = 0.001\,465 \times 10^{15} = 1.47 \times 10^{12} \text{ Hz} \end{aligned} \tag{7.37}$$

(ii) Let the total gate capacitance including both intrinsic and extrinsic components be x. Then, we can write

$$C_{\text{GS(Intrinsic)}} + C_{\text{GD(Intrinsic)}} + C_{\text{GS(Extrinsic)}} + C_{\text{GD(Extrinsic)}} = x \tag{7.38}$$

Since the extrinsic gate capacitance represents 40% of the total gate capacitance,

$$C_{\mathrm{GS(Intrinsic)}} + C_{\mathrm{GD(Intrinsic)}} + 0.4x = x \tag{7.39}$$

or,

$$C_{\mathrm{GS(Intrinsic)}} + C_{\mathrm{GD(Intrinsic)}} = x - 0.4x = 0.6x \tag{7.40}$$

or,

$$x = \frac{C_{\mathrm{GS(Intrinsic)}} + C_{\mathrm{GD(Intrinsic)}}}{0.6} = 1.67\left\{C_{\mathrm{GS(Intrinsic)}} + C_{\mathrm{GD(Intrinsic)}}\right\} \tag{7.41}$$

From equations (7.37)–(7.39) and (7.41),

$$\begin{aligned} f_T &= \frac{g_{m\mathrm{Intrinsic}}}{2\pi x} = \frac{0.082}{2 \times 3.141\,59 \times 1.67 \times (4.61 + 4.3) \times 10^{-15}} \\ &= \frac{0.082}{2 \times 3.141\,59 \times 14.88 \times 10^{-15}} = 0.000\,877 \times 10^{15} = 0.877 \times 10^{12}\ \mathrm{Hz} \end{aligned} \tag{7.42}$$

(iii) We note that

$$\begin{aligned} \mathrm{ON} - \mathrm{resistance} = {} & \text{Total contact resistance of source and drain} \\ & + \text{Access resistance of source and drain} \\ & + \text{Channel resistance} \end{aligned} \tag{7.43}$$

The on resistance is given as 140 Ω – μm. For a device operating in the saturation region, the channel resistance is zero, i.e., $g_{o\mathrm{Intrinsic}} = 0$. The straight line path length in figure 7.4 is 5 + 10 + 5 + 10 + 700/2 – 10 = 370 nm on the source side and an equal length on the drain side; 10 nm is subtracted to account for half the channel length on each side. The total path length is 2 × 370 = 740 nm = 740 × 10^{-9} × 10^{6} μm = 0.74 μm. Neglecting the contact resistances, we can write from equation (7.43)

$$R_{\mathrm{S}} + R_{\mathrm{D}} = 140 \times 0.74 = 103.6\ \Omega \tag{7.44}$$

(iv) In this condition, equation (7.13) becomes

$$\begin{aligned} f_T &= \frac{1}{2\pi} \frac{g_{\mathrm{mIntrinsic}}}{x + g_{\mathrm{mIntrinsic}}\ C_{GD}(R_S + R_D)\left[1 + \{(C_{GS} + C_{GD})\}\dfrac{0}{g_{\mathrm{mintrinsic}}\ C_{GD}}\right]} \\ &= \frac{1}{2\pi} \frac{g_{\mathrm{mIntrinsic}}}{1.67\left(C_{GS(\mathrm{Intrinsic})} + C_{GD(\mathrm{Intrinsic})}\right) + g_{\mathrm{mIntrinsic}}\ C_{GD}(R_S + R_D) + 0} \\ &= \frac{1}{2 \times 3.141\,59} \times \frac{0.082}{1.67(4.61 + 4.3) \times 10^{-15} + 0.082 \times 4.3 \times 10^{-15} \times 103.6} \\ &= \frac{1}{6.283\,18} \times \frac{0.082}{14.88 \times 10^{-15} + 36.529 \times 10^{-15}} \\ &= \frac{1}{6.283\,18} \times \frac{0.082}{51.409 \times 10^{-15}} = 0.000\,253\,9 \times 10^{15} = 0.25 \times 10^{12}\ \mathrm{Hz} \end{aligned} \tag{7.45}$$

(v) From equation (7.15)

$$\tau_{Intrinsic} = \frac{C_{GS(\text{Intrinsic})} + C_{GD(\text{Intrinsic})}}{g_{mintrinsic}} = \frac{(4.61 + 4.3) \times 10^{-15}}{0.082} = \frac{8.91 \times 10^{-15}}{0.082} \quad (7.46)$$
$$= 108.659 \times 10^{-15}\,\text{s} = 0.109\,\text{ps}$$

(vi) From equation (7.41) taking the extrinsic part of capacitance as 40% of the total capacitance,

$$C_{\text{GS(Extrinsic)}} + C_{\text{GD(Extrinsic)}} = 0.4x = 0.4 \times 1.67\left\{C_{\text{GS(Intrinsic)}} + C_{\text{GD(Intrinsic)}}\right\}$$
$$= 0.668\left\{C_{\text{GS(Intrinsic)}} + C_{\text{GD(Intrinsic)}}\right\} \quad (7.47)$$
$$= 0.668(4.61 + 4.3) \times 10^{-15} = 5.952 \times 10^{-15}\,\text{F}$$

Applying equations (7.16), (7.36) and (7.47)

$$\tau_{\text{Extrinsic}} = \frac{C_{\text{GS(Extrinsic)}} + C_{\text{GD(Extrinsic)}}}{g_{m\text{intrinsic}}} = \frac{5.952 \times 10^{-15}}{0.082} \quad (7.48)$$
$$= 72.585 \times 10^{-15}\,\text{s} = 0.0726\,\text{ps}$$

(vii) From equations (7.17), (7.35) and (7.44)

$$\tau_{\text{Parasitic}} \approx C_{\text{GD}}(R_{\text{S}} + R_{\text{D}}) = 4.3 \times 10^{-15} \times 103.6 = 445.8 \times 10^{-15}\,\text{s} \quad (7.49)$$
$$= 0.446\,\text{ps}$$

7.5 InP terahertz HEMT technology

7.5.1 Reasons for interest in InP HEMTs

The wide bandgap of InP (1.34 eV) together with high electron drift velocity $>2 \times 10^5$ m s^{-1} makes InP HEMT technology very attractive. InP HEMTs based on the InP/InAlAs/InGaAs material system utilize the outstanding 2DEG properties of InAlAs/InGaAs heterojunction. As a consequence, development of InP HEMTs capable of working at ever increasing frequencies has engaged the attention of several research groups. The progress from 0.27 to 1 THz InP HEMTs is traced below (Ajayan *et al* 2018, Božanić and Sinha 2019).

7.5.2 High-frequency HEMT amplifiers

35 nm InP HEMT 3-stage IC 0.27 THz amplifier with peak gain 11.6 dB and positive forward voltage gain up to 0.34 THz was reported by Deal *et al* (2007) while 35 nm InP HEMT IC amplifier with gain 4.4 dB at 0.308 THz and projected maximum gain >7 dB at 0.3 THz was described by Mei *et al* (2007). Subsequently, sub 50 nm-InP HEMTs with estimated $f_{\text{max}} > 1$ THz were used in a 5-stage 0.48 THz amplifier with peak gain 11.7 dB (Deal *et al* 2010). Following this achievement, high

f_{max} 30 nm InP HEMTs were used in a 5-stage coplanar waveguide IC to provide low noise amplification at 0.67 THz with gain >7 dB at this frequency; the gain was 30 dB for a 10-stage version (Deal *et al* 2011). With further refinements, packaged 0.85 THz InP HEMT amplifier was fabricated in a 10-stage coplanar waveguide IC showing gain 13.6 dB and output power up to 0.93 mW (Leong *et al* 2015). In sequel, 25 nm InP HEMT amplifier with gain 9 dB at 1 THz was demonstrated for the first time. The maximum available gain was 3.5 dB at 1 THz and 1.5 THz projected f_{max} (Mei *et al* 2015).

7.5.3 HEMT terahertz emitters

Recall discussions in chapter 6 on the use of MOSFETs as terahertz emitters. The HEMT is also a field-effect device and behaves like a MOSFET providing terahertz emission (Dyakonov and Shur 1996).

Voltage-tunable terahertz emission in the frequency range from 0.4–1.0 THz is observed in lattice-matched InGaAs/AlInAs HEMTs of gate length 60 nm on InP substrates (Knap *et al* 2004). When the source–drain voltage is increased, the maximum emission frequency changes from 0.42 to 1.0 THz. The radiation power is in nW range.

Emission of terahertz waves is demonstrated over the temperature range from 4.2K–300K from two types of nanometer HEMTs (Dyakonova *et al* 2006): (i) InAlAs/InGaAs/InP HEMT has a gate length of 50 nm and the electron concentration in the channel is 2.4×10^{-12} cm^{-2}. (ii) AlGaN /GaN HEMT of gate length of 150 nm has an electron concentration in the channel = 1.2×10^{-13} cm^{-2}.

Square source–drain pulses are used to excite the terahertz emission. At 4.2K, the InP-based HEMT gives several-fold higher emission than GaN-based HEMT. The asymmetric spectra have a maximum at 0.5–1 THz with a tail at higher frequencies. At room temperature, the integrated power from GaN-based HEMT is 250 times that of InP-based HEMT, which becomes very low for measurements. The emission peak of GaN-based HEMT is shifted to 1.5–2 THz. The highest power emitted by a single HEMT device is 1 μW (Dyakonova *et al* 2006).

7.6 Discussion and conclusions

The key to enhancing the speed of HEMTs lies in the reduction of travelling time of electrons through the devices together with increasing the velocity of electron motion. The former is attainable by decreasing the gate length using high-resolution lithographic techniques while the latter depends on the proper choice of the semiconductor material system to support high travelling speed of electrons. The combination of high gain at high frequencies with low noise and low power dissipation has provided great impetus to InP HEMT technologies. Investments of research efforts have culminated in the development of amplifiers working at oscillation frequencies over 1 THz.

References

Abdulwahid O S, Sexton J, Kostakis I, Ian K and Missous M 2018 Physical modelling and experimental characterization of InAlAs/InGaAs avalanche photodiode for 10 Gb/s data rates and higher *IET Optelectron.* **12** 5–10

Ajayan J, Nirmal D, Ravichandran T, Mohankumar P, Prajoon P, Arivazhagan L and Sarkar C K 2018 InP high electron mobility transistors for submillimetre wave and terahertz frequency applications: a review *Int. J. Electron. Commun.* **94** 199–214

Božanić M and Sinha S 2019 Emerging transistor technologies capable of terahertz amplification: a way to re-engineer terahertz radar sensors *Sensors* **19** 32

Datta S, Shi S, Roenker K P, Cahay M M and Stanchina W E 1998 Simulation and design of InAlAs/InGaAs pnp heterojunction bipolar transistors *IEEE Trans. Electron Devices* **45** 1634–43

Deal W R *et al* 2007 Demonstration of a 270-GHz MMIC amplifier using 35-nm InP HEMT technology *IEEE Microwave Wirel. Compon. Lett.* **17** 391–3

Deal W R *et al* 2010 Demonstration of a 0.48 THz amplifier module using InP HEMT transistors *IEEE Microwave Wirel. Compon. Lett.* **20** 289–91

Deal W R *et al* 2011 Low noise amplification at 0.67 THz using 30 nm InP HEMTs *IEEE Microwave Wirel. Compon. Lett.* **21** 368–70

Del A J A 2011 The high-electron mobility transistor at 30: Impressive accomplishments and exciting prospects *Proc. of the 2011 Int. Conf. on Compound Semiconductor Manufacturing Technology (Indian Wells, CA, May 16–19May 16–19)* (Beaverton, OR: CS ManTech) 6pp

Dyakonova N *et al* 2006 Room-temperature terahertz emission from nanometer field-effect transistors *Appl. Phys. Lett.* **88** 141906-1–3

Dyakonov M I and Shur M S 1996 Plasma electronics: novel terahertz devices using two-dimensional electron fluid *IEEE Trans. Electron Devices* **43** 1640–5

Gao W, Berger P R, Hunsperger R G, Zydzik G, Rhodes W W, O'Bryan H M, Sivco D and Cho A Y 1995 Transparent and opaque Schottky contacts on undoped $In_{0.52}Al_{0.48}As$ grown by molecular beam epitaxy *Appl. Phys. Lett.* **66** 3471–3

Islam R, Shine G and Saraswat K C 2014 Schottky barrier height reduction for holes by Fermi level depinning using metal/nickel oxide/silicon contacts *Appl. Phys. Lett.* **105** 182103-1–4

Kim D-H, Brar B and del Alamo J A 2011 f_T = 688 GHz and f_{max} = 800 GHz in L_g = 40 nm $In_{0.7}Ga_{0.3}As$ MHEMTs with g_{m_max} > 2.7 mS/μm *IEEE International Electron Devices Meeting (December 5–7, 2011)* (Washington DC: IEDM Tech. Digest) 13.6.1–4

Knap W, Lusakowski J, Parenty T, Bollaert S, Cappy A, Popov V V and Shur M S 2004 Terahertz emission by plasma waves in 60 nm gate high electron mobility transistors *Appl. Phys. Lett.* **84** 2331–3

Lee D S, Liu Z and Palacios T 2014 GaN high electron mobility transistors for sub-millimeter wave applications *Jpn. J. Appl. Phys.* **53** 100212-1–10

Leong K M K H *et al* 2015 A 0.85 THz low noise amplifier using InP HEMT transistors *IEEE Microwave Wirel. Compon. Lett.* **25** 397–9

Mei X B *et al* 2007 35-nm InP HEMT SMMIC amplifier with 4.4-dB gain at 308 GHz *IEEE Electron Device Lett.* **28** 470–2

Mei X B, Yoshida W, Zhou Z, Lange M, Lee J, Liu P H, Leong K, Lai R and Deal W R 2015 25nm InP HEMT TMIC process with 1 THz amplifier circuit gain *Extended Abstracts of the 2015 Int. Conf. on Solid State Devices and Materials (Sapporo, Japan, Sept 27–Sept 30, 2015)* pp 1034–5

Mei X *et al* 2015 First demonstration of amplification at 1 THz using 25-nm InP high electron mobility transistor process *IEEE Electron Device Lett.* **36** 327–9

Moloi S J and McPherson M 2013 Capacitance–voltage behaviour of Schottky diodes fabricated on p-type silicon for radiation-hard detectors *Radiat. Phys. Chem.* **85** 73–82

Movassaghi Y, Fathipour V, Fathipour M and Mohseni H 2017 Analytical and numerical evaluation of electron-injection detector optimized for SWIR photon detection *J. Appl. Phys.* **121** 084501-1–12

Nirmal D and Ajayan J (ed) 2019 *Handbook for III–V High Electron Mobility Transistor Technologies* (Boca Raton, FL: CRC Press) 1–96 pp

Ono H, Taniguchi S and Suzukil T-k 2004 High-frequency characteristics and saturation electron velocity of InAlAs/InGaAs metamorphic high electron mobility transistors at high temperatures *Int. Conf. on Indium Phosphide and Related Materials, Conf. Proc. 16th IPRM*, 31 May–4 June, 2004 Kagoshima, Japan, pp 288–91

Song B *et al* 2014 Effect of fringing capacitances on the RF performance of GaN HEMTs with T-gates *IEEE Trans. Electron Devices* **61** 747–54

Tung R T 2014 The physics and chemistry of the Schottky barrier height *Appl. Phys. Rev.* **1** 011304-1–54

Umansky V, Heiblum M, Levinson Y, Smet J, Nübler J and Dolev M 2009 MBE growth of ultra-low disorder 2DEG with mobility exceeding 35×10^6 cm^2/Vs *J. Cryst. Growth* **311** 1658–61

Yeon S-J, Park M, Choil J and Seo K 2007 600 GHz InAlAs/In0.75GaAs metamorphic HEMTs with an ultra-short 15-nm-gate *International Electron Devices Meeting* (Washington DC: IEDM Technical Digest) pp 613–6

Part II

Vacuum electronic devices

Chapter 8

Travelling wave tubes and backward wave oscillators

Constructional features of travelling wave tubes (TWTs) and backward wave oscillators (BWOs) are described. The similarities and differences between the two tubes are indicated. For closer inspection of TWT and BWO operation, the concepts of phase and group velocities in general are reviewed, and particularly in a waveguide. Decrease of phase velocity in a slow-wave structure is explained. Electric field distribution in a slow-wave structure is studied and the mechanism of Cherenkov radiation is discussed. From the standpoint of phase/group velocity, the differences between TWT and BWO are elaborated. Other topics addressed include the operation and applications of TWTs and BWOs. The present status of terahertz TWTs and BWOs is surveyed.

After surveying the portfolio of solid-state devices in foregoing chapters (Part I), we now divert our attention to vacuum electronic devices (Part II). The vacuum electronic device collection exhibits wide-ranging diversity, both in frequency and power capability. The variation range is incredible in both parameters. In power it extends over 12 orders of magnitude from milliwatts (10^{-3} W) to gigagwatts (10^{9} W). In frequency, it encompasses greater than two orders of magnitude from less than 0.1 THz to more than 10 THz (Booske *et al* 2011).

The fundamental difference between solid-state and vacuum-electronic devices is the medium of electron transport, which is reflected in their power handling capabilities. In solid-state devices, the electrons suffer frequent scattering while moving through the transport medium. Electron current is a collisional conduction current. In vacuum-electronic devices, the electron transport medium is vacuum. So the electron current is a collision-less convection current resembling ballistic transport. The heat produced by electron collisions is easily dissipated

doi:10.1088/978-0-7503-3171-5ch8

making the vacuum-electronic devices superior to solid-state devices for high power, high-frequency operation.

In this chapter we shall study two commonly used electron tubes. These tubes are: the linear-beam TWT and the BWO. The TWT appears to be identical in construction to the BWO. Although the TWT shares several commonalities with its look-alike vacuum tube, the BWO, the two devices differ profoundly in their operation.

8.1 General constructional features of TWTs and BWOs

8.1.1 Main components

The main components of a TWT/BWO are (figures 8.1 and 8.2): a cathode and an anode collector at the opposite ends of a vacuum tube. There is a slow wave structure in the middle, usually in the form of a helix or a planar grating fabricated by micromachining techniques such as UV-LIGA and deep reactive ion etching (DRIE) (Feng *et al* 2011, Karetnikova *et al* 2018). A magnetic field is applied over the electron beam with the help of an axial solenoid surrounding the tube. In place of the bulky solenoid, a permanent magnet is often used in low power tubes.

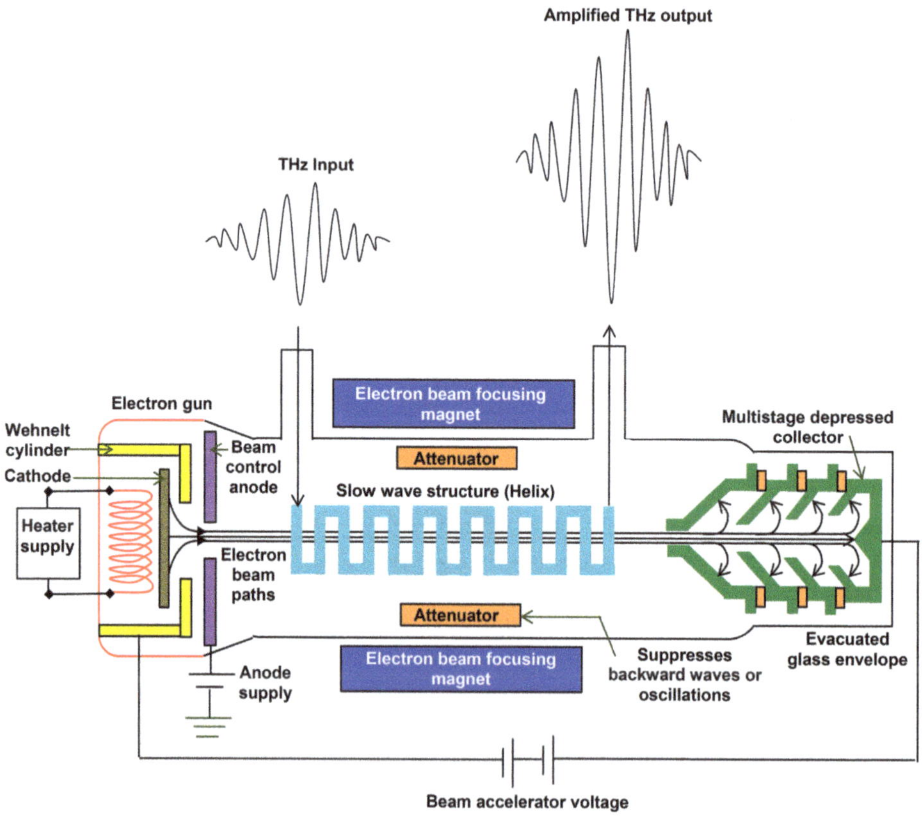

Figure 8.1. Functional diagram of a TWT amplifier.

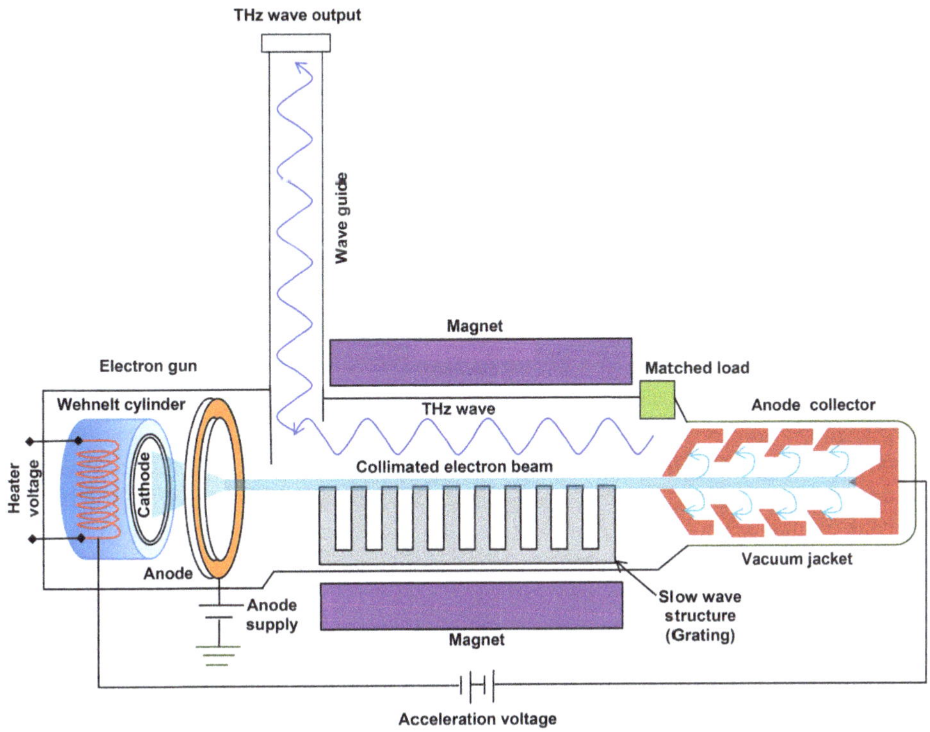

Figure 8.2. Schematic diagram of a BWO.

It is especially befitting to applications such as satellite communications because in these applications, the power consumption and payload must be minimized. The cathode supplies a rectilinear electron beam.

8.1.2 Problems faced with pencil-beams

The difficulty faced with the conventional pencil-like beam at terahertz frequencies is the requirement of high current density ~ 1 kA cm^{-2} together with accurate guidance of this beam in vicinity of the slow wave structure for efficient interaction with the electromagnetic wave travelling in the structure. Due to inaccuracies of the beam guiding optics and the high fields, undesirable current flow and heat loading of the miniaturized components of the slow wave structure (SWS) is inevitable besides ohmic losses.

8.1.3 Performance improvement with planar sheet beams

Thermal performance of the electron tube is improved by using single/multiple planar sheet electron beam configuration of large width instead of a pencil-like beam (Ryskin *et al* 2015, Karetnikova *et al* 2016, Fedotov *et al* 2018, Ploskih *et al* 2019). Due to enlargement of the area of cross-section, the sheet beam decreases the necessary current density but increases the electron current. The average power output of a planar beam TWT/BWO is superior to that of the pencil beam device owing to smaller power loss per unit area and larger heat dissipation area (Shu *et al* 2018). Alternatively, a hollow cylindrical beam is preferred in which electrons are concentrated in proximity of the SWS. When a hollow beam is used, the diameter of the beam tunnel can be significantly increased (Bratman *et al* 2014, 2017). As a result, the current density and dangerous overheating of the fragile slow wave structure are reduced. This allows raising the average radiation at high frequencies.

8.1.4 Functions of electric and magnetic fields

The main tasks performed by these fields are acceleration of electrons, making their motion aligned and parallel to a particular direction, and reducing the spatial cross-section of the electron beam. The electric field applied between the anode and the cathode provides the accelerating voltage for electron motion. The magnetic field serves to collimate the electron beam. It prevents the electrons from spreading apart through electrostatic repulsion. Any electrons diverging radially outwards from their trajectories are sent back towards the axis. In this way, the electron beam is maintained at the required diameter throughout its length.

8.1.5 Function of the slow wave structure

In a waveguide, the phase velocity of electromagnetic waves surpasses the velocity of light in vacuum. How to make the phase velocity less than velocity of light? As its name suggests, the slow wave structure is a contrivance that reduces the phase velocity of electromagnetic waves below the velocity of light. This is necessary in order to ensure that the electrons moving sluggishly than light can get sufficient time for proper interaction with the electromagnetic waves. The fact that the slowing down of the electromagnetic wave below the velocity of light in vacuum is an essential prerequisite to TWT/BWO operation makes them members of the family of devices called slow wave devices. This is an important category of vacuum electronic tubes, the other category is fast wave devices.

8.1.6 Constructional dissimilarities/similarities between TWT and BWO

(i) In a TWT, the input signal is fed near the cathode. The TWT amplifies the input signal injected from near the cathode side while the output signal is derived near the anode side. It is essentially an amplifier. In a BWO, the output signal is extracted from near the cathode side while the anode side terminates in a matched impedance. In this terminating impedance, any forward moving signal is dissipated.

(ii) The TWT has an attenuator which serves to suppress any backward moving signal such as reflected waves from a mismatched load. It prevents such signals from reaching the input where they can cause unwanted oscillations. The BWO has no attenuator.

(iii) In both the TWT and BWO, the spent electron beam is received on the anode collector.

8.2 Closer examination of working of TWT/BWO

After introducing the TWT/BWO superficially, we peep deeper into the functioning of TWT/BWO. Questions that immediately need answering are: the meanings of phase and group velocities in general and particularly their relative magnitudes in a waveguide, and the functioning of slow wave structure, which is an essential component of TWT/BWO. Another question concerns the generation of Cherenkov electromagnetic radiation in a slow wave structure. Many of these concepts are confusing to students. We shall first elaborate on these concepts and then try to understand the operation of the BWO.

8.2.1 Phase velocity (v_p) and group velocity (v_g) concepts

We shall introduce the phase and group velocity concepts in an elementary manner. One complete cycle of a sinusoidal wave starting from zero to peak value in positive direction, back to zero, then up to peak value in the opposite direction and once again to zero, can be related to a angular motion of 2π radians (figure 8.3). For this wave, two angular parameters are defined. The number of radians covered per unit time at a fixed position is the angular frequency ω while the number of radians covered per unit distance at a fixed time is the wave number k. The equation for the wave moving in the positive X-direction can be written using the parameters ω and k as

$$E(x, t) = E_0\cos(kx - \omega t) \tag{8.1}$$

where E is the variable quantity of interest which is a function of space and time, and E_0 is the maximum value or amplitude of the wave. The sine function can be used in place of the cosine function. The only difference from the wave given by cosine function is in phase.

The speed of movement of the shape of the wave or any point of fixed phase is called the phase velocity v_p of the wave (figure 8.4). In other words, it is the velocity at which the loci of constant phase are propagated. The phase velocity

$$\begin{aligned} v_p &= \text{Number of radians of the wave passing a fixed position per unit time} \\ &\quad \times \text{the spatial length of the wave per radian} = \omega \times \frac{1}{k} = \frac{\omega}{k} \end{aligned} \tag{8.2}$$

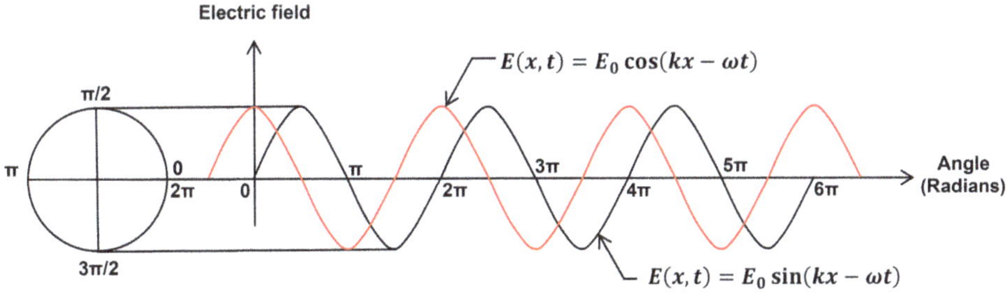

Figure 8.3. Sinusoidal and cosinusoidal waveforms.

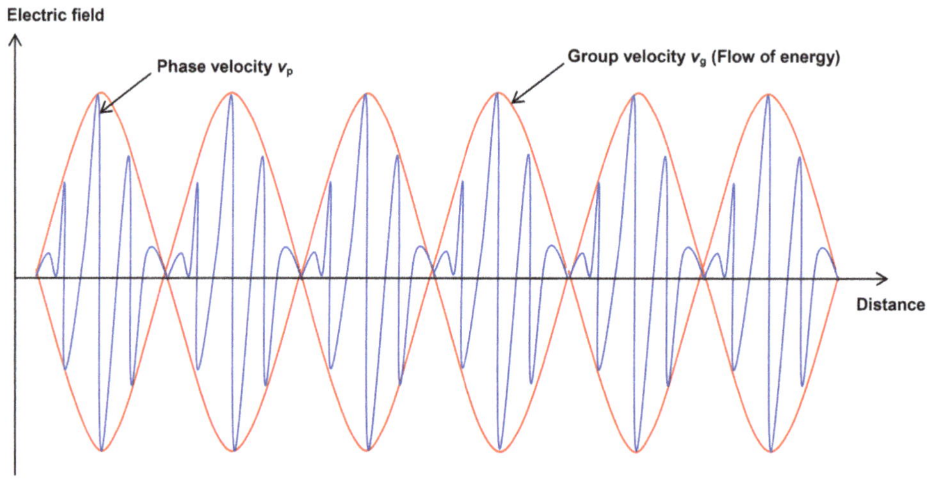

Figure 8.4. Phase and group velocities of a wave.

Since a single wave is not an embodiment of the causal flow of physical effects, the phase velocity is not bounded by any upper limit. This does not contradict Einstein's theory of relativity that messages cannot travel faster than the velocity of light in vacuum.

Energy or information is conveyed in the form of a change in wave property such as a change in amplitude by amplitude modulation. Hence, the speed of information/energy content is expressed as the ratio of change in angular frequency ($\Delta\omega$)/change in wave number (Δk). This is the velocity of amplitude wave. But since each amplitude wave comprises an ensemble of internal waves, the so-called wave packet, its velocity is known as group velocity. So the group velocity of a wave is written as a ratio of changes:

$$v_g = \frac{d\omega}{dk} \tag{8.3}$$

The refractive index (n) is an important property of a transparent optical medium. Also called index of refraction, it is a dimensionless number representing the ratio of two velocities. It is defined as the fraction obtained by dividing the velocity of light in vacuum (c) by the phase velocity of light in the medium under consideration (v). This fraction is given by

$$n = \frac{c}{v_p} = \frac{c}{\frac{\omega}{k}} = \frac{ck}{\omega} \tag{8.4}$$

by applying equation (8.2) for v_p. Equation (8.4) can be re-arranged as

$$\omega = \frac{ck}{n} \tag{8.5}$$

Since the refractive index n is a function of frequency ω, the derivative of ω with respect to k is obtained as

$$\begin{aligned}\frac{d\omega}{dk} &= \frac{d}{dk}\left(\frac{ck}{n}\right) = \frac{c}{n}\frac{dk}{dk} + k\frac{d}{dk}\left(\frac{c}{n}\right) = \frac{c}{n} + k \times \frac{n \times \frac{dc}{dk} - c\frac{dn}{dk}}{n^2} \\ &= \frac{c}{n} + k \times \frac{n \times 0 - c\frac{dn}{dk}}{n^2} = \frac{c}{n} + k \times \frac{-c\frac{dn}{dk}}{n^2} = \frac{c}{n} - \frac{kc}{n^2}\left(\frac{dn}{dk}\right) \\ &= \frac{c}{n} - \frac{c}{n} \times \frac{k}{n}\left(\frac{dn}{dk}\right) = \frac{c}{n}\left\{1 - \frac{k}{n}\left(\frac{dn}{dk}\right)\right\} = v_{\mathrm{p}}\left\{1 - \frac{k}{n}\left(\frac{dn}{dk}\right)\right\}\end{aligned} \tag{8.6}$$

where $v_{\mathrm{p}} = c/n$ has been put from equation (8.4). Combining equations (8.3) and (8.6), we may write

$$v_{\mathrm{g}} = v_{\mathrm{p}}\left\{1 - \frac{k}{n}\left(\frac{dn}{dk}\right)\right\} \tag{8.7}$$

For a medium of constant refractive index,

$$\frac{dn}{dk} = 0 \tag{8.8}$$

$$\therefore v_{\mathrm{g}} = v_{\mathrm{p}}\left(1 - \frac{k}{n} \times 0\right) = v_{\mathrm{p}} \tag{8.9}$$

For a medium in which

$$\frac{dn}{dk} \neq 0 \tag{8.10}$$

$$v_{\mathrm{g}} < v_{\mathrm{p}} \tag{8.11}$$

The possibility exists for the physically irrelevant phase velocity to exceed the velocity of light but the physically meaningful group velocity cannot surpass this barrier.

8.2.2 Phase and group velocities in a waveguide

A waveguide is a hollow metallic tube in which electromagnetic waves propagate by total internal reflection from the opposite walls. The walls prevent the waves from spreading out. Unlike transmission by cables, two separate conductors are not necessary for direct and return paths. Also, only waves having a frequency above a certain frequency known as the cut-off frequency ω_0, are conveyed by the waveguide. The cut-off frequency depends on the geometrical parameters of the waveguide. Imposition of cut-off condition can be expressed by writing the equation of the wave number k of the dominant wave as

$$k = \frac{1}{c}\sqrt{\omega^2 - \omega_0^2} \tag{8.12}$$

Then the phase velocity in the waveguide having the cut-off frequency ω_0 is

$$v_p = \frac{\omega}{k} = \frac{\omega}{\frac{1}{c}\sqrt{\omega^2 - \omega_0^2}} = \frac{c}{\frac{\sqrt{\omega^2 - \omega_0^2}}{\omega}} = \frac{c}{\sqrt{\frac{\omega^2 - \omega_0^2}{\omega^2}}} = \frac{c}{\sqrt{1 - \frac{\omega_0^2}{\omega^2}}} \tag{8.13}$$

This equation implies that as long as $\omega > \omega_0$, c is divided by a fraction <1 so that $v_p > c$. When $\omega = \omega_0$,

$$v_p = \frac{c}{\sqrt{1 - 1}} = \frac{c}{0} = \infty \tag{8.14}$$

When $\omega < \omega_0$, the denominator of the phase velocity equation becomes imaginary and the waveguide is unable to sustain wave motion.

Differentiating the equation for k, equation (8.12), with respect to ω, we get

$$\begin{aligned}\frac{dk}{d\omega} &= \frac{d}{d\omega}\left(\frac{1}{c}\sqrt{\omega^2 - \omega_0^2}\right) = \frac{1}{c}\frac{d}{d\omega}\left\{(\omega^2 - \omega_0^2)^{\frac{1}{2}}\right\} = \frac{1}{c} \times \frac{1}{2}(\omega^2 - \omega_0^2)^{\frac{1}{2}-1} \\ &\times 2\omega^{2-1} = \frac{\omega}{c}(\omega^2 - \omega_0^2)^{-\frac{1}{2}} = \frac{\omega}{c(\omega^2 - \omega_0^2)^{\frac{1}{2}}} = \frac{\omega}{c\sqrt{\omega^2 - \omega_0^2}}\end{aligned} \tag{8.15}$$

Hence, the group velocity of the wave is

$$v_g = \frac{d\omega}{dk} = \frac{c\sqrt{\omega^2 - \omega_0^2}}{\omega} = c\sqrt{\frac{\omega^2 - \omega_0^2}{\omega^2}} = c\sqrt{1 - \frac{\omega_0^2}{\omega^2}} \tag{8.16}$$

The waveguide transmits electromagnetic waves as long as $\omega > \omega_0$. As c is multiplied by a fraction, $v_g < c$.

Moreover, multiplying together v_p and v_g as given by equations (8.13) and (8.16), we have

$$v_p v_g = \frac{c}{\sqrt{1 - \frac{\omega_0^2}{\omega^2}}} \times c\sqrt{1 - \frac{\omega_0^2}{\omega^2}} = c^2 \tag{8.17}$$

The product of phase and group velocities of the electromagnetic wave equals the square of the velocity of light in vacuum.

A rectangular waveguide is one having a rectangular cross-section. Referring to figure 8.5, let P and Q be two consecutive positions of the crests of the electromagnetic wave. Let t_1 be the time during which the crest shifts from point P to point Q. Then the velocity of the electromagnetic wave is given by

$$c = \frac{\mathrm{PQ}}{t_1} \tag{8.18}$$

The phase velocity of the wave is obtained by drawing a perpendicular at the point Q on the straight line PA to intersect the line PB at R. During the time t_1 when

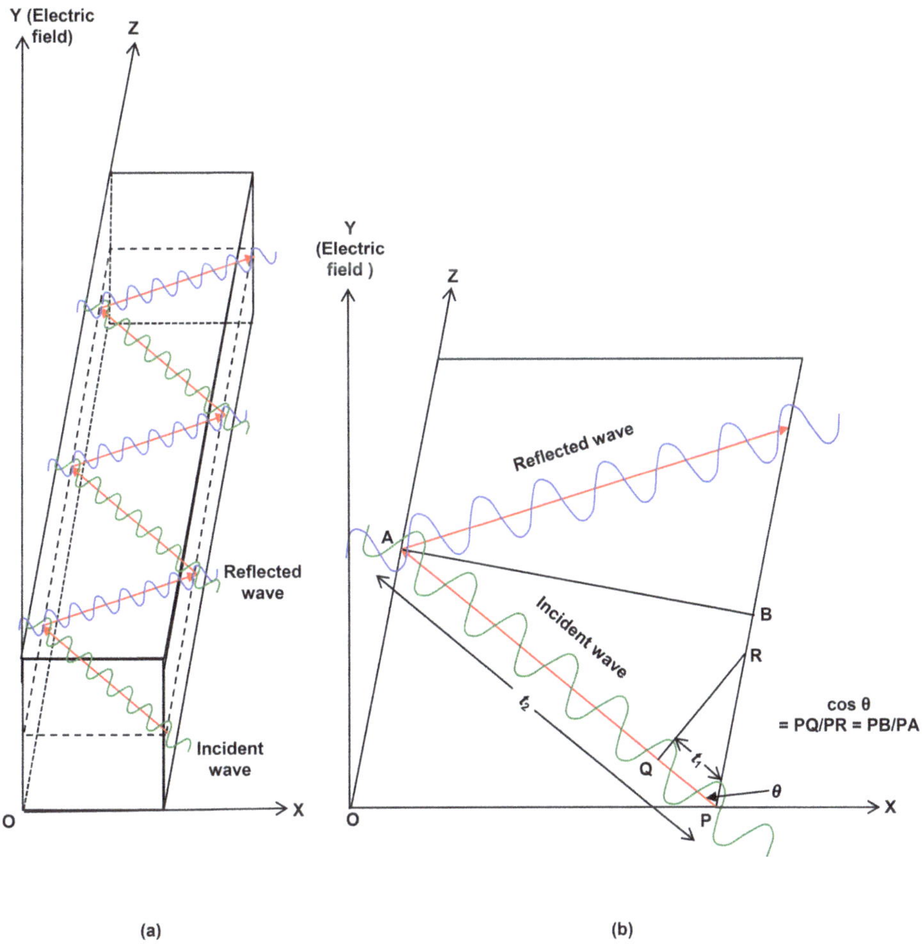

Figure 8.5. Rectangular waveguide: (a) electromagnetic wave propagation in the waveguide by repeated reflections from the walls, and (b) consideration of enlarged view of a portion of the electromagnetic wave for mathematical analysis.

the crest is transferred from P to Q, the point P shifts to the point R along the Z-direction of wave propagation. The phase velocity is written as

$$v_{\mathrm{p}} = \frac{\mathrm{PR}}{t_1} \tag{8.19}$$

Dividing equation (8.19) by equation (8.18),

$$\frac{v_{\mathrm{p}}}{c} = \frac{\frac{\mathrm{PR}}{t_1}}{\frac{\mathrm{PQ}}{t_1}} = \frac{\mathrm{PR}}{t_1} \times \frac{t_1}{\mathrm{PQ}} = \frac{\mathrm{PR}}{\mathrm{PQ}} \tag{8.20}$$

or,

$$v_{\rm p} = c \times \frac{\rm PR}{\rm PQ} \tag{8.21}$$

Looking at figure 8.5, we find that PR > PQ. Hence $v_{\rm p} > c$.
Since from triangle PRQ,

$$\cos\theta = \frac{\rm PQ}{\rm PR} \tag{8.22}$$

Equation (8.21) becomes

$$\frac{v_{\rm p}}{c} = \frac{1}{\cos\theta} \tag{8.23}$$

or,

$$v_{\rm p} = \frac{c}{\cos\theta} \tag{8.24}$$

Since $\cos\theta \leqslant 1$, minimum value of $v_{\rm p} = c$ for $\cos\theta = 1$ and $v_{\rm p}$ can acquire higher values than c for $\cos\theta < 1$.

Like the phase velocity, the group velocity is obtained by drawing a perpendicular from the point A on the waveguide wall to meet the opposite wall at the point B. If t_2 denotes the time during which the wave moves from point P to point A, the velocity of electromagnetic wave is

$$c = \frac{\rm PA}{t_2} \tag{8.25}$$

During this time t_2, the wave moves from P to B along the Z-direction of wave propagation. So, the group velocity of the wave is

$$v_{\rm g} = \frac{\rm PB}{t_2} \tag{8.26}$$

Dividing equation (8.26) by equation (8.25), we have

$$\frac{v_{\rm g}}{c} = \frac{\frac{\rm PB}{t_2}}{\frac{\rm PA}{t_2}} = \frac{\rm PB}{t_2} \times \frac{t_2}{\rm PA} = \frac{\rm PB}{\rm PA} \tag{8.27}$$

or,

$$v_{\rm g} = c \times \frac{\rm PB}{\rm PA} \tag{8.28}$$

Inspection of figure 8.5 shows that PB < PA. Hence, $v_{\rm g} < c$.
From triangle PAB

$$\cos\theta = \frac{\rm PB}{\rm PA} \tag{8.29}$$

Hence, equation (8.28) yields

$$\frac{v_g}{c} = \cos\theta \tag{8.30}$$

or,

$$v_g = c\cos\theta \tag{8.31}$$

Remembering that $\cos\theta \leqslant 1$, maximum value of $v_g = c$ for $\cos\theta = 1$, while $v_g < c$ for $\cos\theta < 1$.

$$v_g\Big|_{\text{Maximum}} = c \tag{8.32}$$

Multiplication of v_p and v_g leads to

$$v_p v_g = \frac{c}{\cos\theta} \times c\cos\theta = c^2 \tag{8.33}$$

which again confirms the truth of equation (8.17). Equations (8.24) and (8.31) are applied.

8.2.3 Reduction of the phase velocity of electromagnetic waves by a slow wave structure

A slow wave structure increases the path length of electromagnetic wave so that a longer time is spent to reach the destination, which is observed as a reduction of velocity of the wave (figure 8.6). Consider two cities A and B located at distance **D** apart, as measured along a straight line (figure 8.6(a)). When a train/car goes from A to B on the straight line route, it moves on the shortest path and reaches early. But a train/car moving on a helical or zigzag route between stations A and B (the path APQRSTUVWXB, suppose) takes a longer time to travel from A to B even if it is moving at the same speed. By virtue of the longer time taken, it can be said that the train/car has slowed down in velocity although it has actually travelled at the same speed. This follows from the fact that the displacement **D** is a vector whose direction is given by the straight line joining the points A and B. The displacement is always **D** irrespective of whether a shorter or longer route is chosen. So the velocity = displacement/time is less for the straight line route. The slow wave structure follows essentially the same principle to make the phase velocity smaller.

Slow wave structures are made in several shapes such as a helix, a spiral, a meander or serpentine geometry. The objective is always to increase the path length. Taking the example of a helical SWS (figure 8.6(b)–(c)), let d be diameter of the coil so that its circumference = πd. If p be the distance between the two adjoining turns of the helix, by applying Pythagora's theorem, the total path length for the electromagnetic wave travelling along the helical path is (Panahi *et al* 2017)

$$P_{\text{Helical}} = \sqrt{p^2 + (\pi d)^2} = \sqrt{p^2 + \pi^2 d^2} \tag{8.34}$$

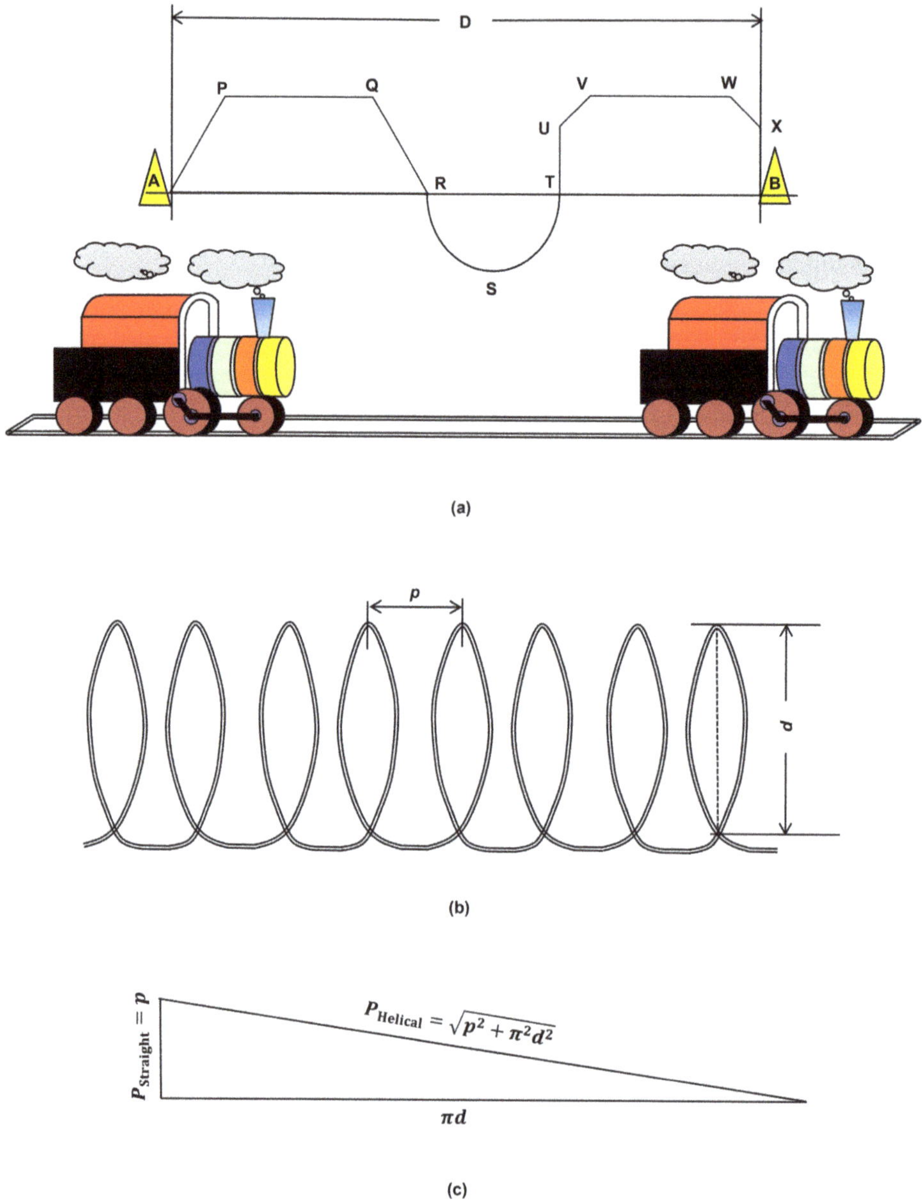

Figure 8.6. Increase in the path length of an electromagnetic wave when travelling along a helical path as compared to the case when it is propagating along a straight line : (a) showing the effect of zigzag path on lengthening the distance between two cities A and B, (b) helix slow wave structure and its dimensional parameters and (c) calculating the path length followed by electromagnetic wave moving along the helix.

while the path length for the electromagnetic wave propagating along the straight line path is

$$P_{\text{Straight}} = p \tag{8.35}$$

The variable p is called the pitch of the helix.
Evidently,

$$\frac{\text{Phase velocity of electromagnetic wave traveling along the straight line path of electron beam } (v_\text{p})}{\text{Phase velocity of electromagnetic wave propagating along the curved length of the helix } (c)} = \frac{\text{Path length for the electromagnetic wave propagating along the straight line path } (P_\text{Straight})}{\text{Path length for the electromagnetic wave traveling along the helical path } (P_\text{Helical})} \tag{8.36}$$

or,

$$\frac{v_\text{p}}{c} = \frac{P_\text{Straight}}{P_\text{Helical}} = \frac{p}{\sqrt{p^2 + \pi^2 d^2}} \tag{8.37}$$

$$v_\text{p} = c \times \frac{p}{\sqrt{p^2 + \pi^2 d^2}} = c \times \frac{1}{\sqrt{1 + \dfrac{\pi^2 d^2}{p^2}}} = \frac{c}{\sqrt{1 + \dfrac{\pi^2 d^2}{p^2}}} \tag{8.38}$$

The phase velocity can be approximated as

$$v_\text{p} \approx \frac{c}{\sqrt{\dfrac{\pi^2 d^2}{p^2}}} = c \times \frac{p}{\pi d} = c \times \frac{\text{Pitch of the helix}}{\text{Circumference of the helix turn}} \tag{8.39}$$

A p/d ratio of 0.25 decreases the phase velocity by 1/(0.25/3.141 59) = 12.57 times below the velocity of light in vacuum.

8.2.4 Floquet's theorem and electric field distribution in the slow-wave structure

The slow wave structure is a periodic structure consisting of a repetition of unit cells. It is described by Floquet's theorem. This theorem states that: the steady-state solutions for the electromagnetic fields of a single propagating mode in a periodic structure have the property that the fields within the adjacent cells vary only by a complex constant, which is the same for all pairs of adjacent cells. The theorem is expressed as

$$E(x,\ y,\ z - L) = E(x,\ y,\ z)\exp(jk_{z0}L) \tag{8.40}$$

where $E(x, y, z)$ is a periodic function of z with period L. The symbol k_{z0} stands for the wave number in the axial direction. The axial direction is the direction of electron velocity in the slow-wave structure.

Let us assume that the solution to Maxwell's equation in a periodic structure can be written as

$$E(x,\ y,\ z) = \phi(x,\ y,\ z)\exp(-jk_{z0}z) \tag{8.41}$$

where $\phi\ (x, y, z)$ is a periodic function of z with period L, the period of the slow wave structure. This equation can be rewritten by replacing z by z-L because the structure is periodic. By doing so, we obtain

$$E(x,\ y,\ z - L) = \phi(x,\ y,\ z - L)\exp\{-jk_{z0}(z - L)\} \tag{8.42}$$

Noting that $\phi\ (x, y, z)$ is a periodic function having the period L, we may write

$$\phi(x,\ y,\ z - L) = \phi(x,\ y,\ z) \tag{8.43}$$

Substituting for $\phi(x, y, z - L)$ from equation (8.43) in equation (8.42) we get

$$E(x,\ y,\ z - L) = \phi(x,\ y,\ z)\exp\{-jk_{z0}(z - L)\} = \phi(x,\ y,\ z)\exp(-jk_{z0}z)\exp(jk_{z0}L) \tag{8.44}$$

Substitution for $\phi(x, y, z)$ from equation (8.41) in equation (8.44) yields

$$E(x,\ y,\ z - L) = E(x,\ y,\ z)\exp(jk_{z0}z)\exp(-jk_{z0}z)\exp(jk_{z0}L) = E(x,\ y,\ z)\exp(jk_{z0}L) \tag{8.45}$$

which is the mathematical statement of Floquet's theorem, as can be verified by revisiting equation (8.40).

Since any periodic, single-valued and continuous function can be represented by a Fourier series, the field distribution function $E(x, y, z)$ can be expanded into a Fourier series of period L as

$$\begin{aligned} E(x,\ y,\ z) &= \sum_{l=-\infty}^{\infty} E_l(x,\ y)\exp\left\{-j\left(\frac{2\pi l}{L}\right)z\right\}\exp(-jk_{z0}z) \\ &= \sum_{l=-\infty}^{\infty} E_l(x,\ y)\exp\left\{-j\left(k_{z0} + \frac{2\pi l}{L}\right)z\right\} \end{aligned} \tag{8.46}$$

where

$$\begin{aligned} E_l(x,\ y) &= \frac{1}{L}\int_0^L E(x,\ y,\ z)\exp\left\{j\left(\frac{2\pi l}{L}\right)z\right\}dz = A_l \\ &= \text{Amplitudes of the } n \text{ harmonics} \end{aligned} \tag{8.47}$$

Equation (8.46) may be recast as

$$E(x,\ y,\ z) = \sum_{l=-\infty}^{\infty} A_l\exp(-jk_{zl}z) \tag{8.48}$$

where

$$k_{zl} = k_{z0} + \frac{2\pi l}{L} \tag{8.49}$$

An electromagnetic wave can be represented as a superimposition of spatial harmonics

$$E(x,\ y,\ z,\ t) = \exp(-j\omega t) \sum_{l=-\infty}^{\infty} A_l \exp(-jk_{zl}z) \tag{8.50}$$

where ω is the angular frequency of the electromagnetic wave, l is the harmonic number, A_l the amplitudes, as defined by equation (8.47); k_{zl} are the axial wave numbers given by equation (8.49) with k_{z0} denoting the wave number of the zeroth order spatial harmonic ($-\pi/L<k_{z0}<+\pi/L$).

Equation (8.50) shows that the electric field in a periodic structure can be expanded into an infinite series of waves. All these waves have the same angular frequency ω. However, they have different phase velocities

$$v_{pl} = \frac{\omega}{k_{zl}} = \frac{\omega}{k_{z0} + \dfrac{2\pi l}{L}} \tag{8.51}$$

and the associated group velocity is

$$\begin{aligned} v_g &= \frac{\partial\omega}{\partial k_{zl}} = \frac{\partial\omega}{\partial\left(k_{z0} + \dfrac{2\pi l}{L}\right)} = \left\{\frac{\partial\left(k_{z0} + \dfrac{2\pi l}{L}\right)}{\partial\omega}\right\}^{-1} = \left\{\frac{\partial k_{z0}}{\partial\omega} + \frac{\partial\left(\dfrac{2\pi l}{L}\right)}{\partial\omega}\right\}^{-1} \\ &= \left\{\frac{\partial k_{z0}}{\partial\omega} + 0\right\}^{-1} = \left\{\frac{\partial k_{z0}}{\partial\omega}\right\}^{-1} = \frac{\partial\omega}{\partial k_{z0}} \end{aligned} \tag{8.52}$$

where equation (8.49) for k_{zl} is applied.

Looking at equations (8.51) and (8.52), we note that the phase velocity in the axial Z-direction is a function of l, whereas the group velocity is independent of l. Furthermore, for positive k_{z0} and l, the phase velocity decreases with increasing l. The obvious implication is that in the slow-wave structure, it is possible that the phase velocity diminishes to a value which is smaller than the velocity of light in vacuum. At this decreased phase velocity, interaction can take place between the electric field and the electron beam which travels at a velocity less than that of light. By utilization of this interaction of electron beam with electric field, high-frequency waves can be amplified (Liao 2008).

8.2.5 Emission of Cherenkov radiation

Electromagnetic radiation emitted by electrons in a periodic slow-wave structure is referred to as Smith–Purcell radiation (Gold and Nusinovich 1997). The Smith–Purcell radiation can be understood as a type of Cherenkov radiation. Let us see how this can be done? Cherenkov radiation is the electromagnetic radiation which is emitted when two conditions are met. Firstly, a charged particle such as the electron moves through a medium with refractive index > 1. Secondly, the velocity of motion of the charged particle (here electron) is greater than the phase velocity of propagation of electromagnetic waves in the medium.

Recalling again the definition of refractive index given in equation (8.4), we note that the refractive index tells us how fast light travels in a given medium relative to the velocity c of light in vacuum. Applying this definition to the slow-wave structure,

$$\text{Refractive index of slow wave structure } (n_{\text{Slow wave}}) = \frac{\text{Velocity of electromagnetic wave in vacuum } (c)}{\text{Phase velocity of electromagnetic wave in slow wave structure } (v_{\text{p}})} \tag{8.53}$$

or,

$$n_{\text{Slow wave}} = \frac{c}{v_{\text{p}}} \tag{8.54}$$

Since in a slow-wave structure, from equation (8.39),

$$v_{\text{p}} < c, \therefore n_{\text{Slow wave}} > 1 \tag{8.55}$$

Referring back to equation (8.51), a spatial harmonic with phase velocity

$$v_p = \frac{\omega}{k_{\text{zl}}} \tag{8.56}$$

can be looked upon as a slow wave travelling in a medium with refractive index given by equation (8.54) as

$$n_{\text{Slow wave}} = \frac{c}{\frac{\omega}{k_{\text{zl}}}} = \frac{ck_{\text{zl}}}{\omega} \tag{8.57}$$

Thus in the backward wave oscillator, Cherekov radiation is emitted by the electron beam moving through the slow wave structure in which the phase velocity of electromagnetic waves is less than their velocity in vacuum.

Since the wave number k is related to the axial wave vector k_{zl} and transverse wave vector $k_{\perp l}$

$$k^2 = k_{\text{zl}}^2 + k_{\perp l}^2 \tag{8.58}$$

or,

$$\left(\frac{\omega}{c}\right)^2 = \left\{n_{\text{Slow wave}}\left(\frac{\omega}{c}\right)\right\}^2 + k_{\perp l}^2 \tag{8.59}$$

since

$$c = \frac{\omega}{k} \tag{8.60}$$

and from equation (8.57)

$$k_{zl} = n_{\text{Slow wave}}\left(\frac{\omega}{c}\right) \tag{8.61}$$

Eq. (8.59) gives

$$k_{\perp l}^{2} = \left(\frac{\omega}{c}\right)^{2} - \left\{n_{\text{Slow wave}}\left(\frac{\omega}{c}\right)\right\}^{2} = \left(\frac{\omega}{c}\right)^{2}(1 - n_{\text{Slow wave}}) \tag{8.62}$$

Recalling from equation (8.55) that

$$n_{\text{Slow wave}} > 1 \tag{8.63}$$

it is evident that $k_{\perp l}$ being the square root of a negative quantity is an imaginary number. This hints at localization of the electromagnetic field of the spatial harmonic near the wall of the slow wave structure. Therefore, the electron beam in a backward wave oscillator too must travel close to the wall of the slow wave structure. Then the beam can couple with the electromagnetic field. This is clarified in example 8.1. In this example, we shall show that an imaginary wavenumber indicates wave attenuation.

Example 8.1 Interpretation of imaginary wavenumber. Imaginary numbers are used for describing real phenomena because representation of waves by imaginary/complex numbers considerably simplifies their mathematical solution by Maxwell's equations. The wave equation (8.1) can be written in an alternative form as

$$\begin{aligned} E(x, t) &= |E_0| \cos(kx - \omega t) = \text{Real part of } [|E_0| \exp\{j(kx - \omega t)\}] \\ &= \text{Real}[|E_0| \exp\{j(kx - \omega t)\}] \end{aligned} \tag{8.64}$$

since

$$\exp\{j(kx - \omega t)\} = \cos(kx - \omega t) + j\sin(kx - \omega t) \tag{8.65}$$

Explain the physical significance of replacing a wave number k by a complex number containing real and imaginary parts.

Solution: If k is purely real, then the wave is a purely cosinusoidal travelling wave having the constant amplitude $|E_0|$. But if k is a complex number having real and imaginary parts, i.e.,

$$k = \alpha + j\beta \tag{8.66}$$

then by substitution of the complex form of k in equation (8.64), the wave equation is modified as

$$E(x, t) = \text{Real}[|E_0| \exp\{j(\alpha + j\beta)x - j\omega t\}] \tag{8.67}$$

If $|E_0|$ is real, equation (8.67) can be decomposed into two factors

$$\begin{aligned} E(x, t) &= \text{Real}[|E_0| \exp\{j\alpha x + j^2\beta x - j\omega t\}] \\ &= \text{Real}\left[|E_0| \exp\left\{j\alpha x + (\sqrt{-1})^2\beta x - j\omega t\right\}\right] \\ &= \text{Real}[|E_0| \exp\{j\alpha x - \beta x - j\omega t\}] \\ &= \text{Real}[|E_0| \exp\{-\beta x + j(\alpha x - \omega t)\}] \\ &= \text{Real}[|E_0| \exp(-\beta x) \times \exp j(\alpha x - \omega t)] \\ &= |E_0| \exp(-\beta x) \times \text{Real}\{\exp j(\alpha x - \omega t)\} \end{aligned} \tag{8.68}$$

The second factor inside the real part, namely

$$\text{Real}\{\exp j(\alpha x - \omega t)\} \tag{8.69}$$

represents a cosinusoidal/sinusoidal oscillation while the first factor is the amplitude of the wave:

$$|E_0| \exp(-\beta x) = \text{Amplitude} \tag{8.70}$$

Thus in place of the amplitude $|E_0|$, we have here $|E_0|$ exp ($-\beta x$). Multiplication of $|E_0|$ by exp ($-\beta x$) shows that $|E_0|$ exp ($-\beta x$) decays as one moves in the X-direction. Hence physically speaking, the imaginary wavenumber is an indicator of attenuation of the wave, as already mentioned.

8.3 Difference between a travelling wave tube and backward wave oscillator from phase/group velocity viewpoint

As already mentioned, the TWT and BWO differ vastly in their operation. The difference between TWT and BWO is elucidated with reference to their operating points on the dispersion diagram displaying the dependence of frequency ω on the axial wave number k in the slow wave structure (figure 8.7(a)).

The TWT differs from the BWO in respect of the signs of group velocity of electromagnetic wave at the operating point. The dispersion diagram is an important characteristic of the device from which an idea of frequency of operation can be obtained. The TWT functions in the portion AB of the dispersion diagram. In a TWT, the phase velocity, given by the ratio of the angular frequency ω to the wave number k_z is positive. At any selected operating point, the group velocity of the electromagnetic wave is positive. What is the reason? The group velocity of electromagnetic wave at a chosen point, given by the gradient of the angular frequency ω with respect to the wave number k_z, and obtained from the slope of the straight line joining the origin of the ω– k_z dispersion diagram of the slow wave

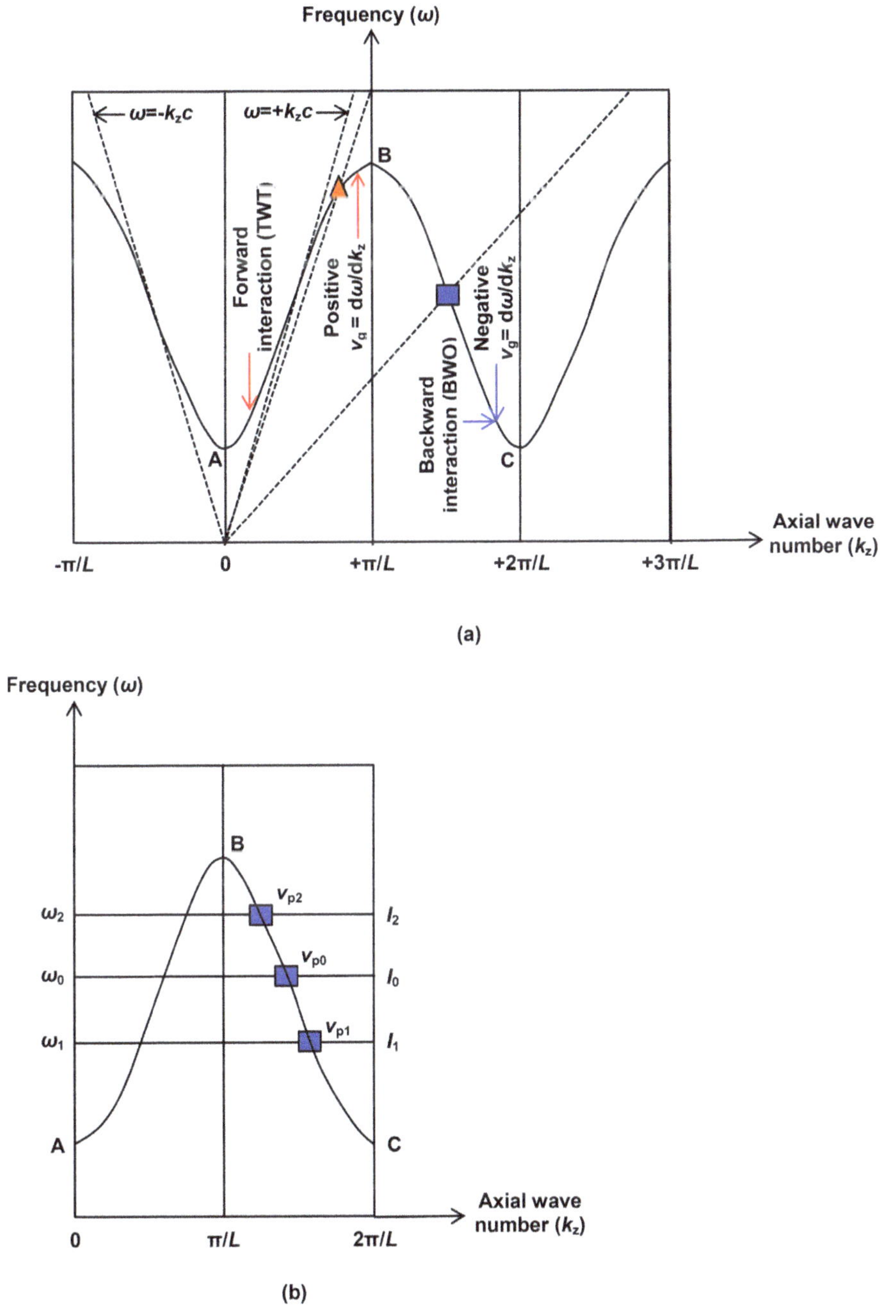

Figure 8.7. Dispersion diagram of the slow wave structure: (a) illustrating the forward and backward interaction regions for TWT and BWO operations in accordance with the positive and negative group velocities of electromagnetic waves revealed by the direct (positive) and inverse (negative) slopes of dispersion curve in the two regions; operating point of TWT is marked with a triangle and that of BWO with a rectangle, and (b) explanation of the frequency tunability of BWO from variation of frequency with operating point corresponding to different phase velocities of electromagnetic waves as controlled by acceleration voltages.

structure to the operating point on this diagram, i.e., $d\omega/dk_z$, is positive because the slope of the joining line is positive.

The BWO works in the portion BC of the dispersion diagram. The portion BC has two distinguishing features: Firstly, the phase velocity of the electromagnetic wave at the selected operating point is positive. Secondly, the group velocity at the marked point is negative because of negative slope of the curve at this point. To reiterate, in both cases, the phase velocities are positive but the group velocity is positive for a TWT while it is negative for a BWO.

Now consider three operating points I_0, I_1 and I_2 on the portion BC of the dispersion diagram for the BWO (figure 8.7(b)). Each point is characterized by a phase velocity v_p of the electromagnetic wave. The point I_0 corresponds to the phase velocity v_{p0} resulting in the operating frequency ω_0. Similarly, the point I_1 pertains to the velocity v_{p1} and frequency ω_1, whereas the point I_2 relates to the velocity v_{p2} and frequency ω_2. This shows that the frequency can be varied from ω_0 to ω_1 and ω_2 by changing the velocity of electron beam to shift the operating point. Thus the BWO works as an oscillator whose frequency is tunable by altering the electron beam velocity through the anode–cathode voltage. Hence, it is a voltage tunable oscillator. A frequency range as large as 2: 1 can be swept by voltage tuning.

8.4 Electron bunching and amplification of the signal in a TWT

The direction of the electric field is the direction in which a positive test charge moves when placed in the field. Considering an electric field produced by a positive charge, the positive test charge will be repelled by the positive field-producing charge. Hence the electric field due to a positive charge is directed away and outward from the positive charge. Similarly, the electric field produced by a negative charge is directed towards and inwards to the negative charge.

Suppose an electron is placed in an electric field due to a positive charge. Then the electron will be attracted towards the positive charge and the direction in which the electron moves is towards the positive charge. But the electric field created by positive charge points in a direction away from the positive charge. So the direction of electron motion is opposite to that of the electric field due to a positive charge.

On the other hand, when the electron is placed in an electric field, due to a negative charge, the electron is repelled away from the negative charge while the direction of the electric field points towards the negative charge. So, in this case also the direction of electron motion is opposite to that of the field. Thus the electron motion always takes place in a direction opposite to that of the electric field.

From such discussions, we can analyse two situations of interest to us, one in which the direction of electron movement is the same as the direction of electric field of a sinusoidal electromagnetic wave, and the other in which these directions are opposite to each other. Consider first the case when the electron and electromagnetic field are moving in the same direction (figure 8.8(a)). We find that during the positive half cycle of the electromagnetic wave, the electron motion is impeded by the electric field of the wave while in the negative half cycle of the wave, it is aided by the electric field of the wave. So during positive half cycle, the electron

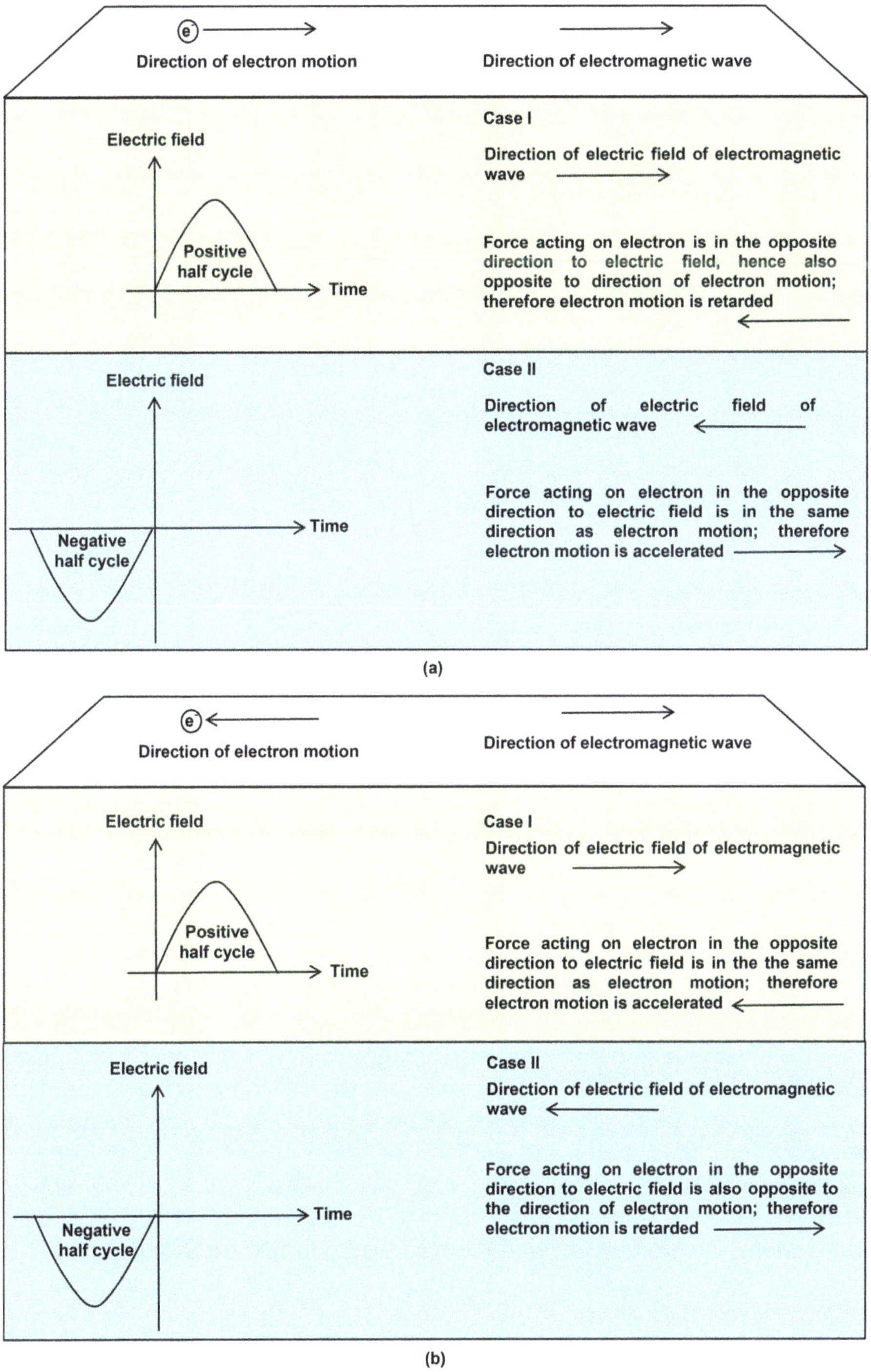

Figure 8.8. Effects of sinusoidal electric field of electromagnetic wave on electron motion when: (a) the electron and the electromagnetic wave are moving in the same direction, and (b) the electron and the electromagnetic wave are moving in opposite directions.

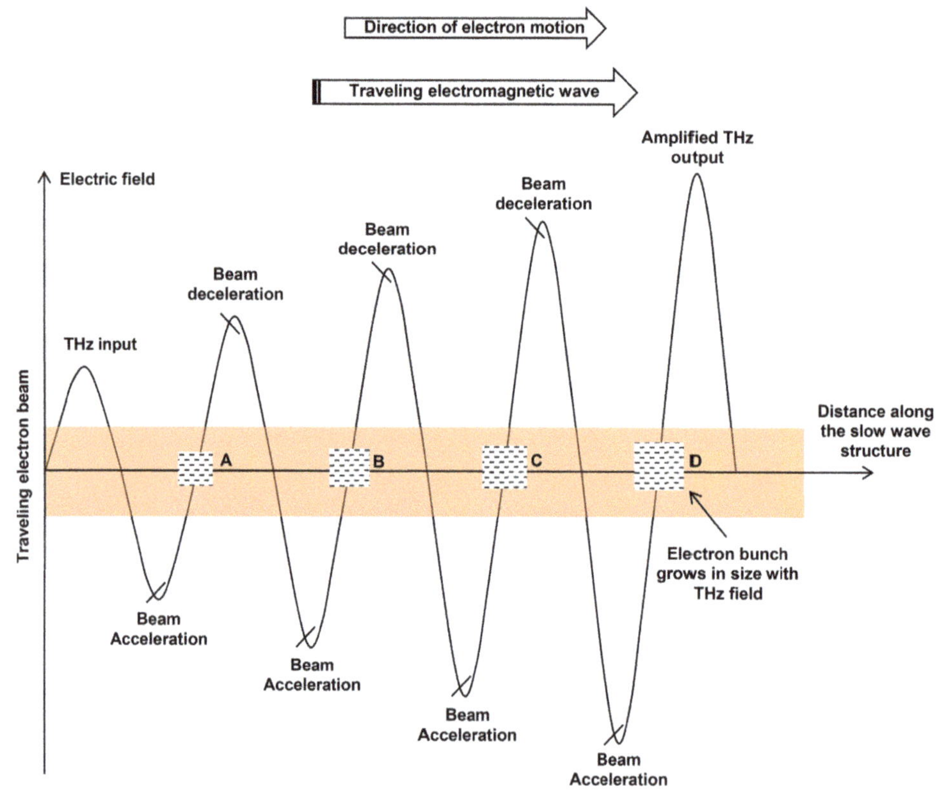

Figure 8.9. Electromagnetic field produced by the SWS in a TWT reduces and increases speed of electrons causing velocity modulation of the electron beam, which is turn leads to its density modulation.

suffers retardation while in the negative half cycle, it is accelerated. The reverse happens when the electron and the electromagnetic wave are travelling in opposite directions (figure 8.8(b)).

When we apply these arguments to the operation of a TWT we note that the electron and electromagnetic wave produced by it propagate in the same direction. During the positive half cycle of the electromagnetic wave when the electric field acts in the same direction as the moving electron, the electrons are decelerated. But during the negative half cycle when the electric field acts in the opposite direction to electron motion, the electrons are accelerated. These deceleration and acceleration regions are indicated in figure 8.9.

We look at electrons behind and after the zero-field point A at any given instant. At this instant, the electrons behind the zero-field point A are travelling faster than average. At the same instant, the electrons ahead of the zero field point A are travelling slower than average. As a consequence, the electrons tend to bunch around point A. In a similar fashion, electrons tend to bunch around the zero field points B, C, D,..., 360° phase away. Since the DC velocity of the electron beam is slightly greater than the phase velocity of electromagnetic wave, a greater number of electrons encounter a decelerating field than those facing an accelerating field. So the electron beam loses energy and the electromagnetic wave gains energy. Hence, the amplitude of the electromagnetic wave increases.

Detailed mathematical analysis of the TWT was performed by Pierce. The output power gain of a helix TWT is defined as (Das and Das 2007)

$$A_{\mathrm{p}} = 10\log_{10}\left|\frac{\text{Output voltage}}{\text{Input voltage}}\right|^2 \text{ dB} \tag{8.71}$$

It is decided by the circuit length N expressed in units of wavelength λ of elctromagnetic waves, given by

$$\begin{aligned} N &= \frac{\text{Physical length } L}{\text{Wavelength of electromagnetic waves } (\lambda)} = \frac{L}{\lambda} \\ &= \text{Number of guided wavelengths along the structure of length } L \end{aligned} \tag{8.72}$$

along with the Pierce gain parameter

$$\begin{aligned} C &= \left(\frac{\text{Beam current } (I_0) \times \text{Impedance of the slow wave structure } (Z_0)}{4 \times \text{Beam voltage } (V_0)}\right)^{\frac{1}{3}} \\ &= \left(\frac{I_0 Z_0}{4V_0}\right)^{\frac{1}{3}} \end{aligned} \tag{8.73}$$

The equation for power gain (in decibels) is

$$A_{\mathrm{p}} = -9.54 + 47.3NC \tag{8.74}$$

The first negative term represents a loss arising from the fact that three forward waves of equal magnitude are set up in the TWT in presence of an electron beam. The first wave is a growing wave with distance. Energy flows from the electron beam to this wave. The second wave is a decaying wave with distance. Energy flows from the wave to the electron beam. The third wave, faster than the first two waves, is an unattenuated wave. No energy exchange occurs between this wave and the electron beam. The fourth backward wave is absorbed by the attenuator.

8.5 Applications of TWTs

TWTs are low noise, reliable, long-life devices with linear characteristics. They are the amplifiers of choice for transmitters in satellite communications. TWT amplifiers provide one to two orders of magnitude higher power than their solid-state counterparts. TWTs are also used in ground-based, ship-borne and airborne radar systems. Video Synthetic Aperture Radar (ViSAR) is an imaging model for performing repeated surveillance of the region of interest by acquisition of a sequence of successive images in real time.

Other possible applications of TWTs include plasma diagnosis of microturbulence in nuclear fusion reactors, imaging, electronic countermeasures, decoys, jamming and processing of materials (Paoloni *et al* 2020).

8.6 Terahertz TWTs

A 0.2145 THz TWT amplifier with microfabricated interaction circuit has been demonstrated by Joye *et al* (2014a, 2014b). The output power at this frequency is 60 W. The beam voltage and the beam current for this tube are 12.1 kV and 118 mA, respectively. The copper interaction circuit of this tube is fabricated by UV LIGA process.

A TWT using folded waveguide structure formed by ion milling with ±5 μm tolerance showed 130 mW maximum power at 0.31824 THz (Hu *et al* 2018). The small signal gain is 19.6 dB.

A 0.670 THz TWT and solid-state amplifier module generated more than 100 mW power while working from 0.640 to 0.685 THz (Tucek *et al* 2012), while a 0.850 THz vacuum electronic amplifier produced 39.4 mW output power at this frequency (Tucek *et al* 2014). Power greater than 50 mW is achieved in the window 0.835–0.842 THz.

A 1.030 THz TWT vacuum electronic power amplifier is a compact amplifier producing 29 mW power at this frequency (Tucek *et al* 2016). It has a gain of 20 dB at saturation. The instantaneous bandwidth is 0.005 THz. It uses a thermionic cathode and a permanent magnet solenoid. For enhancing the efficiency, the electron beam is collected by a depressed collector. The slow wave circuit used in this TWT is a folded waveguide made by microfabrication technology. Deep reactive ion etching of silicon-on-insulator (SOI) wafers is performed. The waveguide is electroplated with OFHC copper. Its dimensions are in the range of tens of microns.

A TWT oscillator has been fabricated with the output fed back to the input employing positive feedback configuration (Gao *et al* 2010). The TWT uses a folded waveguide. The maximum output power is 50 mW at a frequency of 0.656 THz. The frequency is adjusted between 0.607 and 0.675 THz by changing the beam voltage. However, instead of a continuous smoothly changing frequency with beam voltage, step-variations in frequency are observed.

8.7 Operation of the backward wave oscillator

Turning attention to the backward wave oscillator, we are curious to know: how are oscillations produced in a BWO? When the heater is switched on and an electrical voltage is applied between the anode and cathode, an electron beam is set up from cathode to anode. This beam is constrained in its shape and path by the magnetic focusing field. During its flow, it embraces the helix.

Immediately after turn-on, a direct current is flowing through the tube. It is carried by electrons drifting in vacuum. To see how this direct current is converted into oscillations, we note that the electrons produce a noise signal by shot and thermal effects. Shot noise is a form of noise arising from the discrete nature of electric charges, while thermal noise originates from the temperature-induced fluctuations in electron density in a conductor. The noise signals are random in frequency. They comprise signals of a large number of frequencies. Each of these signals has a tendency to develop an electromagnetic wave travelling on the helix towards the cathode. This electromagnetic wave is a Cherenkov radiation. It is generated because the condition for Cherenkov radiation that the phase velocity of

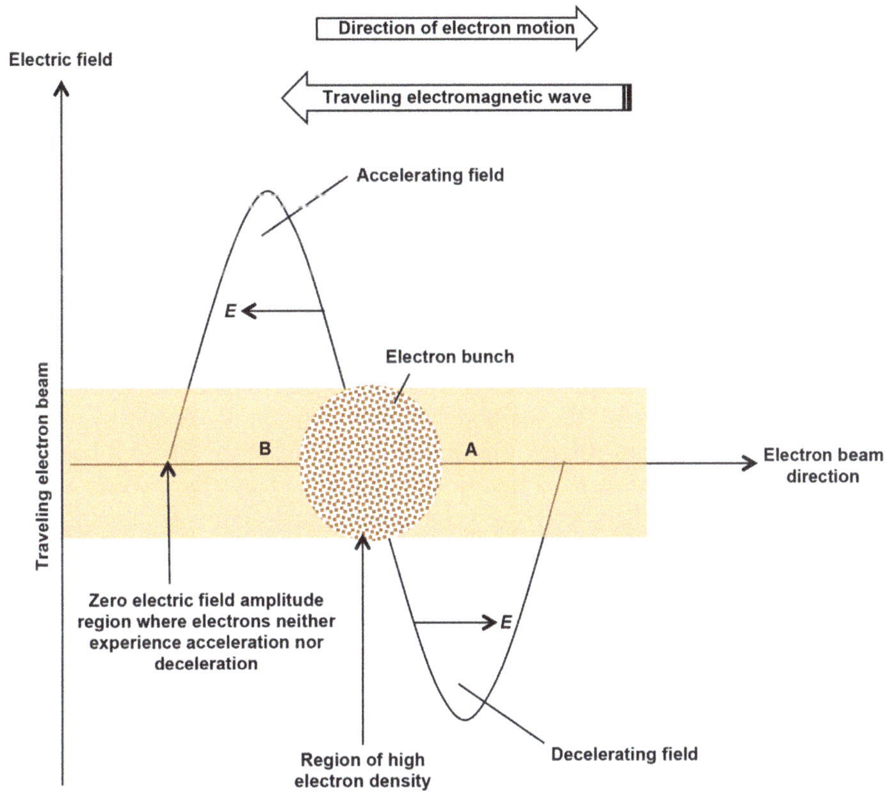

Figure 8.10. Electron bunching caused by velocity modulation produced by axial electric field of the electromagnetic wave along the slow wave structure in a BWO.

the electromagnetic wave is less than the velocity of electron beam, is satisfied by the helical slow-wave structure. Had the electron beam been injected into a smooth wall waveguide, the phase velocity would have been greater than the velocity of light in vacuum and the above condition would have been violated. Then no emission of radiation would have taken place.

Now starting from the electron beam we have come to electromagnetic waves of various frequencies propagating towards the cathode, the so-called noise signals. The electric fields of each of these waves interact with the electron beam on its anode-cathode journey. The helix is made of a cylindrically-wound tape of flat wire. Underneath the helix turns, the electric field is minimum because the helix is made of a conducting material and no field can exist parallel to a conductor. Between any two turns of the helix, there is an electric field. Consider the sinusoidally varying electric field between the helix turns shown in figure 8.10. Electrons at point A, which are ahead on their way towards the anode are decelerated by repulsion from the negative voltage while electrons at point B, which are lagging behind the electrons at point A, are accelerated by the positive voltage. As the velocities of the electrons are being varied, the process is called velocity modulation. By velocity modulations, the group of electrons at point A is brought closer to electron group at point B. Consequently, an electron bunch is formed at the middle point of the decelerating/accelerating fields. Velocity modulation is the process by which the velocity of a stream of electrons is modified by alternately imparting decelerations and accelerations to the electrons causing them to bunch together. At any given instant, velocity modulation results in production of several such electron bunches along the electron beam.

After velocity modulation, we talk about exchange of kinetic energy of the electron beam with the backward moving electromagnetic wave on the helix. Some portion of the electron kinetic energy is delivered to the electromagnetic wave. It is the electron bunch which supplies energy to the electromagnetic wave on the helix moving in the direction of the cathode. Electron beam loses energy while the electromagnetic wave gains energy. The energy supply from the electron beam to the electromagnetic wave serves to replenish the energy lost by the wave in travelling through the helix. It is here that the positive feedback action takes place. If the energy lost by the electromagnetic wave is not recovered in this manner, the wave will soon become weak. But if it is compensated by electron beam, it will not die out.

The interaction between electron bunches and the electromagnetic wave on the slow structure can be described in terms of induced currents. The electron bunches induce electric currents on the helix. These currents move in phase with the electron bunches. The currents induced by the bunches combine with the electromagnetic wave flowing in the helix causing growth of the wave. The growing wave causes formation of more electron bunches. Thus the electromagnetic wave grows exponentially as it travels along the length of the helix.

Apart from the small velocity excursions caused by the electric fields of the electromagnetic wave by velocity modulation *en route* from cathode to anode, the electron beam can move only at one velocity at a time. This velocity is governed by the accelerating voltage between anode and cathode. Since the electron beam can travel with one velocity only, it can support only electromagnetic waves of one specific frequency. As a result, the electromagnetic wave oscillations take place only at one particular frequency and this frequency depends on the velocity of electron beam. Noise signals of remaining frequencies fall off because they are not supported by feedback. So the oscillations stabilize at one frequency and the ultimate control of frequency lies in the hands of the anode-to-cathode accelerating voltage.

From the above discussions, the working of the BWO can be understood with reference to an equivalent circuit made of a chain (figure 8.11(b)) comprising a large number of feedback loops (Department of the Air Force 1963). Each loop has three components: a transmission line, an amplifier and a coupling capacitor (figure 8.11(a)). The amplifier is connected to the input end of the transmission line via the coupling capacitor. The transmission line describes the backward movement of the electromagnetic wave through the helix. For the BWO, it is the forward circuit. The amplifier representing the electron bunching mechanism is the backward circuit providing feedback. The electromagnetic wave weakened by suffering losses in the helix is restored to its original state by the amplifier.

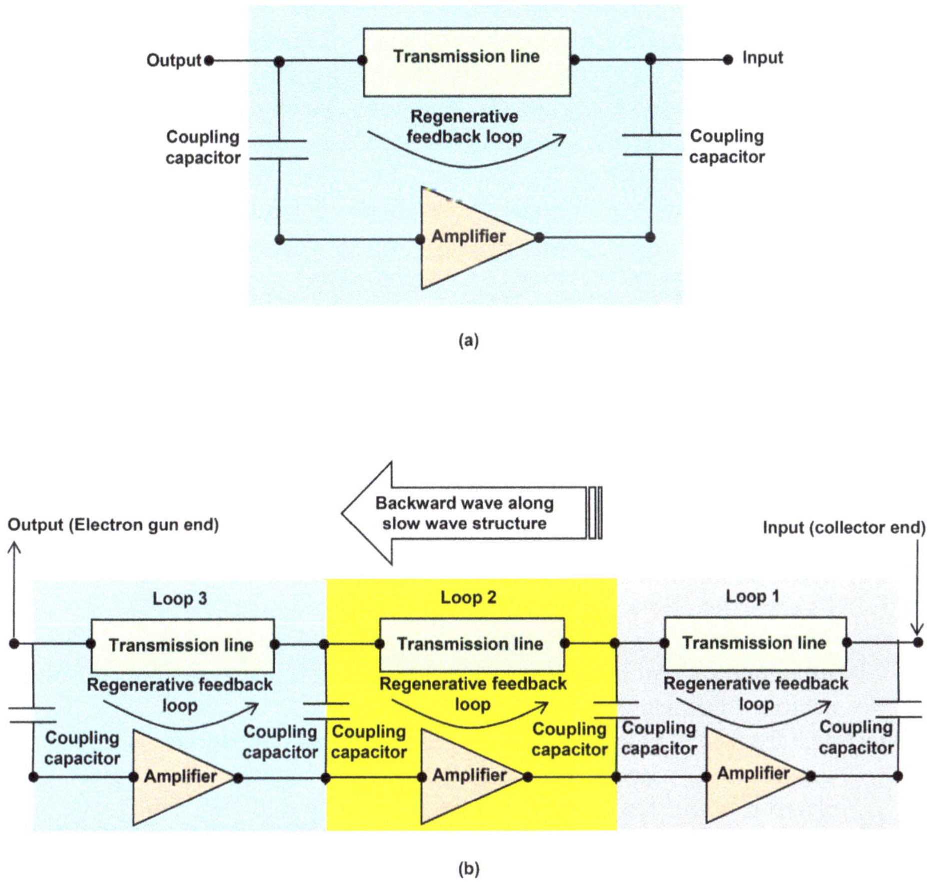

Figure 8.11. Regenerative feedback loops : (a) single loop and (b) chain of loops.

The capacitor helps the amplifier by storing the electron beam space charge. Positive feedback is obtained when the gain of the amplifier is able to overcome the losses in the transmission line. No sooner than the electron beam has the correct velocity, synchronism is achieved between the electron velocity and the backward moving electromagnetic wave, and oscillations occur at a frequency at which the phase delay of each loop = one cycle or an integral number of cycles. The phase delay is changed by changing the electron beam velocity through anode–cathode voltage, thereby altering the frequency of oscillation.

For easy understanding, our explanation is based on the idea of individual feedback loops. In reality, the interaction between the electron beam and the electromagnetic wave is a continuous process. Taking a holistic view, the energy of the electron beam is lost and the spent beam reaches the anode while the energy of the wave is drawn out through a waveguide coupling port. After exiting from the helix, the electron velocities are lowered by a multistage collector. By using this depressed collector arrangement at a lower potential, a large fraction of the energy of the spent electron beam is recovered with a power supply, thereby raising the efficiency of the tube. The electron beam travels from the cathode to anode while the electromagnetic wave propagates in the opposite direction from anode to cathode. Hence the downstream end of the circuit is the region of maximum electron bunching, whereas the upstream end of the circuit is the region where wave amplitude is maximum. Thus the BWO fulfills all the criteria for sustaining oscillations, viz., regenerative feedback, balance of phases and balance of amplitudes.

8.8 Advantages of the backward wave oscillator

The BWO has carved out a coveted niche amongst the wide-ranging spectrum of high frequency generators of coherent radiation. Its main advantages are:

(i) High power terahertz signals with good frequency stability can be produced;
(ii) It is highly adaptive towards variation of the parameters of the electron beam;
(iii) The risetime of oscillations is fast;
(iv) Its compact electrodynamic system is easy to design;
(v) The system can accommodate large stresses of electric field;
(vi) Its frequency can be electronically tuned by extensive margins around an arbitrary frequency;
(vii) The frequency of a BWO is increased by raising the anode voltage so that the electron beam is accelerated to high velocities allowing for faster electron-beam/slow wave structure interaction and electromagnetic wave/electron beam interaction.

8.9 Limitations of the backward wave oscillator

Principal limitations of the BWO are:

(i) Slow wave structures required to build terahertz BWOs are difficult to fabricate because the manufacturing tolerance needs precision up to 1-μm. So, the fabrication processes are costly;
(ii) The BWOs cannot supply high power at high frequencies because the sizes of their components become comparable to wavelength of the waves. Power dissipation becomes a serious issue as the slow wave interaction circuit dimensions shrink at terahertz frequencies. Due to their small sizes, the BWOs are unable to withstand high power;
(iii) The starting oscillation current is raised because of the ohmic losses taking place in the slow wave structure together with the small interaction impedances;
(iv) The electron beam must be small in diameter and high in density, and propagate close to the slow wave structure demanding a high convergence ratio. Consequently, a large spreading of thermal velocities is encountered.

8.10 Frequency/power levels achieved with backward wave oscillators

Due to their shortcomings, the power capabilities of BWOs deteriorate at high frequencies. Nonetheless, the operating frequencies and radiated power levels of BWOs continue to rise with technological innovations.

In a 0.2 THz BWO, the electron beam is generated by pseudospark discharge (He *et al* 2015). The advantage derived from a pseudospark discharge-based electron beam is that such a beam does not need a guide magnetic field. Additionally, the ability to initialize the discharge by a pulsed voltage makes it possible to fit the voltage source in a small volume. So, a portable hand-held THz oscillator can be realized.

The SWS is made with a sinusoidal profile in its inner surface. The beam carries a current of 10 A at a current density of 10^8 Am^{-2}. The beam voltage is swept from 42 to 25 kV and the pulse duration is 25 ns. The frequency is varied from 0.186–0.202 THz giving a maximum power of 20 W.

The 0.179–0.263 THz BWO (© 2019 Insight Product Company) has a power output of 10–30 mW, the 0.258–0.375 THz BWO has a power output 10–20 mW, … the 0.882–0.1111 THz BWO has a power output of 2–8 mW and the 1.034–1.250 THz BWO has a power output of 0.5–2 mW. These sources require a high voltage power supply 1–6 kV at 25–45 mA with heater supply 5.5–7.5 V at 1.4–2.0 A. Water cooling is required at 0.45 L min^{-1}.

The QS1–370–T terahertz generator (Microtech Instruments, Inc. © 2012 a) contains a BWO together with Schottky diode frequency multipliers. The spectral range of its QS1–370 model is 0.22–0.37 THz and the output power is up to 10 mW while the spectral range of QS1–370–1150 model is 0.22–1.15 THz and the output power is up to 0.3 mW.

The QS2–1500 terahertz generator (Microtech Instruments, Inc. © 2012 b) contains a BWO along with Schottky diode frequency multipliers. Its QS2–180 model has a spectral range of 0.1–0.18 THz with output power up to 20 mW; the QS2–350 model has a spectral range 0.2–0.35 THz with output power up to 2.0 mW… the QS2–1500 model has a spectral range 1–1.5 THz with output power up to 0.01 mW.

Let us now solve a few numerical examples on TWTs and BWOs.

Example 8.2 Operational frequency of a TWT is determined by the point of intersection of the electron beam line with first forward space harmonic on the dispersion diagram, and the SWS period is calculated from the associated phase shift (Srivastava 2015). If the position of the intersection point lies at the value of normalized wave number $k = k_z d = 3\pi/2$ where k_z is the wave number and d is the period of the planar periodic slow wave structure, and the TWT is designed for operation with a 10 kV electron beam, calculate the period d of the slow wave structure for the TWT desired to operate at a frequency of 0.22 THz? What is the electron velocity in this structure with respect to velocity of light?

Solution: Given: $k = k_z d = 3\pi/2$, electron beam energy = 10 keV, $f = 0.22$ THz

Since from equation (8.2), the phase velocity of electromagnetic wave

$$v_p = \frac{\omega}{k_z} \tag{8.75}$$

$$\therefore k_z = \frac{\omega}{v_p} \tag{8.76}$$

Also,

$$\omega = 2\pi f \tag{8.77}$$

The phase velocity v_p of electromagnetic wave should be slightly less than the velocity v_e of electron beam for energy transfer from the beam to the wave. As an approximation, we take

$$v_p = v_e \tag{8.78}$$

Combining equations (8.76–8.78),

$$k_z = \frac{2\pi f}{v_e} \tag{8.79}$$

$$\therefore k_z d = \frac{2\pi f d}{v_e} = \frac{3\pi}{2} \tag{8.80}$$

or,

$$d = \frac{3v_e}{4f} \tag{8.81}$$

A relativistic electron beam has a kinetic energy given by the equation

$$\text{Kinetic energy} = \text{Total energy} - \text{Rest mass energy} \tag{8.82}$$

or,

$$E_{\text{Kinetic}} = m_{\text{Relativistic}}c^2 - m_{\text{Rest mass}}c^2 = m_{\text{Relativistic}}c^2 - m_0c^2 \tag{8.83}$$

Under an accelerating potential difference of V Volts, the work done by the electric field is

$$E = eV \tag{8.84}$$

From equations (8.83) and (8.84)

$$eV = m_{\text{Relativistic}}c^2 - m_0c^2 \tag{8.85}$$

The relativistic mass of electron is related to its rest mass by the Lorentz factor γ

$$m_{\text{Relativistic}} = m_0\gamma = m_0 \times \frac{1}{\sqrt{1 - \frac{v_e^2}{c^2}}} \tag{8.86}$$

where m_0 is the rest mass of electron and c is the velocity of light in vacuum.

Substituting for the relativistic mass from equation (8.86) in equation (8.85) we get

$$eV = \frac{m_0c^2}{\sqrt{1 - \frac{v_e^2}{c^2}}} - m_0c^2 = \left(\frac{1}{\sqrt{1 - \frac{v_e^2}{c^2}}} - 1\right)m_0c^2 \tag{8.87}$$

or,

$$\frac{eV}{m_0c^2} = \frac{1}{\sqrt{1 - \frac{v_e^2}{c^2}}} - 1 \tag{8.88}$$

or,

$$1 + \frac{eV}{m_0c^2} = \frac{1}{\sqrt{1 - \frac{v_e^2}{c^2}}} \tag{8.89}$$

Squaring both sides

$$\left(1 + \frac{eV}{m_0c^2}\right)^2 = \frac{1}{1 - \frac{v_e^2}{c^2}} \tag{8.90}$$

or,

$$1 - \frac{v_e^2}{c^2} = \frac{1}{\left(1 + \frac{eV}{m_0c^2}\right)^2} \tag{8.91}$$

or,

$$\frac{v_e^2}{c^2} = 1 - \frac{1}{\left(1 + \frac{eV}{m_0c^2}\right)^2} \tag{8.92}$$

or,

$$v_e^2 = \left\{1 - \frac{1}{\left(1 + \frac{eV}{m_0c^2}\right)^2}\right\}c^2 \tag{8.93}$$

$$\therefore v_e = \sqrt{1 - \frac{1}{\left(1 + \frac{eV}{m_0c^2}\right)^2}} \times c \tag{8.94}$$

From equations (8.81) and (8.94)

$$d = \frac{3v_e}{4f} = \frac{3c}{4f}\sqrt{1 - \frac{1}{\left(1 + \frac{eV}{m_0c^2}\right)^2}}$$

$$= \frac{3 \times 2.9979 \times 10^8}{4 \times 0.22 \times 10^{12}}\sqrt{1 - \frac{1}{\left\{1 + \frac{1.602 \times 10^{-19} \times 10 \times 10^3}{9.109 \times 10^{-31} \times (2.9979 \times 10^8)^2}\right\}^2}}$$

$$= 1.022 \times 10^{-3}\sqrt{1 - \frac{1}{\left\{1 + \frac{1.602 \times 10^{-19} \times 10 \times 10^3}{9.109 \times 10^{-31} \times (2.9979 \times 10^8)^2}\right\}^2}} \tag{8.95}$$

$$= 1.022 \times 10^{-3}\sqrt{1 - \frac{1}{\left(1 + \frac{1.602 \times 10^{-15}}{9.109 \times 10^{-31} \times 8.9874 \times 10^{16}}\right)^2}}$$

$$= 1.022 \times 10^{-3}\sqrt{1 - \frac{1}{\left(1 + \frac{1.602}{9.109 \times 8.9874}\right)^2}}$$

$$= 1.022 \times 10^{-3}\sqrt{1 - 0.96198} = 1.022 \times 10^{-3} \times 0.194987$$

$$= 0.1992767 \times 10^{-3}\,\text{m} = 0.1992767 \times 10^{-3} \times 10^6\,\mu\text{m} = 199.28\,\mu\text{m}$$

The ratio (electron velocity/velocity of light) = the square root factor in equation (8.94) is picked up from the calculation in equation (8.95):

$$v_e = \sqrt{1 - 0.96198} = 0.194987c \tag{8.96}$$

Approximate calculation:

An approximate equation for v_e is derived without taking relativistic effects into account. It is obtained by simply equating the kinetic energy of electrons with their potential energy

$$\frac{1}{2}m_0v_e^2 = eV \tag{8.97}$$

or,

$$v_e^2 = \frac{2eV}{m_0} \tag{8.98}$$

or,

$$v_e = \sqrt{\frac{2eV}{m_0}} = \sqrt{\frac{2e}{m_0}}\sqrt{V} = \sqrt{\frac{2 \times 1.602 \times 10^{-19}}{9.109 \times 10^{-31}}}\sqrt{V} = \sqrt{\frac{2 \times 1.602 \times 10^{12}}{9.109}}\sqrt{V}$$
$$= 5.9308 \times 10^5\sqrt{V} \tag{8.99}$$

From equations (8.81) and (8.99)

$$d = \frac{3v_e}{4f} = \frac{3 \times 5.9308 \times 10^5\sqrt{V}}{4f} = \frac{3 \times 5.9308 \times 10^5\sqrt{10 \times 10^3}}{4 \times 0.22 \times 10^{12}}$$
$$= \frac{3 \times 5.9308 \times 10^5 \times 10^2}{4 \times 0.22 \times 10^{12}} = 20.2186 \times 10^{-5}\,\text{m} \tag{8.100}$$
$$= 20.2186 \times 10^{-5} \times 10^6 \mu\text{m} = 202.186\,\mu\text{m}$$

Also,

$$v_e = 5.9308 \times 10^5\sqrt{V} = 5.9308 \times 10^5\sqrt{10 \times 10^3} = 5.9308 \times 10^7\text{ms}^{-1}$$
$$= \frac{5.9308 \times 10^7\text{ms}^{-1}}{2.9979 \times 10^8\text{ms}^{-1}} = 0.19783c \tag{8.101}$$

Note: From equations (8.96) and (8.101),

$$\frac{v_c|_{\text{Relativistic}}}{v_e|_{\text{Approximate}}} = \frac{0.194987c}{0.19783c} = 0.9856 \tag{8.102}$$

so that the relativistically obtained velocity of electron is less than its approximate velocity.

Example 8.3 The frequency of a helix TWT is 0.22 THz. Its beam voltage is 10 kV and the beam current is 100 mA. If the impedance of the helix is 50 Ω, what is the physical length of the helix for an output power gain of 30 dB?

Solution: Here, $A_p = 30$ dB. In equation (8.72), replacing λ by c/f we get

$$N = \frac{L}{\frac{\text{Velocity of light in vacuum }(c)}{\text{Frequency of oscillation }(f)}} = \frac{Lf}{c} = \frac{L \times 0.22 \times 10^{12}}{2.9979 \times 10^8}$$
$$= \frac{L \times 0.22 \times 10^4}{2.9979} = 733.847L \tag{8.103}$$

From equation (8.73),

$$C = \left(\frac{I_0 Z_0}{4V_0}\right)^{\frac{1}{3}} = \left(\frac{100 \times 10^{-3} \times 50}{4 \times 10 \times 10^3}\right)^{\frac{1}{3}} = (1.25 \times 10^{-4})^{\frac{1}{3}} = 0.05 \tag{8.104}$$

Applying equation (8.74) for power gain A_p

$$30 = -9.54 + 47.3 \times 733.847L \times 0.05 \tag{8.105}$$

or,

$$39.54 = 1735.548L \tag{8.106}$$

$$L = \frac{39.54}{1735.548} = 0.02278 \text{ m} = 0.02278 \times 10^3 = 22.78 \text{ mm} \tag{8.107}$$

Example 8.4

(a) The phase synchronization condition of a BWO is located at the point of intersection of the electron beam line with first backward space harmonic on the dispersion diagram (Xi *et al* 2018). Given that the intersection point lies at the value of normalized wave number k = wave number (k_z) × period (d) of the planar periodic slow wave structure = $k_z d$ = 1.67π (300.6°), and further that the BWO is designed for operation with 25 kV electron beam and its frequency is 1.15 THz. Find the period d of the slow wave structure?

What is the electron velocity in this structure relative to velocity of light?

(b) For the beam voltage of 25 kV and beam current of 3 mA, what is the output power of the 1.15 THz BWO at an efficiency of 0.3%?

(c) Taking the period of slow wave structure as 66 μm, find the frequency of the relativistic BWO, if the beam voltage is raised to 500 kV. What is the velocity of electron beam?

Solution: (a) Here $k = k_z d = 1.67\pi$, Electron beam energy = 25 keV, f = 1.15 THz. From equation (8.79),

$$k_z d = \frac{2\pi f d}{v_e} = 1.67\pi \tag{8.108}$$

or,

$$d = \frac{1.67}{2f} v_e \tag{8.109}$$

Applying equation (8.94),

$$d = \frac{1.67c}{2f}\sqrt{1 - \frac{1}{\left(1 + \frac{eV}{m_0c^2}\right)^2}}$$

$$= \frac{1.67 \times 2.9979 \times 10^8}{2 \times 1.15 \times 10^{12}} \times \sqrt{1 - \frac{1}{\left(1 + \frac{1.602 \times 10^{-19} \times 25 \times 10^3}{9.109 \times 10^{-31} \times (2.9979 \times 10^8)^2}\right)^2}}$$

$$= 2.1767 \times 10^{-4} \times \sqrt{1 - \frac{1}{\left(1 + \frac{1.602 \times 10^{-16} \times 25}{9.109 \times 10^{-31} \times 8.9874 \times 10^{16}}\right)^2}} \quad (8.110)$$

$$= 2.1767 \times 10^{-4} \times \sqrt{1 - \frac{1}{\left(1 + \frac{1.602 \times 10^{-1} \times 25}{9.109 \times 8.9874}\right)^2}}$$

$$= 2.1767 \times 10^{-4} \times \sqrt{1 - 0.9089} = 2.1767 \times 10^{-4} \times 0.3018$$

$$= 0.6569 \times 10^{-4}\ \text{m} = 0.6569 \times 10^{-4} \times 10^6\ \mu\text{m} = 65.69\ \mu\text{m}$$

The (electron velocity/velocity of light) ratio is given by the square root factor in equation (8.110) as

$$v_e = \sqrt{1 - 0.9089} = 0.3018c$$

Approximate calculation:
Putting $v_e = 5.9308 \times 10^5\sqrt{V}$ from equation (8.99) in equation (8.109), we have

$$d = \frac{1.67v_e}{2f} = \frac{1.67 \times 5.9308 \times 10^5\sqrt{V}}{2f} = \frac{1.67 \times 5.9308 \times 10^5\sqrt{25 \times 10^3}}{2 \times 1.15 \times 10^{12}}$$

$$= \frac{1.67 \times 5.9308 \times 10^5 \times 158.114}{2 \times 1.15 \times 10^{12}} = 6.8088 \times 10^{-5}\ \text{m} \quad (8.111)$$

$$= 6.8088 \times 10^{-5} \times 10^6\ \mu\text{m} = 68.09\ \mu\text{m}$$

Also,

$$v_e = 5.9308 \times 10^5\sqrt{V} = 5.9308 \times 10^5\sqrt{25 \times 10^3} = 5.9308 \times 10^5 \times 158.114\ \text{m s}^{-1} = \frac{9.3774 \times 10^7\ \text{m s}^{-1}}{2.9979 \times 10^8\ \text{m s}^{-1}} = 0.3128c \quad (8.112)$$

(b)

$$\text{Output power} = \text{Beam voltage} \times \text{Beam current} \times \text{Efficiency}$$

$$= 25 \times 10^3 \times 3 \times 10^{-3} \times \frac{0.3}{100} = 0.225\ \text{W} \quad (8.113)$$

$$= 0.225 \times 1000\ \text{mW} = 225\ \text{mW}$$

(c) Rewriting the first line of equation (8.110),

$$d = \frac{1.67c}{2f}\sqrt{1 - \frac{1}{\left(1 + \frac{eV}{m_0c^2}\right)^2}} \tag{8.114}$$

or,

$$\begin{aligned} f &= \frac{1.67c}{2d}\sqrt{1 - \frac{1}{\left(1 + \frac{eV}{m_0c^2}\right)^2}} = \frac{1.67 \times 2.9979 \times 10^8}{2 \times 66 \times 10^{-6}} \\ &\times \sqrt{1 - \frac{1}{\left(1 + \frac{1.602 \times 10^{-19} \times 500 \times 10^3}{9.109 \times 10^{-31} \times (2.9979 \times 10^8)^2}\right)^2}} \\ &= \frac{1.67 \times 2.9979 \times 10^{14}}{2 \times 66} \times \sqrt{1 - \frac{1}{\left(1 + \frac{1.602 \times 10^{-19} \times 500 \times 10^3}{9.109 \times 10^{-31} \times 10^{16}(2.9979)^2}\right)^2}} \\ &= 3.793 \times 10^{12} \times \sqrt{1 - \frac{1}{\left(1 + \frac{1.602 \times 10^{-1} \times 500}{9.109 \times (2.9979)^2}\right)^2}} \\ &= 3.793 \times 10^{12} \times \sqrt{1 - \frac{1}{3.9142}} = 3.793 \times 10^{12} \times \sqrt{1 - 0.25548} \\ &= 3.793 \times 10^{12} \times 0.862856 = 3.2728 \times 10^{12}\ \text{Hz} = 3.273\ \text{THz} \end{aligned} \tag{8.115}$$

The velocity of electron beam is obtained from the square root factor in equation (8.115) as

$$v_e = \sqrt{1 - 0.25548} = 0.8629c \tag{8.116}$$

Approximate calculation:

As we got equation (8.111), putting $v_e = 5.9308 \times 10^5\sqrt{V}$ from equation (8.99) in equation (8.109), we have

$$\begin{aligned} f &= \frac{1.67v_e}{2d} = \frac{1.67 \times 5.9308 \times 10^5\sqrt{V}}{2 \times 66 \times 10^{-6}} = \frac{1.67 \times 5.9308 \times 10^5\sqrt{500 \times 10^3}}{2 \times 66 \times 10^{-6}} \\ &= \frac{1.67 \times 5.9308 \times 10^{11} \times 707.1068}{2 \times 66} = 5.3057 \times 10^{12}\ \text{Hz} = 5.3\ \text{THz} \end{aligned} \tag{8.117}$$

Also,

$$v_e = 5.9308 \times 10^5\sqrt{V} = 5.9308 \times 10^5\sqrt{500 \times 10^3}$$
$$= 5.9308 \times 10^5 \times 707.1068 \text{ m s}^{-1} = \frac{4.1937 \times 10^8 \text{ m s}^{-1}}{2.9979 \times 10^8 \text{ m s}^{-1}} = 1.399c \quad (8.118)$$

The approximate calculation totally fails because it gives an electron velocity higher than velocity of light in vacuum.

8.11 Discussion and conclusions

Research on subterahertz/terahertz region devices has received a boost from the easily-handled, lightweight vacuum electronic devices made by microfabrication techniques, the TWT and the BWO. Development of terahertz TWTs/BWOs requires severe precision in dimensional control in the range of tens of microns. In addition, high-density cathodes are to be built. High strength magnetic circuits with minimal drift in field intensity are essential. The alignment of the electron gun with circuit must be accurate with tolerance of only a few microns. Operation of TWT amplifiers with tens of milliwatts of power has been demonstrated at terahertz frequencies. The folded waveguided slow wave structure has been widely used. Apart from its use in amplification, TWT is also used for making oscillators with regenerative feedback.

The BWO is an efficient terahertz vacuum source belonging to the family of TWTs supplying hundreds of milliwatts of power that works on energy transference from electrons to electromagnetic waves propagating in a slow wave structure. An extraordinary stability in frequency is obtained independent of load impedance. Frequency tuning is achieved by changing the energy of electron beam through acceleration voltage (voltage tuning). The beam voltage provides tuning up to wide band to the extent of 50% of frequencies.

References

Booske J H, Dobbs R J, Joye C D, Kory C L, Neil G R, Park G-S, Park J and Temkin R J 2011 Vacuum electronic high power terahertz sources *IEEE Trans. Terahertz Sci. Technol.* **1** 54–75

Bratman V L, Fedotov A E, Makhalov P B and Manuilov V N 2014 Design and numerical analysis of W-band oscillators with hollow electron beam *IEEE Trans. Electron Devices* **61** 1795–9

Bratman V L, Fedotov A E, Makhalov P B and Manuilov V N 2017 Non-relativistic hollow electron beam formation for mm-wave BWO *10th Int. Workshop 2017 Strong Microwaves and Terahertz Waves: Sources and Applications* **149** 1–2

Das A and Das S K 2007 *Microwave Engineering* (New Delhi: Tata McGraw-Hill Publishing Co., Ltd) 348–49

Department of the Air Force 1963 *Electronic Circuit AnalysisAir Force Manual No. 52-8 Vol. II* (Washington DC: Department of the Air Force) pp 12-69–72

Fedotov A E, Bratman V L, Makhalov P B and Manuilov V N 2018 Design of mm-wave slow-wave devices with sheet and hollow electron beams *EPJ Web of Conf.: TERA–2018* **195** 01026

Feng J, Ren D, Li H, Tang Y and Xing J 2011 Study of high frequency folded waveguide BWO with MEMS technology *Terahertz Sci. Technol.* **4** 164–80

Gao P, Booske J H and Yang Z 2010 Frequency step-tuning characteristics of traveling-wave tube regenerative oscillators *IEEE Trans. Electron Devices* **57** 1152–9

Gold S H and Nusinovich G S 1997 Review of high-power microwave source research *Rev. Sci. Instrum.* **68** 3945–74

He W, Zhang L, Bowes D, Yin H, Ronald K, Phelps A D R and Cross A W 2015 Generation of broadband terahertz radiation using a backward wave oscillator and pseudospark-sourced electron beam *Appl. Phys. Lett.* **107** 133501

Hu P, Lei W, Jiang Y, Huang Y, Song R, Chen H and Dong Y 2018 Development of a 0.32-THz folded waveguide traveling wave tube *IEEE Trans. Electron Devices* **65** 2164–9

Joye C D, Cook A M, Calame J P, Nguyen K T, Wright E L, Hanna J M, Pershing D E and Abe D K 2014a Development of a 233 GHz high-gain traveling wave amplifier *IEEE Int. Vacuum Electronics Conf. (Monterey, CA, 22–24 April)*

Joye C D *et al* 2014b Demonstration of a high power, wideband 220-GHz traveling wave amplifier fabricated by UV-LIGA *IEEE Trans. Electron Devices* **61** 1672–8

Karetnikova T A, Rozhnev A G, Ryskin N M, Fedotov A E, Mishakin S V and Ginzburg N S 2018 Gain analysis of a 0.2-THz traveling-wave tube with sheet electron beam and staggered grating slow wave structure *IEEE Trans. Electron Devices* **65** 2129–34

Karetnikova T A, Rozhnev A G, Ryskin N M, Torgashov G V, Sinitsyn N I, Grigoriev Y A, Burtsev A A and Shalaev P D 2016 Modeling a subterahertz traveling-wave tube with a slow-wave structure of the double grating type and a sheet electron beam *J. Commun. Technol. Electron.* **61** 50–5

Liao S Y 2008 *Microwave Devices and Circuits* (New Delhi: Pearson Education, Inc) pp 404–5

Panahi N, Saviz S and Ghorannevis M 2017 The new wave-ring helical (WRH) slow-wave structure for traveling wave tube amplifiers *J. Theor. Appl. Phys.* **11** 269–74

Paoloni C, Gamzina D, Letizia R, Zheng Y and Luhmann N C Jr 2020 Millimeter wave traveling wave tubes for the 21st century *J. Electromagn. Waves Appl.* **35** 567–603

Ploskih A E, Ryskin N M, Burtsev A A, Danilushkin A V and Navrotsky I A 2019 Performance improvement of a sub-THz traveling-wave tube by using an electron optic system with a converging sheet electron beam *Results Phys.* **12** 799–803

Microtech Instruments, Inc. 2012a QS1–370–T http://52ebad10ee97eea25d5e-d7d40819259e7d3022d9ad53e3694148.r84.cf3.rackcdn.com/UK_Z22_Microtech_Terahertz-Source-QS1-370-T_DS.pdf (Accessed on 17th October 2021)

Microtech Instruments, Inc. 2012b QS2–1500 http://52ebad10ee97eea25d5e-d7d40819259e7d3022d9ad53e3694148.r84.cf3.rackcdn.com/UK_Z22_Microtech_Terahertz-Source-QS2-1500_DS.pdf (Accessed on 17th October 2021)

Ryskin N M, Karetnikova T A, Rozhnev A G, Torgashov G V, Bushuev N A and Shalaev P D 2015 Development and modeling of a sheet-beam sub-THz traveling wave tube *2015 IEEE Int. Vacuum Electronics Conf. (IVEC) (Beijing, 27–29 April)*

Shu G X, Liu G and Qian Z F 2018 Simulation study of a high-order mode terahertz radiation source based on an orthogonal grating waveguide and multiple sheet electron beams *Opt. Express* **26** 8040–8

Srivastava A 2015 Microfabricated terahertz vacuum electron devices: technology, capabilities and performance overview *Eur. J. Adv. Eng. Technol.* **2** 54–64

Insight Product Company 2019 Sub-mm sources from 179 to 1250 GHz based on BWO (backward wave oscillators) http://insight-product.com/submmbwo3.htm (Accessed on 17th October 2021)

Tucek J C, Basten M A, Gallagher D A, Kreischer K E, Lai R, Radisic V, Leong K and Mihailovich R 2012 A 100 mW, 0.670 THz power module *Int. Vacuum Electronics Conf. (IVEC) 2012 (Monterey, CA, 24–26 April 2012)* (New York: IEEE) pp 31–2

Tucek J C, Basten M A, Gallagher D A and Kreischer K E 2014 0.850 THz vacuum electronic power amplifier *IEEE Int. Vacuum Electronics Conf. (Monterey, CA, USA, 22–24 April 2014)* (New York: IEEE) 153–4

Tucek J C, Basten M A, Gallagher D A and Kreischer K E 2016 Operation of a compact 1.03 THz power amplifier 2016 *IEEE Int. Vacuum Electronics Conf.* (IVEC) *(Monterey, CA, 08 September 2016)* (New York: IEEE) 1–2

Xi H, Wang J, He Z, Zhu G, Wang Y, Wang H, Chen Z, Li R and Liu L 2018 Continuous-wave Y-band planar BWO with wide tunable bandwidth *Sci. Rep.* **8** 7

IOP Publishing

Practical Terahertz Electronics: Devices and Applications, Volume 1

Solid-state devices and vacuum tubes

Vinod Kumar Khanna

Chapter 9

Gyrotrons

Difficulties encountered in applying conventional vacuum tubes to terahertz frequencies are indicated, viz., size of the resonant cavity, the skin depth issue and the costly tooling requirements. Reasons for ability of fast wave devices to work at higher frequencies and powers than slow wave devices are enumerated. The advantages of gyrotrons for terahertz generation are outlined. Principal components of the gyrotron, namely, the magnetron injection gun, the beam tunnel, the resonant cavity, the electron beam collection system, the quasi-optical mode converter, and the warm and superconducting magnets are described. The cyclotron frequency formula is derived in the non-relativistic limit. Laser/maser action with electrons in place of molecules as gain medium is explained. The phase synchronization between electron motion and electromagnetic waves is applied to obtain expressions for resonance frequency, and gyrotron operation is expounded in terms of relativistic variation of electron mass with velocity. The electron bunching mechanism in the gyrotron is visualized from the three-electron model. Dispersion diagram of the gyrotron is drawn. Trends in gyrotron research are summarized.

The esteemed status of present-day high-power vacuum tubes is the result of a series of innovations which revolutionized conventional vacuum tubes. Technological refinements brought spectacular progress in their wake, and the tubes advanced by leaps and bounds, both in power and frequency capabilities. Let us look at the limitations of traditional tubes.

9.1 Difficulties faced with classical electron tubes in the terahertz range

As one marches to terahertz frequencies, the journey of classical electron tubes is jeopardized with formidable obstacles (Tallerico 1995):

doi:10.1088/978-0-7503-3171-5ch9

(i) Resonant cavity size: The cross-sectional area of the interaction circuit (cavity) in a classical vacuum electronic source is inversely proportional to the square of the operating frequency. So the interaction circuit rapidly decreases in size with increase of frequency. In fact, the size of the circuit is of the order of the small wavelength at the high frequency. Obviously, it is difficult to remove heat effectively from the small circuit. This tells upon the life of the circuit which decreases considerably.

(ii) Skin depth menace: The skin depth of a material (the distance from the surface at which the current density decreases to $1/e$ ~37% of its value at the surface) is frequency dependent. At a frequency f, the skin depth δ in a material of resistivity ρ is given by

$$\delta = \sqrt{\frac{\rho}{\pi\mu_0\mu_r f}} \tag{9.1}$$

where μ_0 is the permeability constant $= 4\pi \times 10^{-7}$ H m^{-1} and μ_r is the relative permeability of the material. It is seen that the skin depth is inversely proportional to the square root of the frequency. Hence, the current tends to be restricted to a thin layer at the surface with rise of frequency. Confinement of current to a thin surface layer by skin effect tantamounts to an increase in resistance. It is accompanied by heat generation and the circuit eventually burns out.

(iii) Uneconomical method: Making small size cavities with required tolerances mandates sophisticated and expensive processing. So it is not cost-effective.

9.2 Periodic beam devices versus periodic circuit devices

In classical electron tubes, a periodic slow wave structure such as a helix is used as a waveguide and the electron beam is uniform. Can the situation be reversed? Yes, the electron beam can be made periodic and the waveguide taken is continuous. But in this kind of waveguide, the phase velocity of electromagnetic waves is greater than velocity of light. So these devices are called fast wave devices and the electrons must be accelerated to relativistic velocities for interaction with electromagnetic waves. By relativistic velocities is meant velocities approaching the velocity of light. As the phase velocity of electromagnetic waves exceeds the velocity of light in vacuum, the condition for Cherenkov radiation is no longer valid. Then how will electromagnetic waves be emitted? We know that a charged particle emits electromagnetic radiation when it undergoes acceleration or deceleration. This radiation is known as bremsstrahlung (braking radiation). So, from slow waves we have changed over to fast waves and from Cherenkov radiation, we move to bremsstrahlung. As an example, electromagnetic waves can be produced by oscillations of electrons in a constant magnetic field.

A periodic electron beam moving through a continuous waveguide produces equivalent effects to a uniform electron beam interacting with a waveguide using a periodic slow wave structure. But the benefits accrued by introducing these changes are enormous. The most eloquent advantage is that the requirement of comparable dimensions of the size of the interaction circuit (cavity) with the wavelength of electromagnetic waves that is mandated by slow wave devices is shattered. The freedom gained in making the interaction circuit larger greatly facilitates heat transfer. By this improvement in heat withdrawal capacity, the fast wave devices can be made to operate at much higher frequencies than their slow wave competitors. So both high frequency and high power can be obtained simultaneously. In this way, the main obstacle of low power at high frequencies faced with slow wave devices is no longer a bottleneck.

9.3 Advantages offered by gyrotron for terahertz generation

The gyrotron is a cardinal member of the family of fast wave tubes. The gyrotron is not burdened with the requirement that the resonant cavity must be of the order of the wavelength of the radiation. The gyrotron cavity can be several wavelengths in length. It is much longer in length and broader in diameter than the cavity of a classical electron tube. The mandatory requirement of coupling into the fundamental cavity mode in slow wave devices is not applicable to gyrotrons. So frequency upscaling can be easily done. Therefore, in a gyrotron, the interaction between the electron beam and the electromagnetic wave can be distributed over a larger length and area as compared to the classical tubes. This distribution of interaction makes heat removal easy and obviates all the disadvantages arising from the small size of classical electron tubes, thereby enabling the efficient production of terahertz waves.

9.4 Components and constructional details of gyrotron

9.4.1 Magnetron injection gun (MIG)

Two versions of the gun (figure 9.1) are commonly used (Pagonakis *et al* 2019):

(i) diode gun containing a heater coil or filament, a conical shaped cathode and anode; and

(ii) triode gun containing a heater coil, a conical shaped cathode, a modulating anode and a control anode. An annular cathode may be used to produce a hollow electron beam. In this cathode, only the peripheral part emits electrons while the central part is non-emitting. The hollow beam helps to reduce the beam current for a given output power.

The modulating anode is kept at an intermediate potential using a separate power supply. It contours the electron beam to be restricted within the defined space. The control anode electrostatically constricts and collimates the beam. The use of two anodes in the triode gun provides better control over the electron beam but this gun also requires two separate power supplies.

Vacuum grade refractory materials like molybdenum, tungsten, and tungsten–rhenium are used for the heater. M-type dispenser cathode consists of a porous tungsten matrix impregnated with an emissive mixture of barium calcium aluminate. The impregnated cathode is coated with osmium–ruthenium or osmium–iridium by sputtering and fired in a hydrogen atmosphere (Cronin 1981). At a temperature >1000 °C, the interdispersed barium migrates to the surface as free barium from

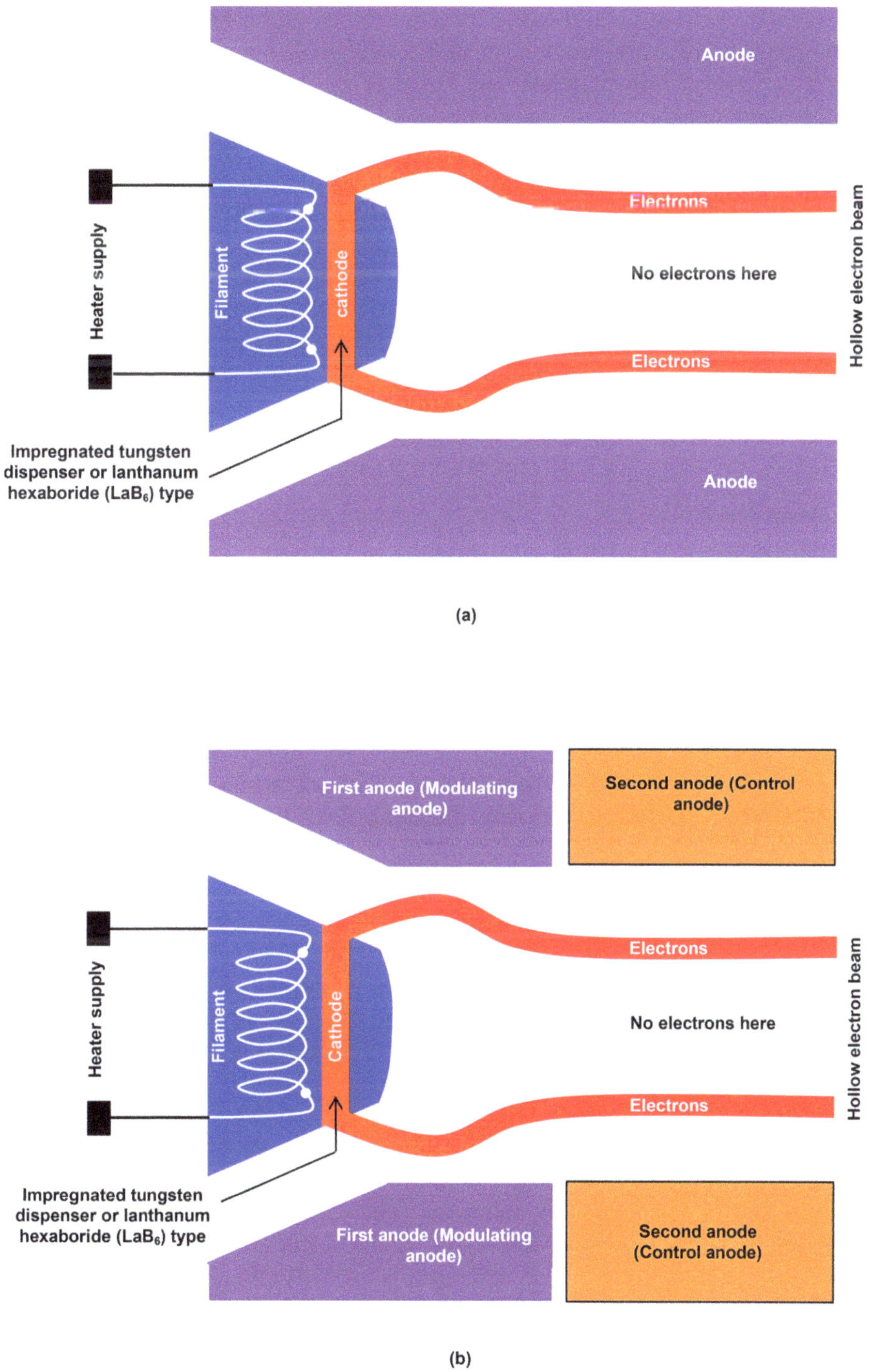

Figure 9.1. Magnetron injection guns: (a) diode gun and (b) triode gun.

where it partly evaporates and partly spreads over the surface of the tungsten matrix, thereby lowering its work function from 4.5 to 2.1 eV. The porous tungsten matrix serves as a reservoir from which the emissive material (barium) diffuses to the surface to form an active layer providing a low work function surface for thermionic emission. Uniformity of electron emission, low rate of barium evaporation, ruggedness towards vibrations, and longevity of cathodes are desired.

Vital parameters to be controlled are electron beam radius, the beam velocity spread and the transverse-to-axial velocity ratio. The temperature-limited region is carefully chosen for operation of the gun. This region provides temperature-limited thermionic emission. In this region, the thermal velocity spread is minimal. The gun is placed in a symmetrical axial magnetic field. By such placement, the gyration of electrons and transmission of gyrating electrons in the interaction region is properly maintained. Further, the magnetic field must be parallel to the cathode surface to ensure quasi-linearity of the electron beam.

9.4.2 Beam tunnel

The beam tunnel (figure 9.2) is the path through which the gyrating hollow electron beam from the magnetron injection gun reaches the interaction cavity. This is a region of inhomogeneous magnetic field with gradually increasing magnetic field to act as an adiabatic compression system in which the average diameter of the electron beam is diminished and its radial (transverse) energy is increased at the cost of axial (longitudinal) energy to provide a ratio of radial-to-axial velocity $\alpha \sim 1.2$, known as the pitch factor. The value of magnetic field reaches its maximum value in the resonance cavity. Magnetic field compression F_{m} is defined as the ratio of magnetic field B_0 at the cathode to the peak magnetic field B_{Peak} at the resonant cavity given by

$$F_{\text{m}} = \frac{B_0}{B_{\text{Peak}}} \tag{9.2}$$

One of the deciding factors of the frequency of emitted electromagnetic waves is the maximum value of magnetic field B_0, the other factor being the accelerating voltage of electrons. The ratio of radial-to-axial velocity α is determined by the ratio of accelerating voltage to the magnetic field upon which the cyclotron frequency is dependent.

The beam tunnel must take care of suppression of spurious oscillations in the compression zone and propagation of parasitic mode excitations from the interaction cavity to the magnetron injection gun, which are identified as backward waves. If this is not ensured, the quality of the beam will be degraded and damage to the cathode of the magnetron injection gun is anticipated.

The beam tunnel is designed in a corrugated structure comprising an alternating stack of oxygen-free high conductive (OFHC) copper and lossy ceramic dielectric (BeO) rings in the axial direction (Schmid *et al* 2011). In this sandwiched structure of copper rings interleaved with BeO rings, the lossy ceramic attenuates the high frequency fields by acting as a damping material. By lowering the quality factor, it

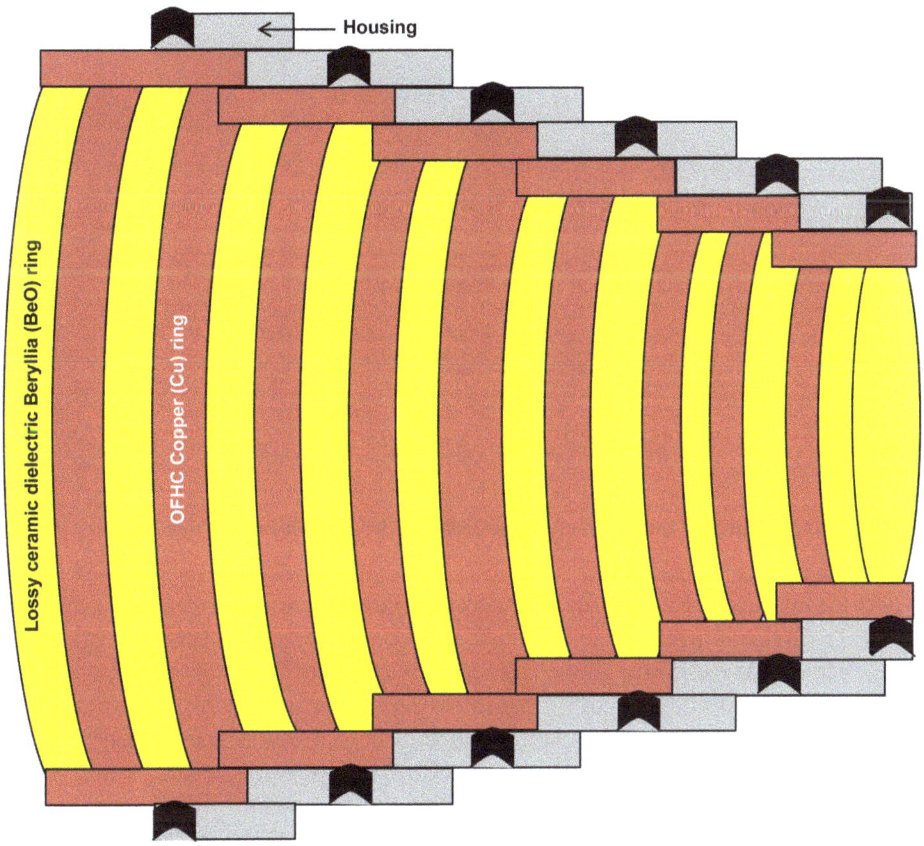

Figure 9.2. A section of beam tunnel of a gyrotron showing the juxtapositioning of alternate rings of OFHC copper and lossy ceramic BeO.

raises the starting current for oscillations leading to their cessation. The copper rings are meant for definition of the electrical potential. Further, the copper rings flow the charges accumulated on the surfaces of ceramic rings to ground. The static charges on the insulators are thereby eliminated.

9.4.3 The resonant cavity

This is the main working component of the gyrotron (Hruszowiec *et al* 2014). It is here that the interaction takes place between the gyrating electron beam and the electromagentic wave. It is a smooth-walled open three-section cylindrical-shaped resonator with a straight central section between tapered sections, one on each side (figure 9.3). The input section has a down taper. It has a high reflection coefficient to prevent backflow of electromagnetic waves to the MIG. The output section has an uptaper. It is a funnel shaped transition section that has a low reflection coefficient to couple the cavity with a launcher or output waveguide. The central section is the place where the beam–wave interactions predominantly take place.

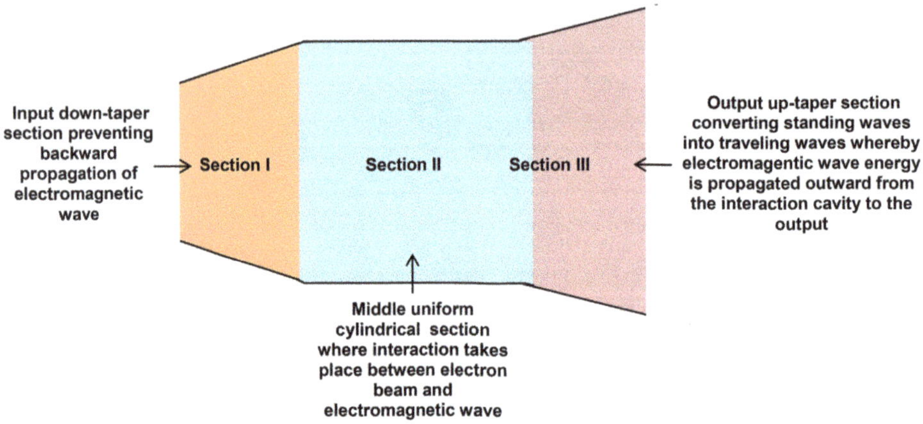

Figure 9.3. Three-section resonant cavity of gyrotron.

9.4.4 Electromagnetic wave output and electron beam collection system

These systems are of two types:

(i) Axial output of both the electromagnetic wave and the electron beam: The resonator output goes to a hollow coaxial region called the collector in which the radial magnetic induction is reduced to almost zero (figure 9.4). The electron beam is not focused by the magnetic field and the electrons are under the influence of spatial charge. Therefore, the walls of the collector capture the expended electron beam. The collector is kept water cooled because a huge power is dissipated on its surface. To prevent damage initiated by any remaining electrons reaching the dielectric window, an extra solenoid generating a crossed magnetic field is frequently used to divert these electrons towards the collector walls.

The difficulty faced with this system is that the accelerator voltage is the same as the collector voltage. So if the collector voltage is decreased, the accelerator voltage will also fall. The reason that we want to reduce the collector voltage is that a collector kept at a lower potential allows partial restoration of energy raising the overall efficiency. It is known as a depressed collector. Moreover, by decreasing the operating voltage, the energy of electron beam declines reducing the intensity of x-ray emission. Such advantages from a depressed collector cannot be availed if the wave and the electron beam are both taken out axially.

(ii) Radial output of the electromagnetic wave beam and axial output of the electron beam

Here, the direction of the electromagnetic wave moving along the axis of the gyrotron is changed to a direction perpendicular to the gyrotron axis by a set of mirrors, altering the multimode signal into a coherent signal (figure 9.5). After passage through the vacuum tight window, the signal enters the receiver. The electron beam

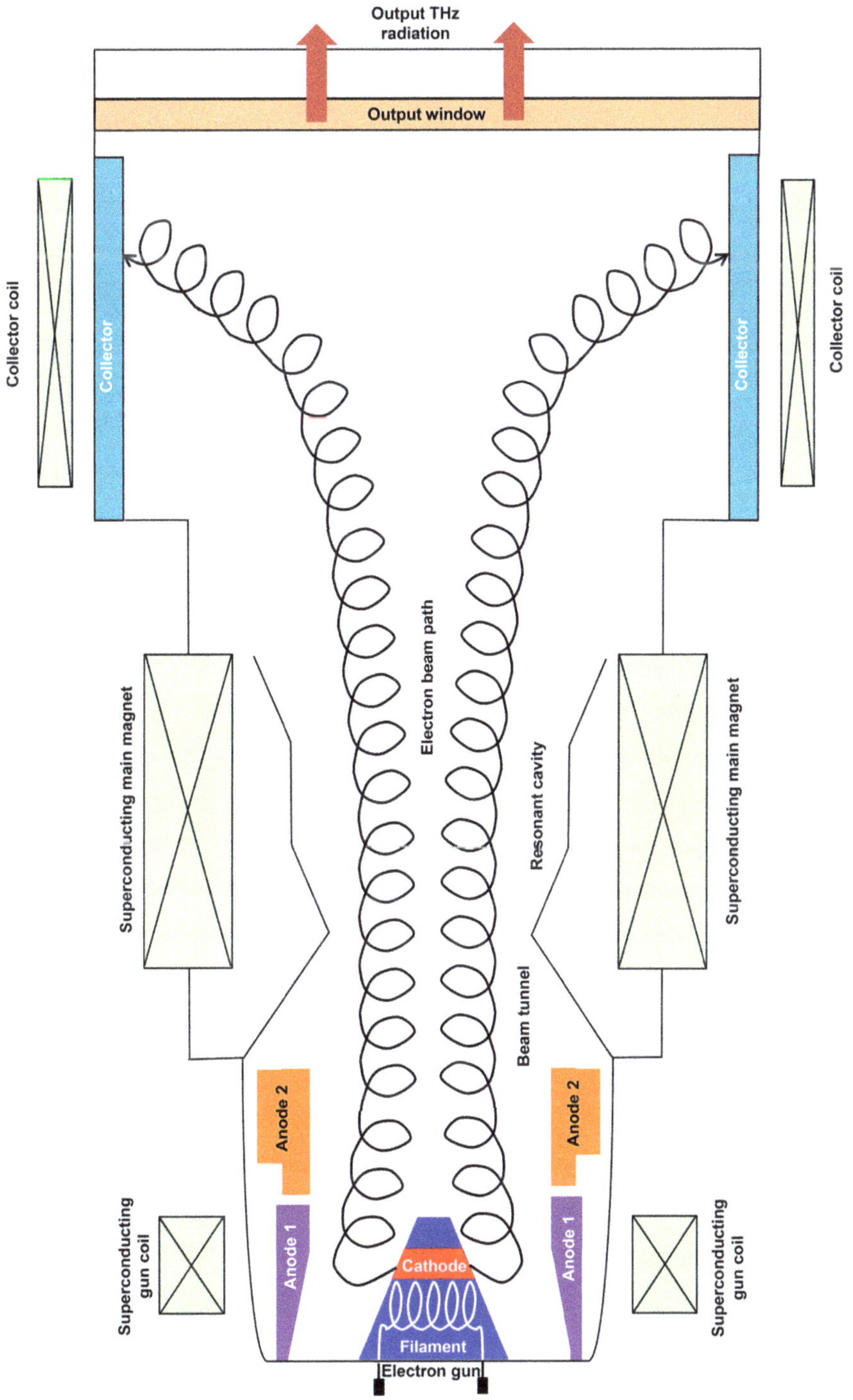

Figure 9.4. Layout of an axial output conventional gyrotron.

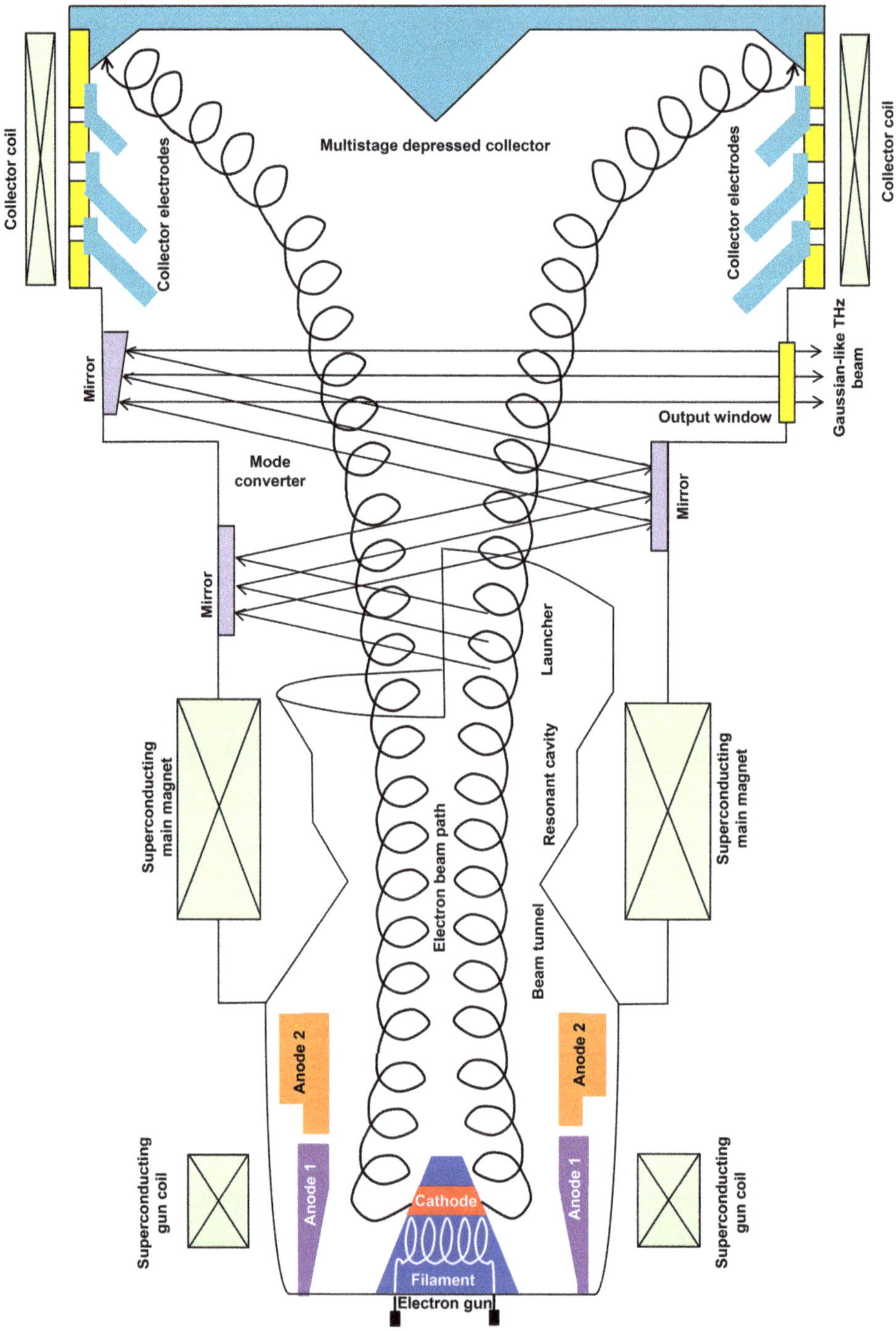

Figure 9.5. Schematic diagram of a radial output quasi-optical gyrotron.

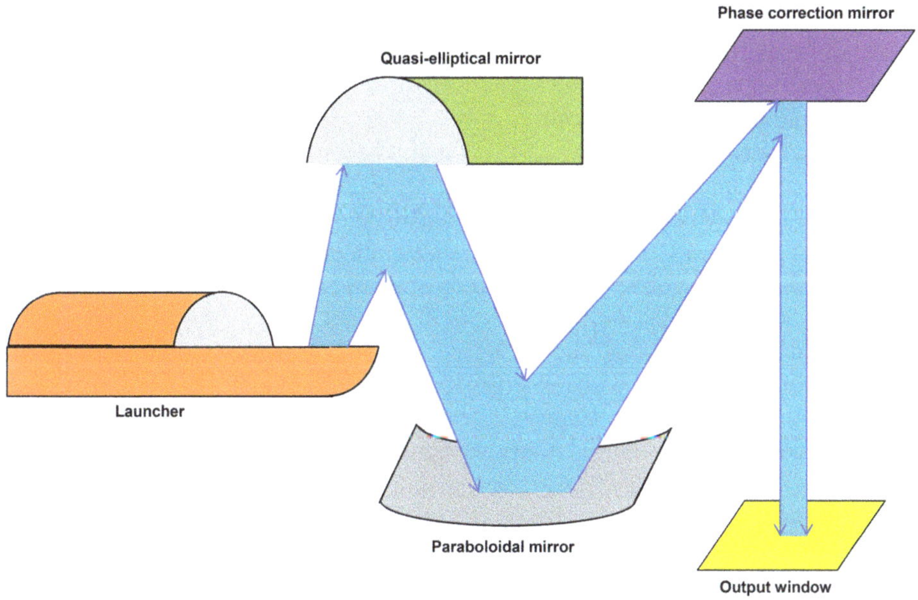

Figure 9.6. Arrangement of the quasi-optical mode converter in a gyrotron.

continues propagating along the axis of the gyrotron and falls on the collector. Thus the electromagnetic wave and the electron beam move in two perpendicular directions. In this way, the accelerator and collector voltages become independent. So, the restriction of choosing different values for these voltages is removed permitting use of a depressed collector. Notwithstanding the benefits derived, the system becomes more complex and also necessitates two power supplies in place of one.

The combined task of providing axial or radial output and converting the mode of the electromagnetic wave to the desired form is performed by the quasi-optical mode converter.

9.4.5 Quasi-optical (Q.O.) mode converter

The high order volume cavity modes in which high power gyrotrons operate are unsuitable for transmitting power in free space owing to the immense diffraction and polarization losses incurred (Litvak *et al* 2001). To circumvent these practical problems, a quasi-optical mode converter is employed for transforming the high-order operating mode into a linearly polarized paraxial beam (figure 9.6). The main components of the Q.O. mode converter located immediately after the interaction circuit of the gyrotron include an irregular cut ending waveguide up-tapering from the cavity (the launcher), along with a set of specially-shaped mirrors. Quasi-elliptic and parabolic reflectors direct the beam onto a phase correction mirror and then to output. The mirrors guide the output electromagnetic wave through a vacuum tight window either in a direction orthogonal to the axis of the cavity in the radial type gyrotron, or in a direction parallel to the cavity axis in an axial gyrotron configuration. Thus the sequential arrangement is:

Mode generator (TE_{mn})→Launcher→Quasi-Optical mirror→Window→TEM_{00} Guassian beam. By using the Q.O. mode converter, the ohmic loss of the resonant mode decreases, and thereby a low loss transport of the high frequency power out of the resonant cavity is achieved.

Radial output coupling of power offers many benefits. Some of the advantages of radial output coupling of the high-frequency power are (Jin 2007):

(i) The linearly polarized TEM_{00} mode can be directly utilized for transmission with low losses and for feeding the antenna effectively. Non-requirement of further mode converters is assured. Fundamental Gaussian mode TEM_{00} is the recommended output mode.

(ii) Due to isolation of the high-frequency electromagnetic wave from the electron beam, the wave emerges from the tube in a direction at right angles to the axis of the magnet. Electron beam collection is no longer the responsibility of the output waveguide. The collector can therefore be designed separately, and energy can be recovered using a depressed collector. Further, the detrimental effects of high frequency power reflected from the output window are avoided. The field profile at the window must accommodate the thermal properties of the material used for the window. The edge losses too must be minimized. Diamond is preferred as a window material owing to its high thermal conductivity. Independent solutions of the collector and window problems afford design flexibility.

9.4.6 Warm and superconducting magnets

For gyrotron operation, three types of magnets are required (Pradhan *et al* 2017): warm gun magnet for electron beam triggering with maximal transverse-to-axial velocity ratio, and minimal beam reflection and velocity spread; superconducting cavity magnet providing a homogeneous field up to 20 T (and beyond) for beam–wave interaction; and warm collector magnet(s) for beam power deposition on the collector surface. The warm magnets are made using enameled copper conductor windings. The superconducting magnet uses high current-carrying multi-filamentary NbTi superconductor strands for winding the solenoid. The magnet is housed inside a low-loss cryostat with ports for liquid nitrogen and liquid helium. To reach the working temperature from room temperature, several cooling hours are needed. Large quantities of liquid helium are used. Maintenance of the working temperature is very expensive.

9.5 Cyclotron frequency

Under the combined influence of the anode–cathode acceleration voltage and the gradually increasing strong axial magnetic field, a dense hollow electron beam is injected by the MIG into the resonant cavity of the gyrotron with a velocity **v** (figure 9.7).

The velocity vector **v** has two components: an axial component $v_{||}$ along the axis of the gyrotron and hence parallel to the magnetic field **B**, and a transverse

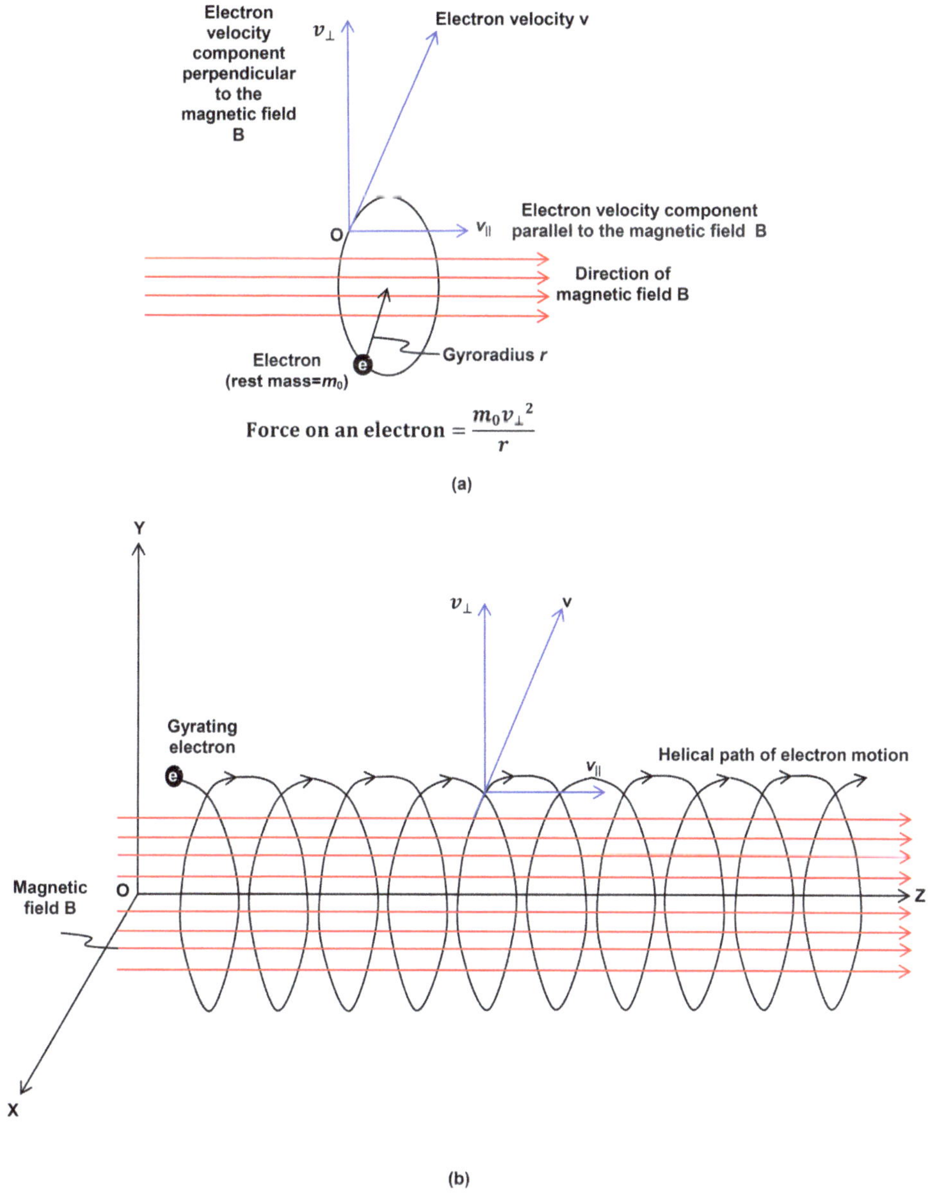

Figure 9.7. Motion of electron in a magnetic field: (a) circular motion when the direction of velocity of electron is perpendicular to that of the magnetic field and (b) helical motion when the two directions are not perpendicular to each other.

component $v_\perp$ at right angles to the gyrotron axis and therefore perpendicular to **B.** The Lorentz force = **v** × **B** from the magnetic field causes the electrons to move in helical paths. The electrons drift under the anode–cathode electric field along the direction of **B** and also rotate about the direction of **B**. Electrons undergo translational motion as well as rotational motion. Cyclotron frequency or gyrofrequency is the frequency of the circular motion of the electron perpendicular to the direction of **B**. For an electron of rest mass m_0 moving in a circle of radius r called the gyroradius, the equation for the balancing of centripetal and magnetic forces is written as

$$\text{Centripetal force} = \text{Magnetic Lorentz force} \tag{9.3}$$

or,

$$\frac{m_0 {v_\perp}^2}{r} = ev_\perp B \sin 90° = ev_\perp B \times 1 = ev_\perp B \tag{9.4}$$

where $v_\perp$ is the perpendicular component of electron velocity $\mathbf{v}$ and e is the electronic charge. The velocity $\mathbf{v}$ acts tangentially to the circular path. The perpendicular component $v_\perp$ is responsible for circular motion of the electron. The other component of velocity of electron $v_{||}$ directed parallel to the magnetic field causes the electron to move along the magnetic field direction, as already said. So the electron path looks like a stretched spring.

From equation (9.4), the linear velocity $v_\perp$ is

$$v_\perp = \frac{eBr}{m_0} \tag{9.5}$$

and the angular velocity ω is related to the linear velocity $v_\perp$ as

$$\omega = \frac{v_\perp}{r} \tag{9.6}$$

Substituting for $v_\perp$ from equation (9.5) in equation (9.6), we get

$$\omega = \frac{\frac{eBr}{m_0}}{r} = \frac{eBr}{m_0} \times \frac{1}{r} = \frac{eB}{m_0} \tag{9.7}$$

which is the cyclotron frequency of rotation given by

$$\Omega_C = \frac{eB}{m_0} \tag{9.8}$$

In the non-relativistic limit, the cyclotron frequency Ω_C is independent of the radius r of circular motion as well as velocity v and therefore kinetic energy of the electron. For an electron moving at relativistic velocities, i.e., velocity close to the velocity of light c in vacuum, equation (9.8) is modified by multiplying m_0 by the Lorentz factor

$$\gamma = \frac{1}{\sqrt{1 - \frac{v^2}{c^2}}} \tag{9.9}$$

to obtain

$$\Omega_C = \frac{eB}{\gamma m_0} \tag{9.10}$$

For the non-relativistic case, $v<<c$

$$\therefore \gamma = \frac{1}{\sqrt{1-0}} = 1 \tag{9.11}$$

so that equation (9.10) reduces to equation (9.8), as expected.

From equation (9.5), the radius r of circular motion called the Larmor radius

$$r = \frac{m_0 v_\perp}{eB} \tag{9.12}$$

for $v<<c$, and

$$r = \frac{\gamma m_0 v_\perp}{eB} = \frac{m_{\text{Effective}} v_\perp}{eB} \tag{9.13}$$

when v approaches c. Also,

$$m_{\text{Effective}} = m_0\gamma = m_0 \times \frac{1}{\sqrt{1-\frac{v^2}{c^2}}} = \text{Effective mass of electron} \tag{9.14}$$

From equations (9.10) and (9.14)

$$\Omega_C = \frac{eB}{m_{\text{Effective}}} \tag{9.15}$$

Example 9.1

(a) If the electron velocity $v = 0.7c$, find the effective mass of electron, Larmor radius of the circular orbit and the cyclotron frequency in a gyrotron with a magnetic field of 17 T? (Assume the pitch factor = 1). (b) Evaluate these parameters when the electron is slowed down to a velocity of $0.33c$. (c) Are Larmor radius, electron beam radius and cavity radius the same in a gyrotron? (d) Find the ratio of cyclotron frequencies per unit Tesla of an electron moving at $v = 0.7c$, by taking into account relativistic effects and by excluding them.

Solution: (a) Equation (9.14) gives

$$m_{\text{Effective}} = \frac{m_0}{\sqrt{1-\frac{v^2}{c^2}}} = \frac{m_0}{\sqrt{1-\frac{(0.7c)^2}{c^2}}} = \frac{m_0}{\sqrt{1-\frac{0.49c^2}{c^2}}} = \frac{m_0}{\sqrt{1-0.49}}$$
$$= \frac{m_0}{0.714} = 1.4m_0 = 1.4 \times 9.109 \times 10^{-31}\,\text{kg} = 1.275 \times 10^{-30}\,\text{kg} \tag{9.16}$$

From equation (9.13)

$$
\begin{aligned}
r &= \frac{m_{\text{Effective}} v_{\perp}}{eB} = \frac{1.275 \times 10^{-30} \times 0.7c}{1.602 \times 10^{-19} \times 17} \\
&= \frac{1.275 \times 10^{-30} \times 0.7 \times 2.9979 \times 10^{8}}{1.602 \times 10^{-19} \times 17} = \frac{1.275 \times 0.7 \times 2.9979 \times 10^{-3}}{1.602 \times 17} \\
&= 0.09825 \times 10^{-3}\,\text{m} = 0.09825 \times 10^{-3} \times 10^{6}\,\mu\text{m} = 98.25\,\mu\text{m}
\end{aligned} \tag{9.17}
$$

and from equation (9.15),

$$
\begin{aligned}
\Omega_C &= \frac{eB}{m_{\text{Effective}}} = \frac{1.602 \times 10^{-19} \times 17}{1.275 \times 10^{-30}} = \frac{1.602 \times 17 \times 10^{11}}{1.275} \\
&= 2.136 \times 10^{12}\,\text{Hz} = 2.136\,\text{THz}
\end{aligned} \tag{9.18}
$$

(b) By equation (9.14)

$$
\begin{aligned}
m_{\text{Effective}} &= \frac{m_0}{\sqrt{1 - \frac{v^2}{c^2}}} = \frac{m_0}{\sqrt{1 - \frac{(0.33c)^2}{c^2}}} = \frac{m_0}{\sqrt{1 - \frac{0.1089c^2}{c^2}}} = \frac{m_0}{\sqrt{1 - 0.1089}} \\
&= \frac{m_0}{0.94398} = 1.0593 m_0 = 1.0593 \times 9.109 \times 10^{-31}\,\text{kg} = 9.649 \times 10^{-31}\,\text{kg}
\end{aligned} \tag{9.19}
$$

Applying equation (9.13)

$$
\begin{aligned}
r &= \frac{m_{\text{Effective}} v_{\perp}}{eB} = \frac{9.649 \times 10^{-31} \times 0.33 \times 2.9979 \times 10^{8}}{1.602 \times 10^{-19} \times 17} = \frac{9.5458 \times 10^{-4}}{27.234} \\
&= 3.5051 \times 10^{-5}\,\text{m} = 3.5051 \times 10^{-5} \times 10^{6}\,\mu\text{m} = 35.05\,\mu\text{m}
\end{aligned} \tag{9.20}
$$

and using equation (9.15)

$$
\begin{aligned}
\Omega_C &= \frac{eB}{m_{\text{Effective}}} = \frac{1.602 \times 10^{-19} \times 17}{9.649 \times 10^{-31}} = \frac{27.234 \times 10^{12}}{9.649} \\
&= 2.8225 \times 10^{12}\,\text{Hz} = 2.8225\,\text{THz}
\end{aligned} \tag{9.21}
$$

(c) No.

$$
\text{Larmor radius} \ll \text{Electron beam radius} \ll \text{Cavity radius} \tag{9.22}
$$

e.g., Larmor radius = 100 μm, electron beam radius = 3 mm and cavity radius = 9 mm (typical values).

(d) From example 9.1(a), equations (9.8) and (9.15), taking $B = 1$ T,

$$(\Omega_C)_{\text{Relativistic}} = \frac{eB}{m_{\text{Effective}}} = \frac{1.602 \times 10^{-19} \times 1}{1.275 \times 10^{-30}} = 1.2565 \times 10^{11}\ \text{Hz T}^{-1}$$
$$= 0.12565 \times 10^{12}\ \text{Hz T}^{-1} = 0.12565\ \text{THz T}^{-1} \tag{9.23}$$

$$(\Omega_C)_{\text{Non-Relativistic}} = \frac{eB}{m_0} = \frac{1.602 \times 10^{-19} \times 1}{9.109 \times 10^{-31}} = 0.17587 \times 10^{12}\ \text{Hz T}^{-1}$$
$$\approx 0.17587\ \text{THz T}^{-1} \tag{9.24}$$

$$\therefore \frac{(\Omega_C)_{\text{Relativistic}}}{(\Omega_C)_{\text{Non-Relativistic}}} = \frac{0.12565\ \text{THz T}^{-1}}{0.17587\ \text{THz T}^{-1}} = 0.7144 \tag{9.25}$$

9.6 Cyclotron resonance maser (CRM)

An electron in an atom may be promoted from a lower energy level to a higher energy level by the action of light (an incoming photon) or heat (phonon). The atom in this condition is said to be in an excited state (figure 9.8). An incident photon of specific frequency may cause the electron to drop from the higher to the lower energy level causing the liberation of a photon having the same frequency, phase, polarization and direction of propagation as the incident electron. This phenomenon is called stimulated emission in contrast to the random phenomenon of spontaneous emission through dropping of electron to the more stable lower energy level.

Normally, a system of atoms contains more atoms in lower energy states than in excited states. By pumping energy, either electrically or optically, the number of atoms in excited states can be made to exceed the number of atoms in lower energy states, bringing about a population inversion. Then one incident photon can activate an avalanche of photons of the same frequency leading to the production of an amplified coherent beam of photons. This is the underlying principle of laser/maser (light/microwave/molecular amplification by stimulation emission of radiation).

While in these lasers/masers, the gain medium comprises atoms or molecules of gases or solids, electrons gyrating in an external magnetic field in vacuum can be exploited for the same purpose. These electrons possess discrete energy states called Landau levels by virtue of their quantized angular momentum (Franck 2017). Here pumping is achieved by continuously supplying higher energy electrons through the cavity resonator. This is the principle of the electron cyclotron maser, which has the additional advantage that its frequency, energy or efficiency is not restricted by material properties of the gain medium. Moreover, the number of rotational states, located at almost equal distances, are much larger than in atomic or molecular lasers.

A net flow of energy from the electrons to the electromagnetic wave is ensured on compliance of a resonance condition between the periodic motion of electrons and the electromagnetic wave in the region of their interaction. This resonance condition is the phase synchronization condition between them, and is met when

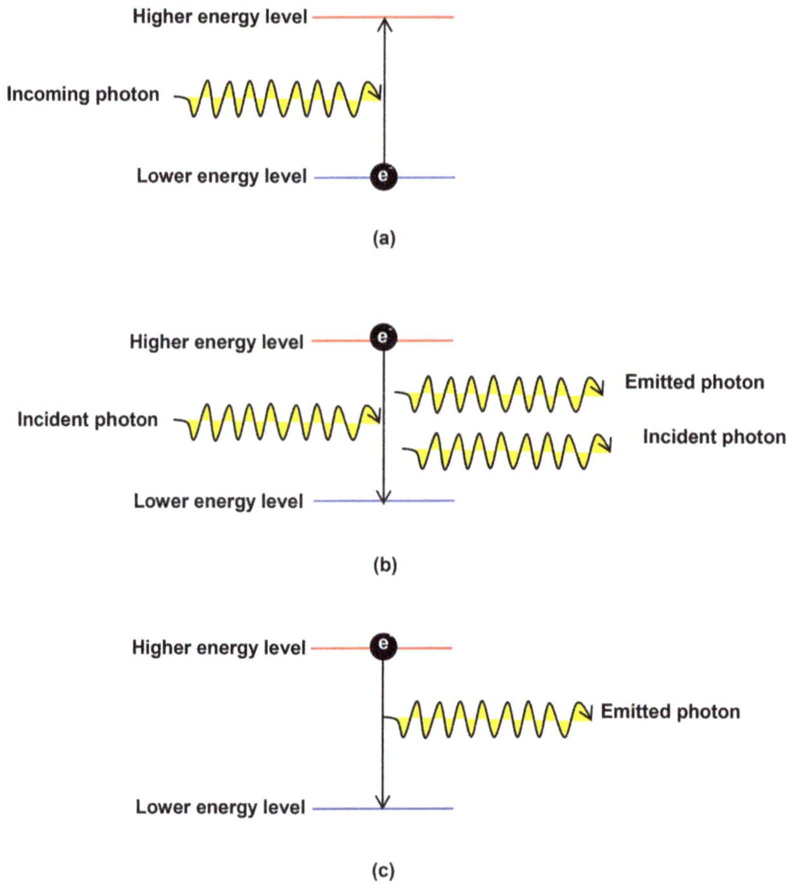

Figure 9.8. Photon absorption and two types of photon emission processes: (a) pumping of atom to excited state by illumination with a photon, (b) stimulated emission of a photon and (c) spontaneous emission of a photon.

$$\begin{aligned}&\text{Angular frequency of the electromagnetic wave } (\omega)\\&\quad - \text{Frequency shift of the electromagnetic wave by relativistic}\\&\quad \text{Doppler effect} = \text{Harmonic number } (s) \times \text{Cyclotron frequency } (\Omega_C)\end{aligned} \tag{9.26}$$

or,

$$\omega - k_z v_z = s\Omega_C \tag{9.27}$$

because

$$\begin{aligned}&\text{Frequency shift of the electromagnetic wave by relativistic Doppler effect}\\&\quad = \text{Axial wave number } (k_z) \times \text{Axial translational velocity of the electron}\\&\quad \text{beam } (v_z) = k_z v_z\end{aligned} \tag{9.28}$$

In a gyrotron,

$$\text{Axial velocity of electron } (v_z) < \text{Transverse velocity of electron } (v_\perp) \tag{9.29}$$

Hence the term $v_z k_z$ is negligibly small so that equation (9.27) reduces to the approximate form

$$\omega \approx s\Omega_C \tag{9.30}$$

Substituting for Ω_C from equation (9.10) in equation (9.30)

$$\omega = s\frac{eB}{\gamma m_0} \tag{9.31}$$

from which

$$\omega = s\frac{1.602 \times 10^{-19}\text{Coulomb} \times B(\text{Tesla})}{\gamma \times 9.109 \times 10^{-31}\text{kilogram}} = 0.17587 \times 10^{12}\frac{sB(\text{Tesla})}{\gamma} \tag{9.32}$$

The linear frequency

$$\nu = \frac{\omega}{2\pi} = \frac{0.176}{2 \times 3.14159} \times 10^{12}\frac{sB(\text{Tesla})}{\gamma} = 0.028011 \times 10^{12}\frac{sB(\text{Tesla})}{\gamma}\text{Hz} \tag{9.33}$$

and wavelength

$$\begin{aligned}\lambda = \frac{c}{\nu} = \frac{2.9979 \times 10^{8}\text{m} - \text{s}^{-1}}{0.028011 \times 10^{12}\frac{sB(\text{Tesla})}{\gamma}\text{Hz}} = 0.0107026\frac{\gamma}{sB(\text{Tesla})}\text{m} \\ = 0.0107026 \times 1000\frac{\gamma}{sB(\text{Tesla})}\text{mm} = 10.7026\frac{\gamma}{sB(\text{Tesla})}\text{mm}\end{aligned} \tag{9.34}$$

If the cathode-to-anode potential difference is V, the work done on the electron beam and hence its kinetic energy is

$$W = -(-eV) = eV \tag{9.35}$$

because the electron moves in a direction opposite to that of the electric field: the negatively charged electron moves from negative to positive potential, whereas the field points from positive to negative potential.

We can write W as

$$W = \frac{1}{2}m_0 v^2 \tag{9.36}$$

From equations (9.35) and (9.36),

$$eV = \frac{1}{2}m_0 v^2 \tag{9.37}$$

or,

$$v^2 = \frac{2eV}{m_0} \tag{9.38}$$

Substituting for v^2 from equation (9.38) in equation (9.9), the equation for the Lorentz factor γ is modified in terms of cathode-to-anode acceleration voltage V as

$$\gamma = \frac{1}{\sqrt{1 - \frac{\frac{2eV}{m_0}}{c^2}}} = \left(1 - \frac{\frac{2eV}{m_0}}{c^2}\right)^{-\frac{1}{2}} \approx 1 - \frac{1}{2} \times \left(-\frac{\frac{2eV}{m_0}}{c^2}\right) = 1 + \frac{1}{2} \times \frac{2eV}{m_0 c^2}$$
$$= 1 + \frac{eV}{m_0 c^2} \tag{9.39}$$

because

$$(1 - x)^{-n} = 1 + nx + \frac{n(n+1)}{2!}x^2 + \ldots \tag{9.40}$$

Equation (9.39) gives

$$\begin{aligned}
\gamma &= 1 + \frac{eV}{m_0 c^2} \\
&= 1 + \frac{1.602 \times 10^{-19}\text{Coulomb} \times V}{9.109 \times 10^{-31}\text{kilogram} \times (2.9979 \times 10^8 \text{m} - \text{s}^{-1})^2} \\
&= 1 + \frac{1.602 \times 10^{-19}}{9.109 \times 10^{-31} \times 8.9874 \times 10^{16}} V \\
&= 1 + \frac{1}{\frac{9.109 \times 10^{-31} \times 8.9874 \times 10^{16}}{1.602 \times 10^{-19}}} V \\
&= 1 + \frac{V}{\frac{9.109 \times 8.9874}{1.602} \times 10^{-31+16+19}} = 1 + \frac{V}{51.10 \times 10^4} \\
&= 1 + \frac{V}{511 \times 10^3} = 1 + \frac{1}{511} \times \frac{V}{10^3} = 1 + \frac{V(kV)}{511}
\end{aligned} \tag{9.41}$$

Combining equations (9.33) and (9.41)

$$\nu = 0.028011 \times 10^{12} \frac{sB(\text{Tesla})}{1 + \frac{V(kV)}{511}} \text{Hz} \tag{9.42}$$

Gyrotron operation is based on the relativistic variation of the mass of electron with velocity. So, gyrotrons are typically designed with a lower limit of cathode voltage = 30 kV and an upper practical limit ~ 100 kV for continuous wave operation.

Example 9.2

(a) A gyrotron is working in the fundamental mode at a cathode-to-anode accelerating voltage of 30 kV. Calculate the magnetic field intensity of the superconducting magnet required for generating 1 THz waves by this gyrotron? (b) What will be field intensity if the accelerating voltage is increased to 50 kV? (c) Repeat the calculation for 100 kV?

Solution: (a) Here $V = 30$ kV, $s = 1$. Equation (9.42) gives

$$1 \times 10^{12}\ \text{Hz} = 0.028011 \times 10^{12} \frac{1 \times B(\text{Tesla})}{1 + \frac{30(kV)}{511}} \text{Hz} = 0.028011 \times 10^{12} \frac{1 \times B(\text{Tesla})}{1 + 0.058708} \text{Hz}$$
$$= 0.028011 \times 10^{12} \frac{1 \times B(\text{Tesla})}{1.058708} \text{Hz} = 0.026458 \times 10^{12} \times B\ \text{Hz} \tag{9.43}$$

giving

$$B = \frac{1 \times 10^{12}}{0.026458 \times 10^{12}} \text{T} = 37.796\ \text{T} \tag{9.44}$$

(b) Here, $V = 50$ kV, $s = 1$. Since

$$1 + \frac{V(kV)}{511} = 1 + \frac{50(kV)}{511} = 1 + 0.09785 = 1.09785 \tag{9.45}$$

Equation (9.42) yields

$$1 \times 10^{12}\ \text{Hz} = 0.028011 \times 10^{12} \frac{1 \times B(\text{Tesla})}{1.09785} \text{Hz} = 0.02551 \times 10^{12} \times B\ \text{Hz} \tag{9.46}$$

So

$$B = \frac{1}{0.02551} \text{T} = 39.2\ \text{T} \tag{9.47}$$

(c) Here, $V = 100$ kV, $s = 1$. Since

$$1 + \frac{V(kV)}{508.8} = 1 + \frac{100(kV)}{511} = 1 + 0.19569 = 1.19569 \tag{9.48}$$

Equation (9.42) yields

$$1 \times 10^{12}\ \text{Hz} = 0.028011 \times 10^{12} \frac{1 \times B(\text{Tesla})}{1.19569} \text{Hz} = 0.02343 \times 10^{12} \times B\ \text{Hz} \quad (9.49)$$

So

$$B = \frac{1}{0.02343} \text{T} = 42.68\ \text{T} \quad (9.50)$$

Note: These high values of magnetic fields are not achievable with present-day superconducting magnets. Therefore, recourse is taken to work with $s = 2$ harmonic when the magnetic field intensities required are considerably reduced falling within the realms of technological realization. Example 10.2 illustrates this aspect.

Example 9.3 Perform the calculations of example 9.2 for the $s = 2$ harmonic of gyrotron.

(a) Solution: (a) From equation (9.44)

$$B = \frac{1 \times 10^{12}}{2 \times 0.026458 \times 10^{12}} \text{T} = 18.898\ \text{T} \quad (9.51)$$

(b) By equation (9.47)

$$B = \frac{1}{2 \times 0.02551} \text{T} = 19.6\ \text{T} \quad (9.52)$$

(c) Applying equation (9.50)

$$B = \frac{1}{2 \times 0.02343} \text{T} = 21.34\ \text{T} \quad (9.53)$$

Example 9.4 A gyrotron is operating with a magnetic field $B = 4.5$ T in the fundamental mode with an electron current of 40 A, beam voltage of 70 kV and efficiency of 40%. Find the output frequency and power.

Solution: Here $s = 1$, $B = 4.5$ T, $V = 70$ kV, $I = 40$ A

From equation (9.42), the output frequency is

$$\nu = 0.028011 \times 10^{12} \frac{1 \times 4.5}{1 + \frac{70}{511}} \text{Hz} = 0.028011 \times 10^{12} \frac{4.5}{1 + 0.13699} \text{Hz}$$
$$= 0.028011 \times 10^{12} \frac{4.5}{1.13699} \text{Hz} = 0.11086 \times 10^{12} \text{Hz} \approx 0.111\ \text{THz} \quad (9.54)$$

The output power is

$$\begin{aligned}\text{Power} &= \text{Electron beam current} \times \text{Beam voltage} \times \text{Efficiency} = 40\text{ A} \\ &\times 70\text{ kV} \times 0.4 = 40\text{ A} \times 70 \times 10^3\text{ V} \times 0.4 = 1120000\text{ W} \\ &= 1.12 \times 10^6\text{ W} = 1.12\text{ MW}\end{aligned} \tag{9.55}$$

Example 9.5 A gyrotron with 30 kV, 250 mA electron beam works in second harmonic mode at an efficiency of 4%. If the gyrotron uses a 20 T superconducting magnet, what is the : (a) frequency achieved, (b) power obtained?

(a) Solution: (a) Here $s = 2$, $V = 30$ kV, $B = 20$ T. From equation (9.42), the frequency is

$$\begin{aligned}\nu &= 0.028011 \times 10^{12} \frac{2 \times 20}{1 + \frac{30}{511}} \text{Hz} = 0.028011 \times 2 \times 20 \times 10^{12} \frac{1}{1 + 0.058708} \text{Hz} \\ &= \frac{1.12044 \times 10^{12}}{1.058708} \text{Hz} = 1.0583 \times 10^{12}\text{ Hz} = 1.0583\text{ THz}\end{aligned} \tag{9.56}$$

(b) Given: $V = 30$ kV, $I = 250$ mA, Efficiency = 4% = 0.04. The power is

$$\begin{aligned}P &= \text{Electron beam current} \times \text{Beam voltage} \times \text{Efficiency} \\ &= 30\text{ kV} \times 250\text{ mA} \times 0.04 = 30 \times 10^3 \times 250 \times 10^{-3} \times 0.04 = 300\text{ W}\end{aligned} \tag{9.57}$$

Note: It is evident that the frequency of a gyrotron oscillator depends mainly on the magnetic field intensity (B) of the superconducting magnet, the harmonic number (s) and weakly on the accelerating voltage V of the electron beam. The harmonic interaction has lower efficiency than the fundamental interaction because of incurrence of higher ohmic losses.

9.7 Explanation of the bunching mechanism of a gyrotron with a simplified three-electron model

The governing equation for the motion of an electron of charge $-e$ in an electromagnetic field is written by using the Lorentz force law:

$$\frac{d\mathbf{p}}{dt} = -e(\mathbf{E} + \mathbf{v} \times \mathbf{B}) \tag{9.58}$$

where $\mathbf{p}$ denotes the momentum of the electron, $\mathbf{v}$ its velocity, $\mathbf{E}$ the applied electric field and $\mathbf{B}$ applied magnetic field. Since the electric force acts along the direction of motion of the electron, it influences its velocity and kinetic energy. The magnetic force acting perpendicularly to the direction of motion of the electron does not affect its kinetic energy but alters its momentum by bending the path of the electron beam.

To find the rate of change of energy of an electron, we find the scalar product of equation (9.58) with velocity **v** obtaining (Sabri 2012)

$$\begin{aligned}\frac{dW}{dt} &= \frac{d\mathbf{p}}{dt}\cdot\mathbf{v} = -e(\mathbf{E}+\mathbf{v}\times\mathbf{B})\cdot\mathbf{v} = -\mathbf{eE}\cdot\mathbf{v} - \mathbf{ev}\times\mathbf{B}\cdot\mathbf{v} \\ &= -\mathbf{eE}\cdot\mathbf{v} - \mathrm{e}\{|\mathbf{v}||\mathbf{B}|\sin 90°\}\{|\mathbf{v}|\cos 90°\} \\ &= -\mathbf{eE}\cdot\mathbf{v} - \mathrm{e}\{|\mathbf{v}||\mathbf{B}|(1)\}\{|\mathbf{v}|(0)\} = -\mathbf{eE}\cdot\mathbf{v} - \mathbf{0} = -\mathbf{eE}\cdot\mathbf{v}\end{aligned} \tag{9.59}$$

For electrons which have a velocity component in the direction of **E**, $dW/dt<0$; hence these electrons lose energy and are decelerated. But for electrons having a velocity component opposite to the direction of **E**, $dW/dt>0$; hence they gain energy and experience acceleration. So we have to look for directions of electron velocities with respect to the electric field to determine whether they will be decelerated or accelerated.

We should also know how deceleration or acceleration of electrons affects their effective mass $m_{\mathrm{Effective}}$ with reference to their rest mass m_0 according to theory of relativity:

$$m_{\mathrm{Effective}} = \frac{m_0}{\sqrt{1-\dfrac{v^2}{c^2}}} \tag{9.60}$$

Electron deceleration implies decrease of electron velocity and thereby the second term under the square root in the denominator of effective mass equation, whereby the denominator increases and effective mass of the electron decreases. Electron acceleration means that electron velocity increases, which is associated with increase in effective mass of electron. From equation (9.13), the Larmor radius of electron is proportional to its effective mass and behaves similarly to the effective mass, decreasing with electron deceleration and increasing with electron acceleration. Due to inverse proportionality of cyclotron frequency with effective mass, as shown by equation (9.15), the cyclotron frequency increases with electron deceleration and decreases with electron acceleration.

At any given instant, let us focus our attention on three electrons (figure 9.9). These electrons are numbered as 1, 2, 3. They have the same initial energies:

$$|V_1| = |V_2| = |V_3| = |V| \tag{9.61}$$

when they are undergoing circular motion about a common center in a uniform magnetic field. As these electrons possess the same energy, their Larmor radii are equal:

$$|r_1| = |r_2| = |r_3| = |r| \tag{9.62}$$

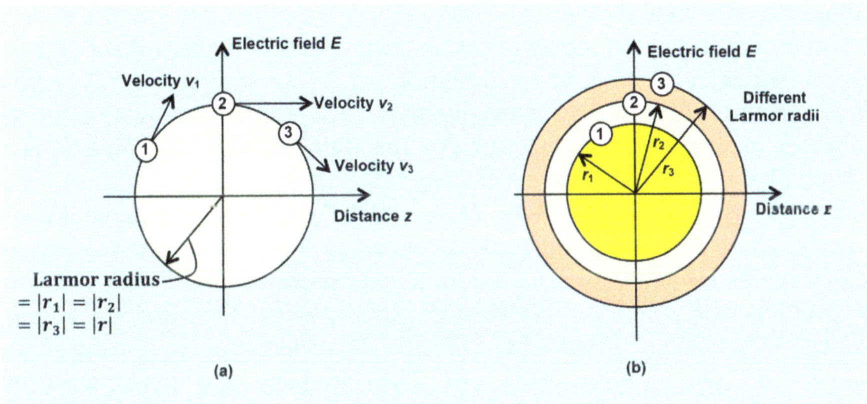

Figure 9.9. Explanation of azimuthal electron bunching mechanism in a gyrotron by three-electron model: (a) starting positions of electrons 1, 2 and 3, and (b) positions of electrons 1, 2, 3 after one period of the cycle of electric field.

The electromagnetic wave in the cavity is associated with a rotating electric field **E**. This electric field is assumed to be spatially uniform. It is also assumed that in accordance with equation (9.27),

$$\begin{aligned}&\text{Angular frequency of rotation of the electric field } (\omega)\\&\geqslant \text{Cyclotron frequency } (\Omega_C)\end{aligned} \tag{9.63}$$

Let the velocity of the electron 1 be $\mathbf{v}_1$, that of electron 2 be $\mathbf{v}_2$ and electron 3 be $\mathbf{v}_3$. If we look at figure 9.9(a), we find that the velocity vectors $\mathbf{v}_1$, $\mathbf{v}_2$, $\mathbf{v}_3$ are oriented at different angles with respect to the electric field **E**.

(i) Velocity $\mathbf{v}_1$ of electron 1 is in the direction of field **E**. For electron 1, $dW/dt<0$; hence this electron suffers a loss of energy, it is decelerated, its effective mass decreases, its Larmor radius decreases and its cyclotron frequency increases.

(ii) Velocity $\mathbf{v}_2$ of electron 2 is perpendicular to the direction of field **E**. For electron 2, $dW/dt = 0$; hence this electron is neither decelerated nor accelerated. Consequently its effective mass, Larmor radius and cyclotron frequency remain unaffected.

(iii) Velocity $\mathbf{v}_3$ of electron 3 is in the opposite direction to field **E**. For electron 3, $dW/dt<0$; hence this electron gains energy. As a result, it is accelerated, its effective mass increases, its Larmor radius increases and its cyclotron frequency decreases.

After one period of the cycle of the field has elapsed, the field again points in the same direction as previously but the positions of electrons 1, 2, 3 have changed because of their interaction with the field during the previous period. The new situation is depicted in figure 9.9(b). Electron 1 has settled down to an orbit of smaller Larmor radius r_1 but has advanced forward because of increase of cyclotron frequency. Electron 2 maintains its relative position with respect to electrons 1, 3. It continues to move in a circular orbit of radius $r_2 = r$. Electron 3 settles down to a larger orbit of radius r_3 but moves backward from its position because of decrease of cyclotron frequency. The necessary condition $\omega \gtrsim \Omega_C$ for stability of energy transfer from the electron beam to the electromagnetic wave ensures that an electron which loses energy exhibits increased Ω_C thereby remaining in the advantageous energy-extraction phase for a longer time with respect to the electromagnetic wave, whereas an electron which gains energy shows decreased Ω_C, therefore decreasing its dwelling time in this phase and preventing removal of energy from the electromagnetic wave. This mechanism leads to accumulation of a large proportion of electrons in similar phase positions with respect to the electromagentic wave. The resultant situation is equivalent to bunching of electrons. Thus the three electrons are bunched together around the positive direction of Y-axis. This bunching is a consequence of the influence of effective mass changes in the relativistic regime on the Larmor radius and cyclotron frequency. The effective mass changes themselves depend on electron energy and velocity. The electron bunching is said to originate from electron cyclotron maser or negative mass instability.

9.8 Dispersion diagram of a gyrotron

The resonator or resonant cavity of the gyrotron located at the maximum value of the static magnetic field contains a section of the cylindrical waveguide of length L and radius r_0. The angular frequency of the electromagnetic wave

$$\omega = 2\pi\nu \tag{9.64}$$

of the transverse electric $\mathrm{TE}_{\mathrm{mnq}}$ mode of this waveguide with mode eigenvalue = χ mn is linked to the axial wave number k_z and transverse wave number $k_\perp$ given by

$$k_z = \frac{q\pi}{L} \ll k_\perp,\ q = \text{an integer};\ k_\perp = \frac{\chi_{mn}}{r_0} = \frac{n^{\text{th}}\ \text{Zero of the derivative of the}\ m^{\text{th}}\ \text{order Bessel function of the first kind}\{J_m'(x)\}}{r_0} \tag{9.65}$$

through the dispersion relation (Sabchevski and Zhelyazkov 2007)

$$k^2 = \frac{\omega^2}{c^2} = k_z^2 + k_\perp^2 \tag{9.66}$$

As we know, the symbol c stands for the velocity of light in vacuum. From equation (9.66)

$$\omega^2 = c^2(k_z^2 + k_\perp^2) \tag{9.67}$$

Equation (9.67) defines the circular waveguide cavity modes. The resonance condition of equation (9.27) can be re-arranged and rewritten as

$$\omega = k_z v_z + s\Omega_C \tag{9.68}$$

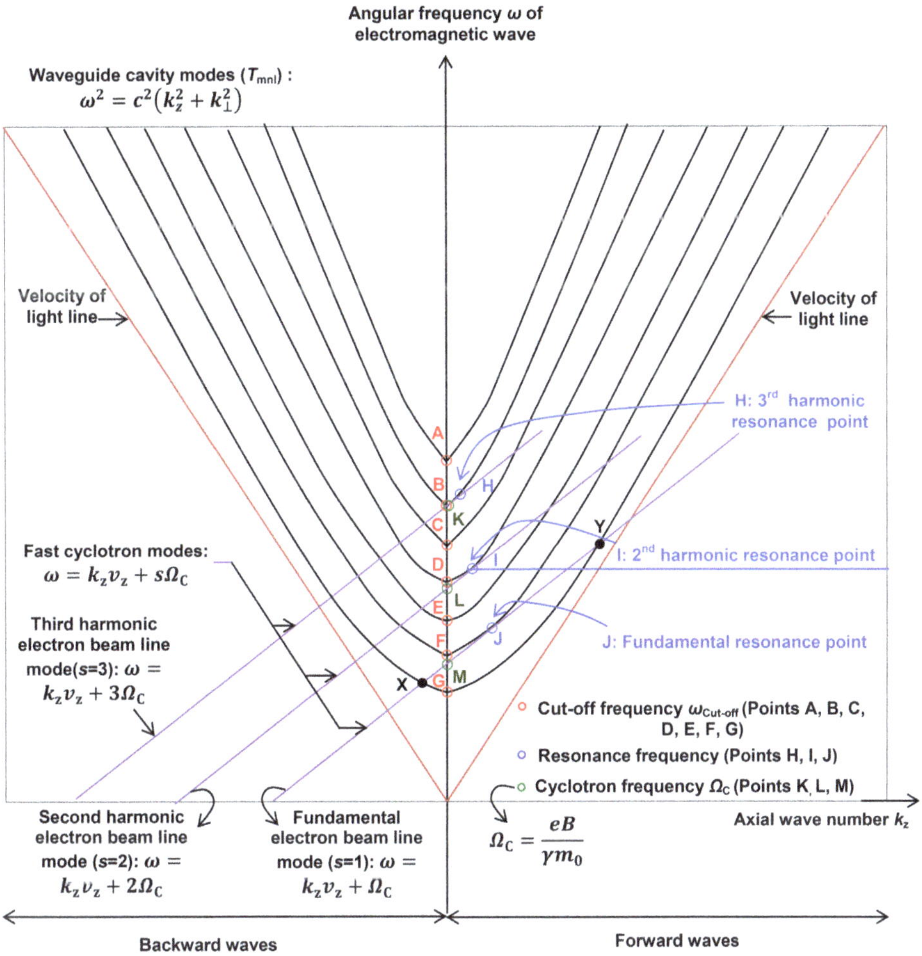

Figure 9.10. Dispersion diagram for cylindrical waveguide cavity gyrotron showing the waveguide cavity modes and the fast cyclotron modes. Cut-off frequency, resonance frequency and cyclotron frequency points are marked on the diagram.

Equation (9.68) defines the cyclotron mode.

The operation of a gyrotron is concisely presented with the help of a diagram obtained by plotting the dispersion relation of the TE waveguide modes in cylindrical waveguide together with the beam wave resonance lines. On the ω–k dispersion diagram for a gyrotron, the waveguide cavity mode hyperbolas and the cyclotron mode straight lines are shown (figure 9.10). The velocity of light lines are also drawn. Points A, B, C, D, E, F, G of intersection of the waveguide modes with the angular frequency ω-axis are marked as the cut-off frequencies. Points H, I, J of intersection of the waveguide with cyclotron modes represent the resonance conditions. Points K, L, M of the intersection of cyclotron modes with the angular frequency ω-axis are the cyclotron frequency Ω_C points.

Focusing attention on points X, J and Y, the interaction points for gyrotron operation, we note that points J and Y represent wave propagation in the forward direction while point X describes backward wave interaction. For points J and Y, $k_z>0$. So from equation (9.68),

$$\omega > s\Omega_C \tag{9.69}$$

while for point X, $k_z<0$. Hence,

$$\omega < s\Omega_C \tag{9.70}$$

Considering point J, the electron beam line is tangential to the hyperbola and the interaction frequency

$$\omega_J \approx \omega_{Cut-off} \tag{9.71}$$

For point Y,

$$\omega_Y > \omega_J \tag{9.72}$$

Even then the mode of interaction is of lower order because it is the fundamental mode.

9.9 Gyrotron research status

Gyrotron FU FIR series is developed (Idehara *et al* 2008a, 2008b, 2009, 2016; Idehara and Sabchevski 2012, 2017) at Research Center for Development of Far Infrared Region, University of Fukui (FIR FU), Japan: FU CW1 provides 2.3 kW at 0.3 THz, FU CW II gives 20–200 W at 0.11–0.44 THz and FU CW III gives 10–220 W at 0.13–1.08 THz, all under continuous wave operation. FU CWI uses a 12 T superconducting magnet, FU CW II uses an 8 T cryo-free magnet while FU CW III uses a 20 T superconducting magnet. Cathode voltage is 30 kV and the gun coil maximum input current is 300 A. Cooling is done by water jackets at the cavity and collector regions.

Pulsed solenoidal coils with magnetic field strength up to 50 T are developed (Glyavin and Luchinin 2009). At a pulse duration of 40 μs, the output power in fundamental mode gyrotron operation is 500 W at 1.3 THz and 5 kW at 1 THz. The electron beam voltage is 30 kV and the beam current is 5 A.

Continuous wave (CW) power output of 100 W is obtained with gyrotrons at frequencies ~0.2–0.5 THz (Glyavin *et al* 2014). In the pulse mode operation with pulse durations of tens of microseconds, the power levels are higher: 200 kW at 0.7 THz and 5 kW at 1 THz.

A coaxial cavity short-pulse gyrotron (pulse length 150 ms) is prototyped using a 6.9 T superconducting magnet (Illy *et al* 2019). At an acceleration voltage of 93 kV and beam current of 80 A, the power output is 2.2 MW at 0.17 THz with an efficiency of 30%. The electron velocity ratio is ~1.3 and the beam radius is 10 mm. The operating mode is $TE_{34, 19}$. The gyrotron activity is being pushed forward targeting a pulse length of 1 s.

9.10 Discussion and conclusions

Gyrotron or electron cyclone maser is a fast-wave vacuum tube for high-power generation (kilowatts of power in terahertz band) working on cyclotron radiation emitted by electrons undergoing rotational motion under the influence of a constant magnetic field of a superconducting magnet. It can operate in pulsed as well as continuous wave regimes. The frequency of a gyrotron oscillator depends on the magnetic field intensity of the superconducting magnet because it is this field intensity which decides the cyclotron resonance frequency. Magnetic field variation provides the largest range of frequency adjustment (magnetic tuning). Frequency tuning is also accomplished by changing the temperature of the cavity (temperature tuning) or magnetron injection gun voltage (voltage tuning).

References

Cronin J L 1981 Modern dispenser cathodes *IEE Proc.* **1** 19–32

Franck J 2017 *Systematic Study of Key Components for a Co-axial Cavity Gyrotron for DEMO* (Karlsruhe: KIT Scientific Publishing, Karsruhe Institute of Technology) p 18

Glyavin M Y and Luchinin A G 2009 Powerful terahertz gyrotrons based on pulsed magnets *Terahertz Sci. Technol.* **2** 150–5

Glyavin M Y, Denisov G G, Zapevalov V E, Kuftin A N, Luchinin A G, Manuilov V N, Morozkin M V, Sedov A S and Chirkov A V 2014 Terahertz gyrotrons: state of the art and prospects *J. Commun. Technol. Electron.* **59** 792–7

Hruszowiec M, Czarczyński W, Pliński E F and Więckowski T 2014 Gyrotron technology *J. Telecommun. Inf. Technol.* **1** 68–76

Idehara T, Ogawa I, Saito T, Mitsudo S, Tatematsu Y, Mori L A H and Kobayashi S 2008a Development of THz gyrotrons and application to high power THz technologies *Terahertz Sci. Technol.* **1** 100–6

Idehara T, Saito T, Ogawa I, Mitsudo S, Tatematsu Y, Mori L A H and Kobayashi S 2008b Development of terahertz FUCW gyrotron series for DNP/NMR *Appl. Magn. Reson.* **34** 265–75

Idehara T, Saito T, Ogawa I, Mitsudo S and Tatematsu Y 2009 THz gyrotrons – FU CW series for high power THz technologies *J. Plasma Fusion Res. Ser.* **8** 1518–21

Idehara T and Sabchevski S P 2012 Development and applications of high-frequency gyrotrons in FIR FU covering the sub-THz to THz range *J. Infrared, Millim. Terahertz Waves* **33** 667–94

Idehara T, Kuleshov A, Khutoryan E, Tatematsu Y, Matsuki Y, Fujiwara T, Asai S, Suehara T, Yamazaki T and Miyazaki A 2016 High power THz technologies opened by high frequency gyrations covering sub-THz to THz region *9th Int. Kharkiv Symp. on Physics and Engineering of Microwaves, Millimeter and Submillimeter Waves (MSMW) (Kharkiv, Ukraine, 21–24 June)* pp 1–3

Idehara T and Sabchevski S P 2017 Gyrotrons for high-power terahertz science and technology at FIR UF *J. Infrared, Millim. Terahertz Waves* **38** 62–86

Idehara T, Ogawa I, Wagner D, Thumm M, Kosuga K and Sabchevski S P 2018 High purity mode CW gyrotron covering the sub-THz to THz range using a 20 T super-conducting magnet *IEEE Trans. Electron Devices* **65** 3486–91

Illy S, Avramidis K A, Gantenbein G, Ioannidis Z, Jin J and Kalaria P C *et al* 2019 Recent status and future prospects of coaxial-cavity gyrotron development at KIT *20th Joint Workshop on*

Electron Cyclotron Emission and Electron Cyclotron Resonance Heating (EC20), Greifswald, Germany, The European Physical Journal (EPJ) Web of Conferences **203** 04005

Jin J 2007 *Quasi-Optical Mode Converter for a Coaxial Cavity Gyrotron* (Karlsruhe: Forschungszentrum Karlsruhe GmbH) pp 1–12

Litvak A G, Alikaev V V, Denisov G G, Kurbatov V I, Myasnikov V E, Tai E M and Zapevalov V E 2001 *Gyrotrons for fusion: status and prospects (IAEA-CSP--8/C)* International Atomic Energy Agency (IAEA) 5 pp

Pagonakis I, Alberti S, Avramidis K, Legrand F, Gantenbein G, Genoud J, Hogge J-P, Illy S, Ioannidis Z and Kalaria P *et al* 2019 Overview on recent progress in magnetron injection gun theory and design for high power gyrotrons *20th Joint Workshop on Electron Cyclotron Emission and Electron Cyclotron Resonance Heating (EC-20), Greifswald, Germany, May 14–17, 2018, The European Physical Journal (EPJ) Web of Conferences* **203** 04011

Pradhan S, Raj P, Prasad U, Ghate M, Khristi Y and Panchal A *et al* 2017 Design, development and functional validation of magnets system in support of 42 GHz gyrotron in India *19th Joint workshop (EC-19) on Electron Cyclotron Emission (ECE) and Electron Cyclotron Resonance Heating (ECRH), April 04–7, 2016, Institute for Plasma Research, Gandhinagar, India, The European Physical Journal (EPJ) Web of Conferences* **147** 04005

Sabchevski S and Zhelyazkov I 2007 Modelling and simulation of new generation powerful gyrotrons for the fusion research *J. Phys. Conf. Ser.* **63** 1–11

Sabri N G 2012 Production of high power by using gyrotron device for electron cyclotron resonance heating in tokamak reactor *Energy Procedia* **18** 944–53

Schmid M, Erckmann V, Gantenbein G, Illy S, Kern S, Liévin C, Samartsev A, Schlaich A, Rzesnicki T and Thumm M 2011 Technical developments at the KIT gyrotron test facility *Fusion Eng. Des.* **86** 518–21

Tallerico P 1995 Ch.4: gyrotrons, magnicons and ubitrons *Handbook of Microwave Technology Vol. 2: Applications* ed T K Ishii (San Diego, CA: Academic) pp 97–119

IOP Publishing

Practical Terahertz Electronics: Devices and Applications, Volume 1
Solid-state devices and vacuum tubes
Vinod Kumar Khanna

Chapter 10

Free electron lasers

Differences of a free electron laser (FEL) from a conventional laser are pointed out. In place of bound atomic or molecular states, the lasing medium in a free electron laser is a relativistic electron beam propagating through a periodic magnetic field; hence it is called by this name. The principal components of a free electron laser are described, viz., the electron injector, the RF linear accelerator, and the undulator. Derivation of the equation of motion of the electron in the undulator is presented. The frequency of terahertz waves emitted by a free electron laser depends on the electron energy, the undulator period and weakly on the strength of the magnetic field. The operating modes of free electron laser such as self-amplified-spontaneous-emission FEL, oscillator and amplifier are described. Endowed with the ability to work over any segment of the electromagnetic spectrum, the free electron lasers can fill any spectral void in which conventional lasers are inoperative.

Although a free electron laser provides a beam of electromagnetic radiation of similar optical properties to a conventional laser, it works on entirely different operating principles.

10.1 Free electron laser versus conventional laser

Free electron laser is a special type of laser which is capable of providing high powers >1 GW with short pulse duration <1 ps. It has a tunable wavelength in a broad range with narrow bandwidth and high transverse and longitudinal coherence (Dattoli *et al* 2012, 2018). Table 10.1 brings out its main features of demarcation from a traditional laser.

10.2 Main components of a free electron laser

10.2.1 Electron injector

To provide high peak current pulses ~100–200 A, electron bunches of low transverse emittance <1 mm-mrad are produced by shining laser beams on the metal or

Table 10.1. Differences between conventional and free electron lasers.

Sl. No.	Feature	Conventional laser	Free electron laser
1.	Electrical nature	Not all-electric laser; uses chemicals	All-electric platform
2.	Gain or lasing medium	Direct bandgap semiconductors (GaAs, GaN); crystals (Nd: YAG, Ti:sapphire); glasses (silicate or phosphate); liquid solutions (dyes); gases (CO_2, HeNe, Ar)	Free electron beam accelerated to relativistic velocity
3.	Condition of electrons in the gain medium	Bound to atoms or molecules	At liberty
4.	Phenomenon/ parameters determining the wavelength or frequency	Electron transitions between quantum states having fixed energy levels	Energy of electron beam, undulator period and magnetic field
5.	Nature of energy transitions	Discrete	Occur in a continuum
6.	Wavelength alteration possibility	Little scope; limited by specificity of electron transitions	Easily tunable continuously from microwave to x-ray wavelengths
7.	Essentiality of resonator cavity	Mandatory	Optional; not required in single-pass FEL; beneficial aspect for VUV and x-ray lasers due to lack of mirrors with adequate reflectivity
8.	Reaching x-ray range	Not possible	Achieved
9.	Ultra-short pulses	Not possible at all wavelengths because mode locking is achievable only in lasers with a broad emission spectrum	Possible at all wavelengths with durations of pico or femtoseconds; provides a variety of pulse formats
10.	Power capability	Low to medium power. Limited by heating and breakdown of lasing medium. Thermal energy is removed at acoustic velocities ~ 10^3 m s^{-1}	High peak and average power; no fundamental limit; thermal energy is carried away at the velocity of light (3×10^8 m s^{-1})
11.	Similarity to a vacuum electronic device like TWT, klystron or gyrotron	No	Yes; it is a short-wavelength analogue of vacuum electronic devices

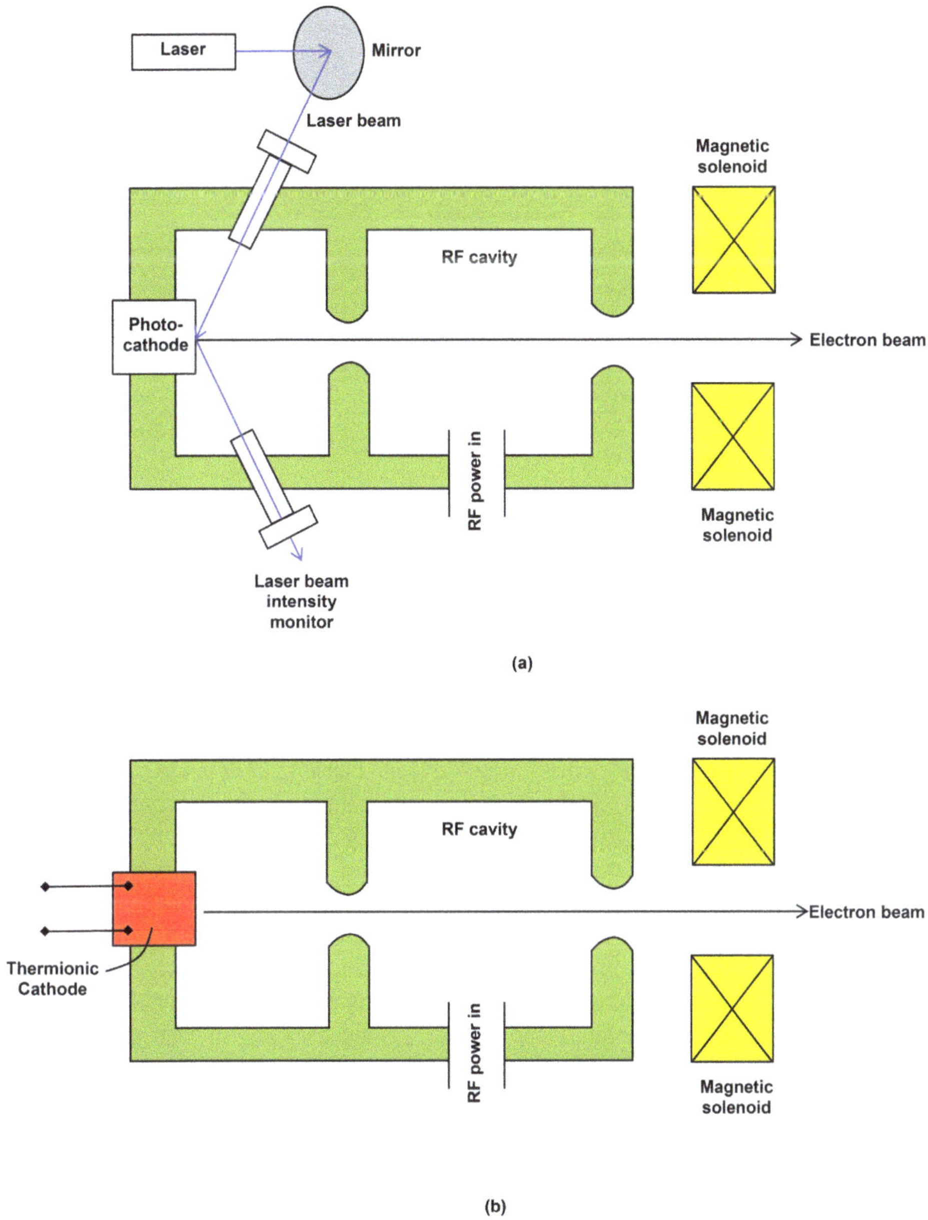

Figure 10.1. RF electron guns: (a) photocathode gun and (b) thermionic gun.

semiconductor photocathode of an RF photocathode gun (figure 10.1(a)). As these guns have short operational lifetimes, thermionic guns are sometimes used (figure 10.1(b)). Besides longer lifetimes, thermionic guns are also cost-effective. RF guns accelerate electrons to very high energies in MeV range over extremely short distances of a few cms. Hence they are preferred over DC guns.

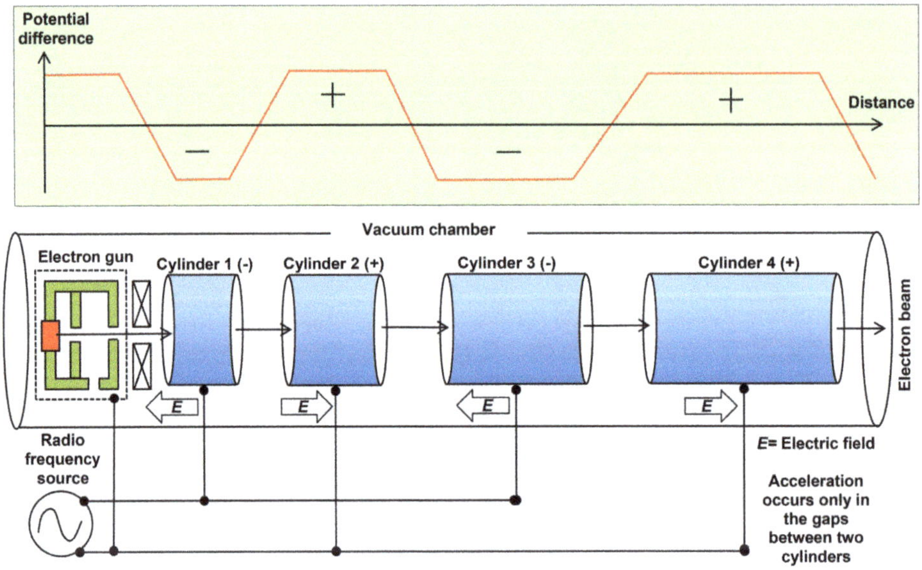

Figure 10.2. Layout of RF linear accelerator.

10.2.2 RF linear accelerator (LINAC)

The accelerator consists of a vacuum chamber in the shape of a hollow pipe (figure 10.2). Extending along the length of the pipe are a series of hollow open-ended cylinders of successively increasing lengths. These cylinders act as the electrodes. The electrons travel along straight line paths through this arrangement under the action of RF potentials produced by an oscillator circuit. After amplification, the potentials are applied to the cylindrical electrodes in a sequence which ensures that at any instant, the potentials applied to any two neighboring electrodes have opposite polarities. Hence the electric field in the gap between the electrodes is of the correct polarity to impart energy to the electrons reaching the gap. In this way, the velocities of electrons are consecutively increased by RF fields oscillating in synchronism with the motion of electrons through high-field regions existing in the gaps between metallic electrodes. The RF fields used have a frequency range of 100 MHz to GHz.

10.2.3 Undulator

Originating from the Latin word 'unda' for wave, undulator means 'wave maker' (Pflueger 2018). The undulator consists of an array of magnets of alternating polarity (figure 10.3). Hence it is a periodic array and the periodicity is spatial only. The magnetic field is constant in time. The magnetic field acts in a direction transverse to the direction of electron beam causing its bending. With reference to the co-ordinate system in figure 10.3, the static magnetic field is parallel to Y-axis direction and is periodic in the direction of Z-axis along which the electron beam propagates. The distance between the center of north pole (or south pole) of a magnet to the center of north pole (or south pole) of the next magnet is called the undulator wavelength λ_u. It is typically a few centimeters in length.

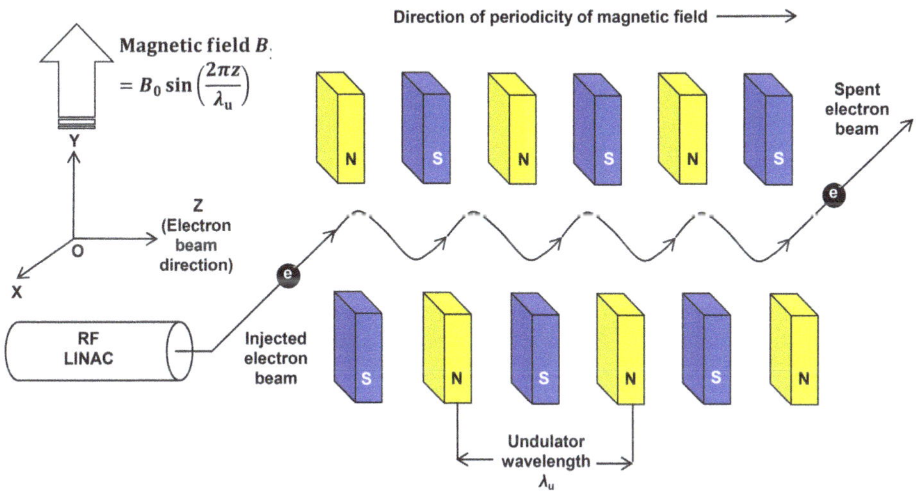

Figure 10.3. The undulator, the co-ordinate system, the electron beam direction, the magnetic field direction and the direction of periodicity of magnetic field of the undulator.

10.3 Equation of motion of the electron in the undulator

Properties of the electromagnetic radiation emitted by electrons moving in the undulator are understood by mathematically analyzing the motion of the electrons in the spatially periodic magnetic field of the undulator (Walker 1996). The analysis is carried out with reference to figure 10.4.

10.3.1 Force acting on the electron

Let e = electron charge, $\mathbf{F}$ = force acting on the electron in the presence of electric field $\mathbf{E}$ and magnetic field strength $\mathbf{B}$, $\mathbf{p}$ = momentum of the electron = electron mass $m \times$ velocity $\mathbf{v}$, c = velocity of light. Then

$$\mathbf{F} = \frac{\mathrm{d}\mathbf{p}}{\mathrm{d}t} = -e(\mathbf{E} + \mathbf{v} \times \mathbf{B}) \tag{10.1}$$

where

$$\mathbf{p} = \gamma m \mathbf{v} \tag{10.2}$$

and the Lorentz factor

$$\gamma = \frac{1}{\sqrt{1 - \frac{v^2}{c^2}}} \tag{10.3}$$

Since the radiated field is weak $\mathbf{E}$~0; hence equation (10.1) reduces to

$$\frac{\mathrm{d}\mathbf{p}}{\mathrm{d}t} \approx -e(\mathbf{v} \times \mathbf{B}) \tag{10.4}$$

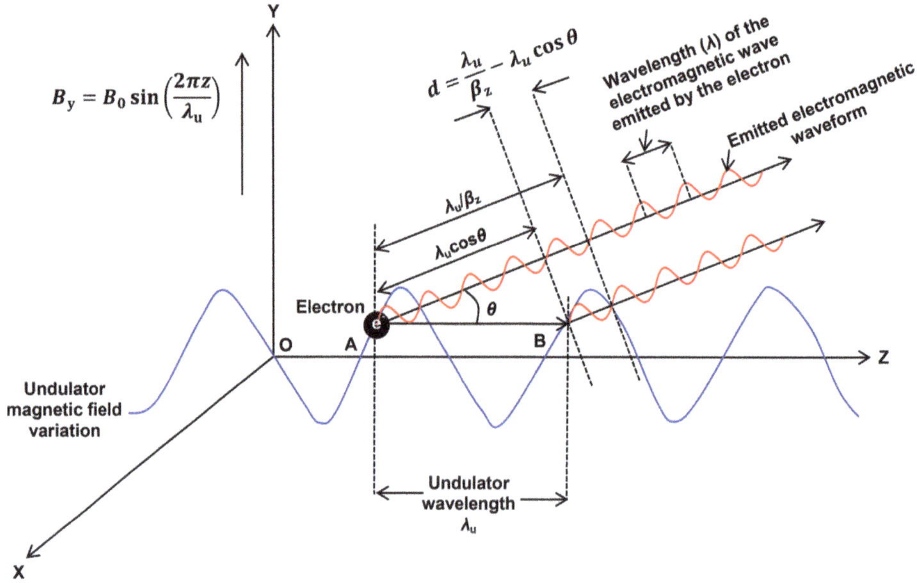

Figure 10.4. Illustration of electron motion in the undulator, electromagnetic wave emission and difference between distances traversed by slow moving electron and the faster electromagnetic wave.

The electron beam advances in the *Z*-direction and simultaneously undergoes periodical deflection in the *X*-direction. So the electron beam has velocity components in both the directions.

10.3.2 Electron velocity component in *x*-direction

If m = mass of the electron, v_x = component of electron velocity in the *X*-direction and v_z = component of electron velocity in the *Z*-direction, B_y = the applied magnetic field in the *Y*-direction, equations (10.2) and (10.4) give

$$\frac{\mathrm{d}(\gamma m \mathrm{v}_x)}{\mathrm{dt}} = -e(\mathbf{v} \times \mathbf{B}) \tag{10.5}$$

or,

$$\gamma m \frac{d\mathrm{v}_x}{dt} = -e\left(\mathrm{v}_z B_y \sin 90^\circ\right) = -e\mathrm{v}_z B_y \times 1 = -e\mathrm{v}_z B_y \tag{10.6}$$

Assuming a sinusoidal magnetic field B_y varying with wavelength λ_u (the undulator wavelength) (Huang and Kim 2007), we can write

$$B_y = B_0 \sin\left(\frac{2\pi z}{\lambda_u}\right) \tag{10.7}$$

where z is the distance in the Z-direction. The magnetic field B_y is oriented in the vertical Y-direction. It varies sinusoidally in the Z-direction. B_0 is the magnetic pole field.

Substituting for B_y from equation (10.7) in equation (10.6)

$$\gamma m\frac{d\mathrm{v}_x}{dt} = -e\mathrm{v}_z B_0 \sin\left(\frac{2\pi z}{\lambda_\mathrm{u}}\right) \tag{10.8}$$

or,

$$\gamma m\frac{d\mathrm{v}_x}{dt} = -e\frac{dz}{dt}B_0 \sin\left(\frac{2\pi z}{\lambda_\mathrm{u}}\right) \tag{10.9}$$

by writing the velocity as a time derivative

$$\mathrm{v}_z = \frac{dz}{dt} \tag{10.10}$$

Equation (10.9) becomes

$$\gamma m d\mathrm{v}_x = -edzB_0 \sin\left(\frac{2\pi z}{\lambda_\mathrm{u}}\right) \tag{10.11}$$

or,

$$\gamma m d\mathrm{v}_x = -eB_0 \sin\left(\frac{2\pi z}{\lambda_\mathrm{u}}\right)dz \tag{10.12}$$

Let us put

$$\frac{2\pi z}{\lambda_\mathrm{u}} = \xi \tag{10.13}$$

Then

$$\gamma m d\mathrm{v}_x = -eB_0 \sin\xi dz \tag{10.14}$$

Also,

$$d\left(\frac{2\pi z}{\lambda_\mathrm{u}}\right) = d\xi \tag{10.15}$$

or,

$$\frac{2\pi}{\lambda_\mathrm{u}}dz = d\xi \tag{10.16}$$

or,

$$dz = \frac{\lambda_\mathrm{u}}{2\pi}d\xi \tag{10.17}$$

Substituting for dz from equation (10.17) into equation (10.12) and applying equation (10.13), we get

$$\gamma m d\mathrm{v}_x = -eB_0 \sin\xi \frac{\lambda_\mathrm{u}}{2\pi} d\xi = -eB_0 \frac{\lambda_\mathrm{u}}{2\pi} \sin\xi d\xi \tag{10.18}$$

Integrating both sides,

$$\int \gamma m d\mathrm{v}_x = \int -eB_0 \frac{\lambda_\mathrm{u}}{2\pi} \sin\xi d\xi \tag{10.19}$$

or,

$$\gamma m \int d\mathrm{v}_x = -\frac{eB_0\lambda_\mathrm{u}}{2\pi} \int \sin\xi d\xi \tag{10.20}$$

or,

$$\gamma m \mathrm{v}_x = -\frac{eB_0\lambda_\mathrm{u}}{2\pi} \times -\cos\xi = \frac{eB_0\lambda_\mathrm{u}}{2\pi} \cos\left(\frac{2\pi z}{\lambda_\mathrm{u}}\right) \tag{10.21}$$

or,

$$\mathrm{v}_x = \frac{eB_0\lambda_\mathrm{u}}{2\pi\gamma m} \cos\left(\frac{2\pi z}{\lambda_\mathrm{u}}\right) \tag{10.22}$$

Let us define

$$\frac{eB_0\lambda_\mathrm{u}}{2\pi mc} \equiv K \tag{10.23}$$

where K is a dimensionless number. It is known as the undulator parameter.

Then

$$\frac{eB_0\lambda_\mathrm{u}}{2\pi\gamma m} = \frac{eB_0\lambda_\mathrm{u}}{2\pi mc} \times \frac{c}{\gamma} = K \times \frac{c}{\gamma} = \frac{Kc}{\gamma} \tag{10.24}$$

Combining equations (10.22) and (10.24),

$$\mathrm{v}_x = \frac{Kc}{\gamma} \cos\left(\frac{2\pi z}{\lambda_\mathrm{u}}\right) \tag{10.25}$$

10.3.3 Electron velocity component in the *z*-direction

Since

$$v^2 = \mathrm{v}_\mathrm{x}^2 + \mathrm{v}_\mathrm{z}^2 \tag{10.26}$$

we can write equation (10.3) as

$$\gamma = \frac{1}{\sqrt{1 - \dfrac{v^2}{c^2}}} = \frac{1}{\sqrt{1 - \dfrac{\mathrm{v}_\mathrm{x}^2 + \mathrm{v}_\mathrm{z}^2}{c^2}}} \tag{10.27}$$

Squaring both sides

$$\gamma^2 = \frac{1}{1 - \dfrac{\mathrm{v}_\mathrm{x}^2 + \mathrm{v}_\mathrm{z}^2}{c^2}} \tag{10.28}$$

or,

$$\frac{1}{\gamma^2} = 1 - \frac{\mathrm{v}_\mathrm{x}^2 + \mathrm{v}_\mathrm{z}^2}{c^2} = 1 - \frac{\mathrm{v}_\mathrm{x}^2}{c^2} - \frac{\mathrm{v}_\mathrm{z}^2}{c^2} \tag{10.29}$$

or,

$$\frac{\mathrm{v}_\mathrm{z}^2}{c^2} = 1 - \frac{1}{\gamma^2} - \frac{\mathrm{v}_\mathrm{x}^2}{c^2} \tag{10.30}$$

From equations (10.25) and (10.30)

$$\frac{\mathrm{v}_\mathrm{z}^2}{c^2} = 1 - \frac{1}{\gamma^2} - \frac{\left\{\dfrac{Kc}{\gamma}\cos\left(\dfrac{2\pi z}{\lambda_\mathrm{u}}\right)\right\}^2}{c^2} = 1 - \frac{1}{\gamma^2} - \frac{K^2}{\gamma^2}\cos^2\left(\frac{2\pi z}{\lambda_\mathrm{u}}\right) \tag{10.31}$$

or,

$$\frac{\mathrm{v}_\mathrm{z}^2}{c^2} = 1 - \frac{1}{\gamma^2} - \frac{1}{2}\frac{K^2}{\gamma^2}\left\{2\cos^2\left(\frac{2\pi z}{\lambda_\mathrm{u}}\right)\right\} = 1 - \frac{1}{\gamma^2} - \frac{1}{2}\frac{K^2}{\gamma^2}\left\{1 + \cos 2\left(\frac{2\pi z}{\lambda_\mathrm{u}}\right)\right\} \tag{10.32}$$

by applying the formula

$$\cos 2A = 2\cos^2 A - 1 \tag{10.33}$$

Equation (10.32) is rewritten as

$$\frac{\mathrm{v}_\mathrm{z}^2}{c^2} = 1 - \frac{1}{\gamma^2} - \frac{1}{2}\frac{K^2}{\gamma^2}\left\{1 + \cos\left(\frac{4\pi z}{\lambda_\mathrm{u}}\right)\right\} = 1 - \left[\frac{1}{\gamma^2} + \frac{1}{2}\frac{K^2}{\gamma^2}\left\{1 + \cos\left(\frac{4\pi z}{\lambda_\mathrm{u}}\right)\right\}\right] \tag{10.34}$$

from which

$$\begin{aligned}\frac{\mathrm{v}_z}{c} &= \sqrt{1 - \left[\frac{1}{\gamma^2} + \frac{1}{2}\frac{K^2}{\gamma^2}\left\{1 + \cos\left(\frac{4\pi z}{\lambda_\mathrm{u}}\right)\right\}\right]} \\ &= 1 - \frac{1}{2}\left[\frac{1}{\gamma^2} + \frac{1}{2}\frac{K^2}{\gamma^2}\left\{1 + \cos\left(\frac{4\pi z}{\lambda_\mathrm{u}}\right)\right\}\right]\end{aligned} \tag{10.35}$$

since

$$(1-x)^{\frac{1}{2}} \approx 1-\frac{1}{2}x \tag{10.36}$$

Equation (10.35) gives

$$\frac{\mathrm{v}_z}{c} = 1-\frac{1}{2\gamma^2}-\frac{1}{4}\frac{K^2}{\gamma^2}-\frac{1}{4}\frac{K^2}{\gamma^2}\cos\left(\frac{4\pi z}{\lambda_\mathrm{u}}\right) \tag{10.37}$$

Hence the relative average velocity in the forward direction is obtained by averaging the velocity component in the Z-direction over a complete cycle or over N full cycles

$$\beta_z = \frac{\bar{\mathrm{v}}_z}{c} \approx 1-\frac{1}{2\gamma^2}-\frac{1}{4}\frac{K^2}{\gamma^2} \tag{10.38}$$

The average value of cosine of an angle over a complete cycle is zero since the two halves of the cosine curve cancel each other out. Hence the last term in equation (10.37) averages to zero. This is shown below for N cycles ($0 \leqslant z \leqslant N\lambda_\mathrm{u}$):

$$\begin{aligned}
\text{Average value of } \cos\left(\frac{4\pi z}{\lambda_\mathrm{u}}\right) &= \left\langle \cos\left(\frac{4\pi z}{\lambda_\mathrm{u}}\right)\right\rangle \\
&= \frac{1}{N\lambda_\mathrm{u}-0}\int_0^{N\lambda_\mathrm{u}} \cos\left(\frac{4\pi z}{\lambda_\mathrm{u}}\right)dz \\
&= \frac{1}{N\lambda_\mathrm{u}}\int_0^{N\lambda_\mathrm{u}} \cos\left(\frac{4\pi z}{\lambda_\mathrm{u}}\right)dz = \frac{1}{N\lambda_\mathrm{u}}\left|\frac{\sin\left(\frac{4\pi z}{\lambda_\mathrm{u}}\right)}{\frac{4\pi}{\lambda_\mathrm{u}}}\right|_0^{N\lambda_\mathrm{u}} \\
&= \frac{1}{N\lambda_\mathrm{u}} \times \frac{\lambda_\mathrm{u}}{4\pi} \times \left|\sin\left(\frac{4\pi z}{\lambda_\mathrm{u}}\right)\right|_0^{N\lambda_\mathrm{u}} \\
&= \frac{1}{4\pi N}\left(\sin\frac{4\pi N\lambda_\mathrm{u}}{\lambda_\mathrm{u}} - \sin\frac{4\pi \times 0}{\lambda_\mathrm{u}}\right) \\
&= \frac{1}{4\pi N}(\sin 4\pi N - \sin 0) = \frac{1}{4\pi N}(0-0) = 0
\end{aligned} \tag{10.39}$$

10.3.4 Times taken by the wave and the electron to cover distance between two points

The electrons travelling through the periodic magnet array of the undulator undergo transverse oscillations. The resulting accelerations experienced by the electrons cause them to emit electromagnetic waves. The electrons and the electromagnetic waves produced by their oscillations co-propagate and interact with each other. By this interaction, the electrons are accelerated and decelerated. Electromagnetic waves are emitted and absorbed by them. The working mechanism of free electron laser is the resonant interaction between the electron beam and the co-propagating electromagentic radiation in the undulator.

The electromagnetic waves and the electrons differ in their velocities of propagation. Let us calculate the time durations taken by the wave and the electron to traverse a fixed distance. We shall calculate the time taken by the wave to traverse a distance from a point A to an equivalent point B. This distance corresponds to one undulator wavelength. The time required by the wave is

$$\Delta t' = \frac{\text{Distance travelled}}{\text{Velocity}} = \frac{\text{Undulator wavelength}}{\text{Electromagnetic wave Velocity}} = \frac{\lambda_u}{c} \tag{10.40}$$

The electron moves slower than the wave with velocity v_z. Therefore, time taken by the electron to traverse the same distance is longer:

$$\Delta t = \frac{\text{Distance travelled}}{\text{Velocity}} = \frac{\text{Undulator wavelength}}{\text{Electron Velocity}} = \frac{\lambda_u}{\bar{v}_z} = \frac{\lambda_u}{c\frac{\bar{v}_z}{c}} = \frac{\lambda_u}{c\beta_z} \tag{10.41}$$

using equation (10.38).

During this time, the wavefront will move from A through a distance

$$s = \Delta t \times \text{Wave velocity} = \Delta t \times \mathrm{c} = \frac{\lambda_u}{c\beta_z} \times c = \frac{\lambda_u}{\beta_z} \tag{10.42}$$

It will be ahead of the wavefront emitted at the point B by a distance d given by

$$d = \frac{\lambda_u}{\beta_z} - \lambda_u \cos\theta \tag{10.43}$$

where θ is the angle of emission of the wavefront with respect to the axis of the electron beam.

10.3.5 Condition for interference of wavefronts emitted by the same electron

Fractions in equations (10.40) and (10.41) differ in the velocities in the denominators. In order that constructive interference takes place between the wavefronts emitted by the same electron at different points in the magnet, the condition that must be satisfied is that the electron must slip back by an integral number of wavelengths over the duration of one period, i.e.,

$$d = n\lambda \tag{10.44}$$

where λ is the wavelength of the wave to be distinguished from the undulator wavelength λ_u and n is the harmonic number.

Combining equations (10.43) and (10.44), we have

$$n\lambda = \frac{\lambda_u}{\beta_z} - \lambda_u \cos\theta \tag{10.45}$$

10.3.6 The undulator equation

Substituting for β_z from equation (10.38) in equation (10.45),

$$n\lambda = \frac{\lambda_u}{1 - \frac{1}{2\gamma^2} - \frac{1}{4}\frac{K^2}{\gamma^2}} - \lambda_u \cos\theta = \frac{\lambda_u}{\left\{1 - \left(\frac{1}{2\gamma^2} + \frac{1}{4}\frac{K^2}{\gamma^2}\right)\right\}} - \lambda_u \cos\theta$$
$$= \lambda_u\left\{1 - \left(\frac{1}{2\gamma^2} + \frac{1}{4}\frac{K^2}{\gamma^2}\right)\right\}^{-1} - \lambda_u \cos\theta = \lambda_u\left\{1 + \left(\frac{1}{2\gamma^2} + \frac{1}{4}\frac{K^2}{\gamma^2}\right)\right\} - \lambda_u \cos\theta \tag{10.46}$$

since

$$(1 - x)^{-1} \approx 1 + x \tag{10.47}$$

Further,

$$1 - \cos\theta = 2\sin^2\left(\frac{\theta}{2}\right) \tag{10.48}$$

or,

$$\cos\theta = 1 - 2\sin^2\left(\frac{\theta}{2}\right) \tag{10.49}$$

From equations (10.46) and (10.49),

$$n\lambda = \lambda_u\left\{1 + \left(\frac{1}{2\gamma^2} + \frac{1}{4}\frac{K^2}{\gamma^2}\right)\right\} - \lambda_u\left\{1 - 2\sin^2\left(\frac{\theta}{2}\right)\right\}$$
$$= \lambda_u + \lambda_u\left(\frac{1}{2\gamma^2} + \frac{1}{4}\frac{K^2}{\gamma^2}\right) - \lambda_u$$
$$+ 2\lambda_u \sin^2\left(\frac{\theta}{2}\right) = \lambda_u\left(\frac{1}{2\gamma^2} + \frac{1}{4}\frac{K^2}{\gamma^2}\right) + 2\lambda_u \sin^2\left(\frac{\theta}{2}\right) \tag{10.50}$$
$$= \lambda_u\left\{\left(\frac{1}{2\gamma^2} + \frac{1}{4}\frac{K^2}{\gamma^2}\right) + 2\sin^2\left(\frac{\theta}{2}\right)\right\}$$

In the approximation of assumption of a small angle θ

$$\sin\left(\frac{\theta}{2}\right) = \frac{\theta}{2} \tag{10.51}$$

$$\therefore \sin^2\left(\frac{\theta}{2}\right) = \frac{\theta^2}{4} \tag{10.52}$$

Equations (10.50) and (10.52) lead to

$$\begin{aligned} n\lambda &= \lambda_u\left\{\left(\frac{1}{2\gamma^2} + \frac{1}{4}\frac{K^2}{\gamma^2}\right) + 2 \times \frac{\theta^2}{4}\right\} = \lambda_u\left\{\left(\frac{1}{2\gamma^2} + \frac{1}{4}\frac{K^2}{\gamma^2}\right) + \frac{\theta^2}{2}\right\} \\ &= \lambda_u\left(\frac{1}{2\gamma^2} + \frac{1}{4}\frac{K^2}{\gamma^2} + \frac{\theta^2}{2}\right) = \frac{\lambda_u}{2\gamma^2}\left(1 + \frac{K^2}{2} + \theta^2\gamma^2\right) \end{aligned} \tag{10.53}$$

which is the undulator equation. It describes the production of electromagnetic radiation by a relativistic electron beam traversing a periodic magnet structure. For the first harmonic $n = 1$. The equation is written in approximate form as

$$\lambda \approx \frac{\lambda_u}{2\gamma^2}\left(1 + \frac{K^2}{2}\right) \tag{10.54}$$

The equation tells us that the wavelength and therefore the frequency of the emitted radiation is determined by three parameters: the undulator wavelength λ_u (depending on the geometry), the Lorentz factor γ (electron velocity and hence electron energy) and the undulator parameter K (and hence the magnetic field).

Example 10.1

(a) Calculate the undulator parameter K per unit peak magnetic field intensity B_0 (T) per unit undulator spatial period λ_u (m); T here means 'Tesla'. (b) In the undulator of a free electron laser, the amplitude of magnetic field B_0 is 0.5 T. If the undulator spatial period λ_u is 20 mm, find the undulator constant K.

Solution: (a) From equation (10.23)

$$\begin{aligned} K &= \frac{eB_0\lambda_u}{2\pi m_0 c} = \frac{e}{2\pi m_0 c}B_0\lambda_u = \frac{1.602 \times 10^{-19}}{2 \times 3.14159 \times 9.109 \times 10^{-31} \times 2.9979 \times 10^{8}}B_0\lambda_u \\ &= 0.0093367 \times 10^4 B_0\lambda_u = 93.37B_0(\mathrm{T})\lambda_u(\mathrm{m}) \end{aligned} \tag{10.55}$$

Hence,

$$K = 93.37\ \mathrm{T}^{-1}\,\mathrm{m}^{-1} \tag{10.56}$$

(b) From equation (10.55), for $B_0 = 0.5$ T and $\lambda_u = 20$ mm $= 20 \times 10^{-3}$ m $= 2 \times 10^{-2}$ m, $K = 93.37 \times 0.5 \times 2 \times 10^{-2} = 0.9337$.

Example 10.2

The undulator of a free electron laser employs a magnetic field of amplitude $B_0 = 0.5$ T. The undulator period λ_u is 2.1 cm. If the Lorentz factor γ is 17, find the frequency of emitted electromagnetic radiation. Also calculate the velocity and energy of electron beam.

Solution: For $B_0 = 0.5$ T and $\lambda_u = 2.1$ cm $= 2.1 \times 10^{-2}$ m

Applying equation (10.55)

$$K = 93.37B_0 \times 0.5 \times 2.1 \times 10^{-2} = 0.98 \tag{10.57}$$

By equation (10.54),

$$\lambda \approx \frac{\lambda_u}{2\gamma^2}\left(1 + \frac{K^2}{2}\right) = \frac{2.1 \times 10^{-2}}{2(17)^2}\left\{1 + \frac{(0.98)^2}{2}\right\} = 3.633 \times 10^{-5} \times 1.4802$$
$$= 5.3776 \times 10^{-5}\ \mathrm{m} = 5.3776 \times 10^{-5} \times 10^{6}\ \mu\mathrm{m} = 53.776\ \mu\mathrm{m} \quad (10.58)$$

The frequency of radiation emitted is

$$f = \frac{c}{\lambda} = \frac{2.9979 \times 10^{8}}{5.3776 \times 10^{-5}} = 0.5575 \times 10^{13}\ \mathrm{Hz} = 5.575 \times 10^{12}\ \mathrm{Hz}$$
$$= 5.58\ \mathrm{THz} \quad (10.59)$$

From equation (10.3)

$$\frac{1}{\sqrt{1 - \left(\frac{v}{c}\right)^2}} = 17 \quad (10.60)$$

or,

$$\frac{1}{1 - \left(\frac{v}{c}\right)^2} = (17)^2 = 289 \quad (10.61)$$

or,

$$289 - 289\left(\frac{v}{c}\right)^2 = 1 \quad (10.62)$$

or,

$$289\left(\frac{v}{c}\right)^2 = 288 \quad (10.63)$$

$$\therefore \frac{v}{c} = \sqrt{\frac{288}{289}} = 0.9982684 \quad (10.64)$$

or,

$$v = 0.9982684c \quad (10.65)$$

and applying the formula for relativistic energy

$$E = \gamma m_0 c^2 = 17 \times 9.109 \times 10^{-31} \times (2.9979 \times 10^{8})^2\ \mathrm{J} = 1.3917 \times 10^{-12}\ \mathrm{J}$$
$$= \frac{1.3917 \times 10^{-12}}{1.602 \times 10^{-19}}\mathrm{eV} = 0.8687 \times 10^{7}\ \mathrm{eV} = 8.687 \times 10^{6}\ \mathrm{eV} = 8.687\ \mathrm{MeV} \quad (10.66)$$

Example 10.3

The velocity of electrons in a linear particle accelerator is 0.9999995c. The undulator spatial period of free electron laser built with this accelerator is 50 mm. Find the: (a)

Lorentz factor and (b) the energy of the electron in MeV. (c) Argue using the Lorentz contraction and relativistic longitudinal Doppler effect to calculate: the wavelength of radiation and linear frequency of radiation. Looking at the electromagnetic spectrum, find out the portion of the spectrum in which the calculated wavelength/frequency is located? (d) What electron velocity and energy of electron are required to produce a radiation of 9 THz for a free electron laser of undulator spatial period 25 mm?
Solution: (a) From equation (10.3), the Lorentz factor is

$$\gamma = \left(1 - \frac{v^2}{c^2}\right)^{-\frac{1}{2}} = \left\{1 - \left(\frac{v}{c}\right)^2\right\}^{-\frac{1}{2}} = \left\{1 - (0.9999995)^2\right\}^{-\frac{1}{2}} = 1000 \quad (10.67)$$

(b) The energy of the electron is

$$\begin{aligned} E = \gamma m_0 c^2 &= 1000 \times 9.109 \times 10^{-31} \times \left(2.9979 \times 10^8\right)^2 \text{ J} = 81.866 \times 10^{-12} \text{ J} \\ &= 81.866 \times 10^{-12} \times 6.242 \times 10^{18} \text{ eV} = 511.0076 \times 10^6 \text{ eV} \approx 511 \text{ MeV} \end{aligned} \quad (10.68)$$

(c) Let λ_u be the period of the magnet arrangement in the undulator. According to the theory of relativity, the wavelength of the undulator measured by an observer in a co-ordinate system moving with the same average velocity as the electron beam, i.e., the moving electron's frame of reference will appear to be shortened with respect to the length measured by a stationary observer. This shortening is known as Lorentz–FitzGerald contraction or Lorentz contraction. The wavelength after length contraction is expressed by Lorentz transformation as

$$\lambda_u^* = \frac{\lambda_u}{\gamma} \quad (10.69)$$

The angular frequency of electron oscillation will be correspondingly higher and given by

$$\omega^* = \frac{2\pi c}{\lambda_u^*} = \frac{2\pi c}{\frac{\lambda_u}{\gamma}} = \gamma \frac{2\pi c}{\lambda_u} \quad (10.70)$$

using equation (10.69).

The oscillating electrons emit radiation. By relativistic longitudinal Doppler effect, the wavelength of radiation is altered from λ_u^* in a moving frame of reference to a value λ in the laboratory frame of reference, i.e., the frame of reference of a stationary observer, and the wavelengths λ, λ_u^* after and before alteration are related as

$$\begin{aligned} \lambda &= \lambda_u^* \times \sqrt{\frac{1 - \frac{v}{c}}{1 + \frac{v}{c}}} = \sqrt{\frac{1 - \frac{v}{c}}{1 + \frac{v}{c}}} \times \frac{\sqrt{1 + \frac{v}{c}}}{\sqrt{1 + \frac{v}{c}}} \lambda_u^* = \frac{\sqrt{1 - \left(\frac{v}{c}\right)^2}}{1 + \frac{v}{c}} \lambda_u^* \\ &= \frac{1}{\gamma\left(1 + \frac{v}{c}\right)} \lambda_u^* = \frac{1}{\gamma(1 + 1)} \lambda_u^* = \frac{\lambda_u^*}{2\gamma} \end{aligned} \quad (10.71)$$

in the limit

$$v \approx c \tag{10.72}$$

Equation (10.3) for γ has been applied.

From equations (10.71) and (10.69),

$$\lambda = \frac{\lambda_u^*}{2\gamma} = \frac{\frac{\lambda_u}{\gamma}}{2\gamma} = \frac{\lambda_u}{\gamma} \times \frac{1}{2\gamma} = \frac{\lambda_u}{2\gamma^2} \tag{10.73}$$

Applying equations (10.70) and (10.71), the angular frequency of radiation in the laboratory system is

$$\omega = \frac{2\pi c}{\lambda} = \frac{2\pi c}{\lambda_u^*} \times \frac{\lambda_u^*}{\lambda} = \omega^* \times 2\gamma = 2\gamma\omega^* = 2\gamma \times \gamma\frac{2\pi c}{\lambda_u} = \gamma^2\frac{4\pi c}{\lambda_u} \tag{10.74}$$

From equation (10.73), the wavelength of radiation is

$$\begin{aligned} \lambda &= \frac{\lambda_u}{2\gamma^2} = \frac{50 \times 10^{-3}}{2(1000)^2} = \frac{5}{2} \times 10^{-2} \times 10^{-6} = 2.5 \times 10^{-8}\ \text{m} \\ &= 2.5 \times 10^{-8} \times 10^{9}\ \text{nm} = 25\ \text{nm} \end{aligned} \tag{10.75}$$

Equation (10.74) gives the linear frequency as

$$\begin{aligned} f &= \frac{\omega}{2\pi} = \frac{\gamma^2\frac{4\pi c}{\lambda_u}}{2\pi} = \gamma^2\frac{4\pi c}{\lambda_u} \times \frac{1}{2\pi} = \gamma^2\frac{2c}{\lambda_u} = (1000)^2\frac{2 \times 2.9979 \times 10^8}{50 \times 10^{-3}} \\ &= \frac{2 \times 2.9979}{5} \times 10^{16} = 1.1992 \times 10^{16}\ \text{Hz} \end{aligned} \tag{10.76}$$

which pertains to the ultraviolet region (8×10^{14} to 3×10^{16} Hz) of the electromagnetic spectrum.

The correctness of this equation is verified by straightaway calculating frequency from wavelength (25 nm) as obtained in equation (10.75)

$$f = \frac{c}{\lambda} = \frac{2.9979 \times 10^8}{2.5 \times 10^{-8}} = 1.1992 \times 10^{16}\ \text{Hz} \tag{10.77}$$

(d) Here $\lambda_u = 25\ \text{mm} = 25 \times 10^{-3}$ m and $f = 9 \times 10^{12}$ Hz; from equation (10.76)

$$\begin{aligned} f &= \gamma^2\frac{2c}{\lambda_u} = \gamma^2\frac{2 \times 2.9979 \times 10^8}{25 \times 10^{-3}} \\ &= \gamma^2\frac{2 \times 2.9979 \times 10^{11}}{25} = \gamma^2 0.2398 \times 10^{11} = 2.398 \times 10^{10}\gamma^2 \end{aligned} \tag{10.78}$$

$$\therefore 2.398 \times 10^{10}\gamma^2 = 9 \times 10^{12}$$

or,

$$\gamma^2 = \frac{9 \times 10^{12}}{2.398 \times 10^{10}} = \frac{9}{2.398} \times 10^2 = 3.753 \times 10^2 = 375.3 \tag{10.79}$$

$$\therefore \gamma = \sqrt{375.3} = 19.37 \tag{10.80}$$

From equation (10.3)

$$\frac{1}{\sqrt{1 - \left(\frac{v}{c}\right)^2}} = 19.37 \tag{10.81}$$

or,

$$\frac{1}{1 - \left(\frac{v}{c}\right)^2} = (19.37)^2 = 375.197 \tag{10.82}$$

or,

$$375.197\left\{1 - \left(\frac{v}{c}\right)^2\right\} = 1 \tag{10.83}$$

or,

$$375.197 - 375.197\left(\frac{v}{c}\right)^2 = 1 \tag{10.84}$$

or,

$$-375.197\left(\frac{v}{c}\right)^2 = 1 - 375.197 = -374.197 \tag{10.85}$$

or,

$$\left(\frac{v}{c}\right)^2 = \frac{374.197}{375.197} \tag{10.86}$$

or,

$$\frac{v}{c} = \sqrt{\frac{374.197}{375.197}} = 0.998666 \tag{10.87}$$

$$\therefore v = 0.998666c \tag{10.88}$$

and

$$\begin{aligned} E &= \gamma m_0 c^2 = 19.37 \times 9.109 \times 10^{-31} \times (2.9979 \times 10^8)^2 \text{ J} \\ &= 1.5857 \times 10^{-12} \text{ J} = \frac{1.5857 \times 10^{-12}}{1.602 \times 10^{-19}} \text{eV} = 0.9898 \times 10^7 \text{ eV} \\ &= 9.898 \times 10^6 \text{ eV} = 9.898 \text{ MeV} \end{aligned} \tag{10.89}$$

Example 10.4

A THz free electron laser works at average beam current of 4 mA. The electron beam energy is 10 MeV? The average power output is 500 W and the pulsed power output is 7.5×10^5 W. Calculate the efficiency of conversion of electron beam energy into laser beam energy, and also the pulsed beam current. Assume the output coupling efficiency to be unity.

Solution: The efficiency of a free electron laser is defined as

$$\begin{aligned} \eta_{\text{FEL}} &= \frac{\text{Output power}}{\text{Input power}} \times 100\% \\ &= \frac{\text{Power of emitted radiation}}{\text{Power of injected electron beam}} \times 100\% \end{aligned} \tag{10.90}$$

The power output of a free electron laser is given by (Benson 1999)

$$\begin{aligned} \text{Power } (P_{\text{FEL}}) = {} & \text{Electron beam energy } (E_{\text{EB}}) \\ & \times \text{Average electron beam current } (I_{\text{EB}}) \\ & \times \text{Electron-to-optical conversion efficiency } (\eta_{\text{FEL}}) \\ & \times \text{Output coupling efficiency of the optical cavity } (\eta_{\text{OPT}}) \end{aligned} \tag{10.91}$$

or,

$$P_{\text{FEL}} = E_{\text{EB}} I_{\text{EB}} \eta_{\text{FEL}} \eta_{\text{OPT}} = E_{\text{EB}} I_{\text{EB}} \eta_{\text{FEL}} \tag{10.92}$$

taking

$$\eta_{\text{OPT}} = 1 \tag{10.93}$$

From equation (10.92),

$$\begin{aligned} \eta_{\text{FEL}} &= \frac{P_{\text{FEL}}}{E_{\text{EB}} I_{\text{EB}}} = \frac{500}{10 \times 10^6 \times 4 \times 10^{-3}} = \frac{5}{4} \times 10^{-2} = 1.25 \times 10^{-2} \\ &\approx 0.0125 = 0.0125 \times 100\% = 1.25\% \end{aligned} \tag{10.94}$$

For calculating the electron beam current in pulse mode (I_{PEB}), equation (10.92) is written in terms of pulsed power output P_{PFEL} as

$$P_{\text{PFEL}} = E_{\text{EB}} I_{\text{PEB}} \eta_{\text{FEL}} \tag{10.95}$$

from which

$$I_{PEB} = \frac{P_{PFEL}}{E_{EB}\eta_{FEL}} = \frac{7.5 \times 10^5}{10 \times 10^6 \times 0.0125} = \frac{7.5}{0.0125} \times 10^{-2} \tag{10.96}$$
$$= 600 \times 10^{-2} = 6 \text{ A}$$

10.4 Operating modes of the free electron laser

10.4.1 Self-amplified-spontaneous-emission (SASE) FEL

In this mode, the FEL works without any resonant cavity built with mirrors and without using any seed laser (figure 10.5).

The relativistic electron beam from the accelerator travels in the undulator along the Z-direction. Under the influence of the magnetic field of the undulator acting in the Y-direction, the electron beam undergoes periodical deflection in the X-direction making it trace out a sinusoidal path in the Z-direction.

10.4.1.1 Opening angle for emitted radiation

The deflection of the relativistic electron beam by the Lorentz force of the magnetic field results in its acceleration transverse to the direction of motion of the electrons. A charged particle does not radiate energy under normal conditions when it is in uniform motion. But when accelerated, it radiates electromagnetic energy in the forward direction in a narrow cone with a small opening angle (figure 10.6). This solid angle is given by

$$\psi = \frac{1}{\gamma} = \frac{m_0 c^2}{E_b} \tag{10.97}$$

where m_0 is the rest mass of electron, c is the velocity of light and E_b is the energy of electron beam. Equation (10.66) is applied.

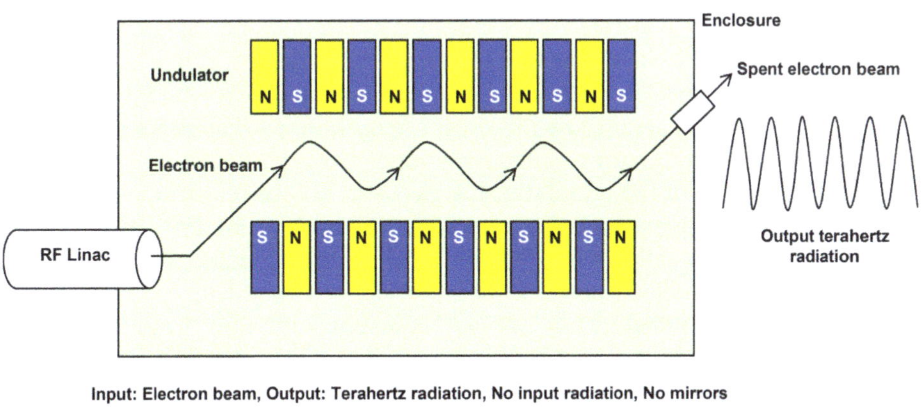

Figure 10.5. Self-amplified spontaneous emission (SASE) free electron laser.

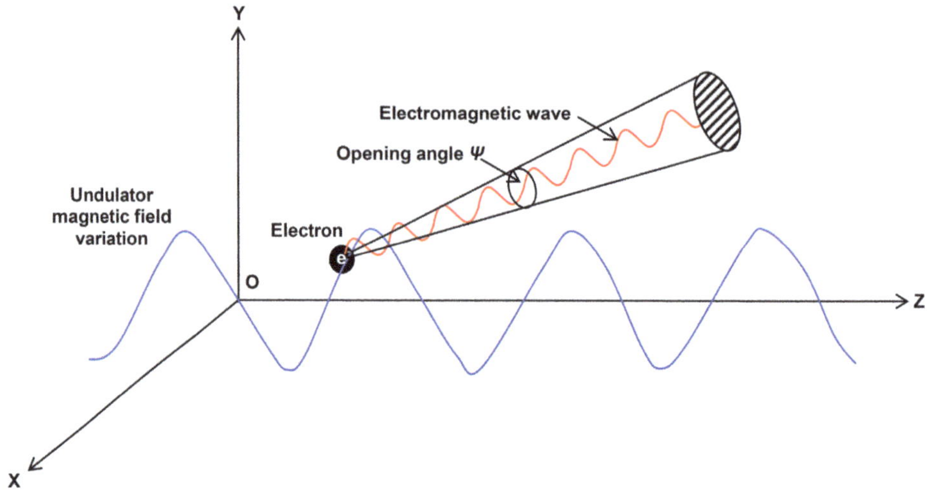

Figure 10.6. The opening angle of the emitted electromagnetic wave.

Example 10.5

What is the opening angle for the radiation emitted by electron beam energy of 10 MeV?

Applying equation (10.97)

$$\begin{aligned}\psi &= \frac{m_0c^2}{E_b} = \frac{9.109 \times 10^{-31} \times (2.9979 \times 10^8)^2}{10 \times 10^6 \times 1.602 \times 10^{-19}} \\ &= \frac{9.109 \times (2.9979)^2 \times 10^{-3}}{1.602} = 51.1025 \times 10^{-3}\,\text{rad} \\ &= 51.1025 \times 10^{-3} \times 10^3\text{mrad} = 51.1\,\text{mrad}\end{aligned} \tag{10.98}$$

10.4.1.2 Mechanism of radiation amplification

In the SASE FEL, we have two mobile entities: the electron beam and the electromagnetic wave generated by electron acceleration, i.e., the electron beam and the co-propagating wave. The first section of the undulator is marked by chaotic distribution of electrons. The phases of electromagnetic waves emitted by different electrons are randomly arranged with respect to each other. The radiation is largely incoherent. But the undulator is sufficiently long.

In the next section of the undulator, the electrons and the electromagnetic waves interact. As a consequence of this interaction, the electrons are organized into collections or clusters called microbunches. The electromagnetic waves emitted by these microbunches of electrons have their phases correlated with each other. Also, the radiation is amplified. Thus we get coherent radiation from the starting incoherent wave. Evidently, the spontaneous emission in the first section of the undulator is progressively amplified in its ensuing sections. This phenomenon is called self-amplified spontaneous emission.

Basically SASE FEL is a power amplifier, which amplifies the shot noise in the electron beam propagating through a long undulator. The exponential growth of shot noise power of spontaneous radiation P_n with the distance z along the undulator is described by the equation (see example 10.6):

$$P = \alpha P_n \exp\left(\frac{z}{L_G}\right) \tag{10.99}$$

where α is the coupling coefficient = the efficiency of coupling of noise power with gain process = the fraction of noise power coupled into the dominant growing mode, L_G is the power gain length = the distance at which the power grows to e times its starting value P_n, i.e., the distance at which the radiation intensity multiplies e-fold. If the power saturates at a level $P = P_{Sat}$ in a length L_{Sat}, we can write the saturation power from equation (10.99) as

$$P_{Sat} = \alpha P_n \exp\left(\frac{L_{Sat}}{L_G}\right) \tag{10.100}$$

or,

$$\frac{P_{Sat}}{\alpha P_n} = \exp\left(\frac{L_{Sat}}{L_G}\right) \tag{10.101}$$

or,

$$\ln\left(\frac{P_{Sat}}{\alpha P_n}\right) = \frac{L_{Sat}}{L_G} \tag{10.102}$$

so that the saturation length L_{Sat} (the length along the undulator to attain the maximum output power, i.e., the length at which microbunching is maximum), is

$$L_{Sat} = L_G \ln\left(\frac{P_{Sat}}{\alpha P_n}\right) \tag{10.103}$$

Example 10.6

Exponential growth law for power: For an SASE amplifier, given that the rate of growth of intensity of the wave (I_{Wave}) is proportional to the intensity itself (Margaritondo and Ribic 2011), show that the intensity growth in the amplifier occurs exponentially.

Solution: Here,

$$\frac{dI_{Wave}}{dt} = AI_{Wave} \tag{10.104}$$

where A is a constant. From equation (10.104),

$$\frac{dI_{Wave}}{I_{Wave}} = Adt \tag{10.105}$$

On integration

$$\int_{I_{\text{Wave}(0)}}^{I_{\text{Wave}}} \frac{dI_{\text{Wave}}}{I_{\text{Wave}}} = A\int_0^t dt \tag{10.106}$$

or,

$$|\ln I_{\text{Wave}}|_{I_{\text{Wave}(0)}}^{I_{\text{Wave}}} = A\ |t|_0^t \tag{10.107}$$

or,

$$\ln\left(I_{\text{Wave}}\right) - \ln\left(I_{\text{Wave}(0)}\right) = \ln\left(\frac{I_{\text{Wave}}}{I_{\text{Wave}(0)}}\right) = A(t-0) \tag{10.108}$$

or,

$$\frac{I_{\text{Wave}}}{I_{\text{Wave}(0)}} = \exp(At) \tag{10.109}$$

or,

$$I_{\text{Wave}} = I_{\text{Wave}(0)}\exp(At) \tag{10.110}$$

To put it in a form similar to the power growth equation (10.99), we set

$$A = \frac{v}{L_{\text{G}}} \tag{10.111}$$

where

$$v \approx c \tag{10.112}$$

is the electron velocity. Then

$$I_{\text{Wave}} = I_{\text{Wave}(0)}\exp\left(\frac{vt}{L_{\text{G}}}\right) = I_{\text{Wave}(0)}\exp\left(\frac{z}{L_{\text{G}}}\right) \tag{10.113}$$

since $vt = z$. Hence the intensity growth occurs exponentially.

10.4.1.3 One-dimensional model of SASE FEL

We consider the simplest one-dimensional model and the ideal case, which assumes that the electron beam is spatially distributed uniformly in the transverse direction, and its emittance and energy spread are both zero. Diffraction and time dependence of energy are neglected. Then the various parameters of equations (10.99), (10.103) are expressed by the equations (Xie 1995)

$$\alpha = \text{Coupling coefficient} = \frac{1}{9} \tag{10.114}$$

$$P_n = \text{Starting value of power} = \text{Effective input noise power} \approx \frac{\rho^2 c E_{EB}}{\lambda} \quad (10.115)$$

$$L_G = \text{Gain length} = \frac{\lambda_u}{4\pi\sqrt{3}\rho} \quad (10.116)$$

$$L_{Sat} = \text{Saturation length} \approx \frac{\lambda_u}{\rho} \sim 20 L_G \quad (10.117)$$

$$P_{Sat} = \text{Saturation value of power} \approx \rho P_{EB} \quad (10.118)$$

$$\rho = \text{Pierce parameter} - \frac{1}{2\gamma}\left(\frac{I_{EB(Peak)}}{I_A}\right)^{\frac{1}{3}}\left[\frac{K\lambda_u\{JJ(\xi)\}}{2\sqrt{2}\pi\sigma_x}\right]^{\frac{2}{3}} \quad (10.119)$$

$$\begin{aligned} P_{EB} &= \text{Electron beam power} \\ &= \text{Electron beam energy} \times \text{Electron beam current} = E_{EB} I_{EB} \\ \text{Peak electron beam power} &= P_{EB(Peak)} \\ &= \frac{\gamma m_0 c^2}{\text{Electronic charge}} \text{Peak electron beam current} \\ &= \frac{\gamma m_0 c^2}{e} I_{EB(Peak)} \end{aligned} \quad (10.120)$$

where equation (10.66) is applied.

$$\gamma = \frac{E_{EB}}{m_0 c^2} = \text{Electron beam energy normalized to } m_0 c^2 \quad (10.121)$$

$$\begin{aligned} I_A &= \text{Alfvén current} = \frac{4\pi\varepsilon_0 m_0 c^3}{e} \\ &= \frac{4 \times 3.14159 \times 8.854 \times 10^{-12} \times 9.109 \times 10^{-31} \times (2.9979 \times 10^8)^3}{1.602 \times 10^{-19}} \\ &= \frac{4 \times 3.14159 \times 8.854 \times 9.109 \times (2.9979)^3}{1.602} \times \frac{10^{-12} \times 10^{-31} \times 10^{24}}{10^{-19}} \\ &= 17045.46\text{A} = \frac{17045.46}{1000}\text{kA} = 17.0455\text{kA} \approx 17\text{ kA} \end{aligned} \quad (10.122)$$

$$\xi = \frac{K^2}{4 + 2K^2} \quad (10.123)$$

$$\begin{aligned} JJ(\xi) &= \text{Reduction factor of FEL interaction in the planar undulator} \\ &= J_0\left(\frac{K^2}{4 + 2K^2}\right) - J_1\left(\frac{K^2}{4 + 2K^2}\right) \end{aligned} \quad (10.124)$$

For the undulator parameter $K \leqslant 1$,

$$JJ(\xi) = 1 - \frac{\xi^2}{4} - \frac{\xi}{2} \tag{10.125}$$

The Alfvén current I_A is the limiting current at which the self-magnetic field of an electron beam is adequate to reverse the directions of paths of the electrons near the outer edge of the beam. It is the maximum current that can propagate up to an infinite distance and for an infinite time from a source in a charge-neutral electron beam. The undulator parameter K is given by equation (10.23). J_0 and J_1 are the zeroth order and first order Bessel functions of first kind; a planar undulator is considered. The parameter σ_x is the electron beam RMS transverse dimension defined by the emittance and Twiss coefficients:

$$\sigma_x = \sqrt{\text{Emittance}(\varepsilon_{n,x}) \times \text{Twiss factor}\left(\beta_x\right)} \tag{10.126}$$

The dimensionless number ρ called the FEL parameter or Pierce parameter represents the efficiency of conversion from FEL electron beam power to electromagnetic radiation power. It is the ratio of maximum radiation power (at saturation) to electron beam power. It is a pivotal parameter in FEL analysis playing a vital role in FEL design calculations.

For a planar undulator, the small-signal gain is given by (Nguyen *et al* 2009)

$$\begin{aligned} g_{ss} &= \left(\frac{2\pi N_u}{\gamma}\right)^3 \left\{\frac{I_{EB(Peak)}}{I_A}\right\} \left[\frac{K\lambda_u\{JJ(\xi)\}}{2\sqrt{2}\,\pi\sigma_x}\right]^2 \\ &= \{2(2\pi N_u)\}^3 \times \left(\frac{1}{2\gamma}\left(\frac{I_{EB(Peak)}}{I_A}\right)^{\frac{1}{3}}\left[\frac{K\lambda_u\{JJ(\xi)\}}{2\sqrt{2}\,\pi\sigma_x}\right]^{\frac{2}{3}}\right)^3 \\ &= \{2(2\pi N_u)\}^3 \times \rho^3 = \{2(2\pi N_u\rho)\}^3 \end{aligned} \tag{10.127}$$

showing that the small-signal gain is directly proportional to the cube of the Pierce parameter ρ. The symbol N_u denotes the number of periods in the undulator.

Example 10.7

(a) Calculate the Pierce parameter, the gain length, the saturation length and the small-signal gain for a THz FEL designed with the following parameters: electron beam energy = 7 MeV, peak electron beam current = 16 A, undulator spatial period = 1.8 cm, undulator parameter = 0.83, number of undulator periods = 10, and beam size = 370 μm. Find also the wavelength and frequency of radiation generated.

(b) What are the values of Pierce parameter and gain length when the electron beam energy is changed to 1.74 GeV, the peak electron beam current to 160 kA, the undulator spatial period to 5 mm and the beam size to 30 μm, other variables remaining same? What is the effect on the frequency of radiation emitted?

Solution: (a) Here, $E_{EB} = 7$ MeV $= 7 \times 10^6$ eV, $I_{EB(PEAK)} = 16$ A, $K = 0.83$, $\lambda_u = 1.8$ cm $= 1.8 \times 10^{-2}$ m, $N_u = 10$, $\sigma_x = 370$ μm $= 370 \times 10^{-6}$ m. From equation (10.121)

$$\gamma = \frac{E_{\rm EB}}{m_0c^2} = \frac{7\times10^6\times1.602\times10^{-19}}{9.109\times10^{-31}\times(2.9979\times10^8)^2}\frac{\rm J}{\rm kg(m\ s^{-1})^2}$$
$$= \frac{7\times1.602}{9.109\times(2.9979)^2}\times10^{6-19+31-16}\frac{\rm kg\ m^2\ s^{-2}}{\rm kg\ m^2\ s^{-2}} \tag{10.128}$$
$$= 0.136979\times10^2 = 13.698$$

The Xi parameter of equation (10.123) is calculated as

$$\frac{K^2}{4+2K^2} = \frac{(0.83)^2}{4+2(0.83)^2} = \frac{0.6889}{4+2\times0.6889} = \frac{0.6889}{5.3778} = 0.1281 \approx 0.1 \tag{10.129}$$

We shall apply equations (10.123) and (10.124):

$$J_0\left(\frac{K^2}{4+2K^2}\right) = J_0(0.1) = 0.9975 \tag{10.130}$$

$$J_1\left(\frac{K^2}{4+2K^2}\right) = J_1(0.1) = 0.0499 \tag{10.131}$$

$$\therefore J_0\left(\frac{K^2}{4+2K^2}\right) - J_1\left(\frac{K^2}{4+2K^2}\right) = 0.9975 - 0.0499 = 0.9476 \tag{10.132}$$

Equations (10.119) and (10.122) yield

$$\rho = \frac{1}{2\times13.698}\left(\frac{16}{17.0455\times10^3}\right)^{\frac{1}{3}}\left(\frac{0.83\times1.8\times10^{-2}\times0.9476}{2\times3.14159\times1.414\times370\times10^{-6}}\right)^{\frac{2}{3}}$$
$$= \frac{1}{27.396}(9.386642\times10^{-4})^{\frac{1}{3}}(4.3067)^{\frac{2}{3}} \tag{10.133}$$
$$= 0.0365\times0.0979122\times2.647 = 0.00946$$

From equation (10.116)

$$L_{\rm G} = \frac{\lambda_{\rm u}}{4\pi\sqrt{3}\rho} = \frac{1.8\times10^{-2}}{4\times3.14159\times1.732\times0.00946} = 8.7423\times10^{-2}\ {\rm m}$$
$$= 8.7423\times10^{-2}\times10^2\ {\rm cm} = 8.74\ {\rm cm} \tag{10.134}$$

From equation (10.117)

$$L_{\rm Sat} \approx \frac{\lambda_{\rm u}}{\rho} = \frac{1.8\times10^{-2}}{0.00946} = 190.275\times10^{-2}\ {\rm m} = 1.9\ {\rm m} \tag{10.135}$$

From equation (10.127)

$$g_{\rm ss} = \{2(2\pi N_{\rm u}\rho)\}^3 = \{2(2\times3.14159\times10\times0.00946)\}^3 = 1.68 \tag{10.136}$$

From equation (10.54)

$$\lambda \approx \frac{\lambda_u}{2\gamma^2}\left(1 + \frac{K^2}{2}\right) = \frac{1.8 \times 10^{-2}}{2(13.698)^2}\left\{1 + \frac{(0.83)^2}{2}\right\} = 4.7965 \times 10^{-5}(1 + 0.6889)$$
$$= 4.7965 \times 10^{-5} \times 1.6889 = 8.1008 \times 10^{-5}\,\text{m} \tag{10.137}$$
$$= 8.1008 \times 10^{-5} \times 10^{6}\mu\text{m} = 81.008\,\mu\text{m}$$

$$f = \frac{c}{\lambda} = \frac{2.9979 \times 10^8}{8.1008 \times 10^{-5}} = 0.37 \times 10^{13}\text{Hz} = 3.7 \times 10^{12}\text{ Hz} = 3.7\text{ THz} \tag{10.138}$$

(b) Here $E_{\text{EB}} = 1.74$ GeV $= 1.74 \times 10^9$ eV, $I_{\text{EB(PEAK)}} = 160$ kA, $K = 0.83$, $\lambda_u = 5$ mm $= 5 \times 10^{-3}$ m, $N_u = 10$, $\sigma_x = 30$ μm $= 30 \times 10^{-6}$ m. From equation (10.121)

$$\gamma = \frac{E_{\text{EB}}}{m_0c^2} = \frac{1.74 \times 10^9 \times 1.602 \times 10^{-19}}{9.109 \times 10^{-31} \times (2.9979 \times 10^8)^2} = \frac{1.74 \times 1.602}{9.109 \times (2.9979)^2}$$
$$\times 10^{9-19+31-16} = 0.034049 \times 10^5 = 3404.9 \tag{10.139}$$

Equations (10.119) and (10.122) give

$$\rho = \frac{1}{2 \times 3404.9}\left(\frac{160}{17.0455}\right)^{\frac{1}{3}}\left(\frac{0.83 \times 5 \times 10^{-3} \times 0.9476}{2 \times 3.14159 \times 1.414 \times 30 \times 10^{-6}}\right)^{\frac{2}{3}}$$
$$= \frac{1}{6809.8}(9.386642)^{\frac{1}{3}}(14.754)^{\frac{2}{3}} \tag{10.140}$$
$$= 0.0001468 \times 2.10945 \times 6.0155 = 0.00186$$

From equation (10.116)

$$L_G = \frac{\lambda_u}{4\pi\sqrt{3}\rho} = \frac{5 \times 10^{-3}}{4 \times 3.14159 \times 1.732 \times 0.00186} = 123.509 \times 10^{-3}\,\text{m} \tag{10.141}$$
$$= 0.123509\text{ m} = 0.123509\text{ m} \times 10^2\text{ cm} = 12.351\text{ cm}$$

From equation (10.54)

$$\lambda \approx \frac{\lambda_u}{2\gamma^2}\left(1 + \frac{K^2}{2}\right) = \frac{5 \times 10^{-3}}{2(3404.9)^2}\left\{1 + \frac{(0.83)^2}{2}\right\} = 2.156409 \times 10^{-10}$$
$$\times 1.34445 = 2.89918 \times 10^{-10}\text{ m} \approx 2.9 \times 10^{-10} \times 10^9\text{ nm} = 0.29\text{ nm} \tag{10.142}$$

$$f = \frac{c}{\lambda} = \frac{2.9979 \times 10^8}{2.9 \times 10^{-10}} = 1.03376 \times 10^{18}\text{ Hz} \tag{10.143}$$

The wavelength and frequency of emitted radiation are in the x-ray region (10^{-11} m-10^{-8} m, 3×10^{16}–3×10^{19} Hz).

Example 10.8

A table-top terahertz FEL is designed with the parameters: Energy of electron beam = 5 MeV, peak electron beam current = 500 mA, transverse beam size = 437 μm, undulator parameter = 1.2, undulator spatial period = 26 mm, number of undulator periods = 30. Find the Pierce parameter, gain length, gain of the laser and the wavelength and frequency of emitted radiation.

Solution: Here, $E_{EB} = 5$ MeV $= 5 \times 10^6$ eV, $I_{EB(PEAK)} = 500$ mA $= 500 \times 10^{-3}$ A, $K = 1.2$, $\lambda_u = 26$ mm $= 26 \times 10^{-3}$ m, $N_u = 30$, $\sigma_x = 437$ μm $= 437 \times 10^{-6}$ m.

$$\text{Parameters to be determined} = \rho, L_G, g_{ss}, \lambda, \nu \tag{10.144}$$

From equation (10.121)

$$\begin{aligned}\gamma &= \frac{E_{EB}}{m_0c^2} = \frac{5 \times 10^6 \times 1.602 \times 10^{-19}}{9.109 \times 10^{-31} \times (2.9979 \times 10^8)^2} \\ &= \frac{5 \times 1.602}{9.109 \times (2.9979)^2} \times 10^{6-19+31-16} = 0.09784 \times 10^2 = 9.784\end{aligned} \tag{10.145}$$

The Xi parameter of equation (10.123) is

$$\frac{K^2}{4 + 2K^2} = \frac{(1.2)^2}{4 + 2(1.2)^2} = \frac{1.44}{4 + 2 \times 1.44} = \frac{1.44}{6.88} = 0.2093 \approx 0.2 \tag{10.146}$$

Since

$$J_0\left(\frac{K^2}{4 + 2K^2}\right) = J_0(0.2) = 0.9900 \tag{10.147}$$

$$J_1\left(\frac{K^2}{4 + 2K^2}\right) = J_1(0.2) = 0.0995 \tag{10.148}$$

$$\therefore J_0\left(\frac{K^2}{4 + 2K^2}\right) - J_1\left(\frac{K^2}{4 + 2K^2}\right) = 0.9900 - 0.0995 = 0.8905 \tag{10.149}$$

From equations (10.119) and (10.149)

$$\begin{aligned}\rho &= \frac{1}{2 \times 9.784}\left(\frac{500 \times 10^{-3}}{17.0455 \times 10^3}\right)^{\frac{1}{3}}\left(\frac{1.2 \times 26 \times 10^{-3} \times 0.8905}{2 \times 3.14159 \times 1.414 \times 437 \times 10^{-6}}\right)^{\frac{2}{3}} \\ &= \frac{1}{19.568}(2.933 \times 10^{-5})^{\frac{1}{3}}(7.156)^{\frac{2}{3}} = 0.0511 \times 0.030839 \times 3.713 \\ &= 0.00585\end{aligned} \tag{10.150}$$

From equation (10.116)

$$L_G = \frac{\lambda_u}{4\pi\sqrt{3}\rho} = \frac{26 \times 10^{-3}}{4 \times 3.14159 \times 1.732 \times 0.00585} = 204.2 \times 10^{-3}\ \text{m} \tag{10.151}$$
$$= 0.2042\ \text{m} = 0.2042 \times 10^2\ \text{cm} = 20.42\ \text{cm}$$

From equation (10.117)

$$L_{Sat} \approx \frac{\lambda_u}{\rho} = \frac{26 \times 10^{-3}}{0.00585} = 4444.44 \times 10^{-3}\ \text{m} = 4.44\ \text{m} \tag{10.152}$$

From equation (10.127)

$$g_{ss} = \{2(2\pi N_u \rho)\}^3 = \{2(2 \times 3.14159 \times 30 \times 0.00585)\}^3 = 10.7265{\sim}10.73 \tag{10.153}$$

From equation (10.54)

$$\lambda \approx \frac{\lambda_u}{2\gamma^2}\left(1 + \frac{K^2}{2}\right) = \frac{26 \times 10^{-3}}{2(9.784)^2}\left\{1 + \frac{(1.2)^2}{2}\right\} = 0.1358 \times 10^{-3} \times 1.72 \tag{10.154}$$
$$= 2.3358 \times 10^{-4}\ \text{m} = 2.3358 \times 10^{-4} \times 10^6\ \mu\text{m} = 233.6\ \mu\text{m}$$

$$f = \frac{c}{\lambda} = \frac{2.9979 \times 10^8}{2.3358 \times 10^{-4}} = 1.2835 \times 10^{12}\ \text{Hz} \tag{10.155}$$
$$= 1.28 \times 10^{12}\ \text{Hz} = 1.28\ \text{THz}$$

Example 10.9

A free electron laser is designed with $K = 1.85$, electron beam energy of 0.5 GeV, peak electron beam current of 20 kA, undulator spatial period of 2.18 cm and transverse beam size of 60 μm. Find the Pierce parameter and gain length. Does this free electron laser emit at terahertz frequency?

Solution: Here $K = 1.85$, $E_{EB} = 0.5$ GeV $= 0.5 \times 10^9$ eV, $I_{EB(Peak)} = 20$ kA $= 20 \times 10^3$ A, $\lambda_u = 2.18$ cm $= 2.18 \times 10^{-2}$ m,, $\sigma_x = 60$ μm $= 60 \times 10^{-6}$ m.

To apply equation (10.124), we shall determine J_0 and J_1:

$$J_0\left\{\frac{(1.85)^2}{4 + 2 \times 1.85^2}\right\} = J_0\left(\frac{3.4225}{10.845}\right) = J_0(0.3156) \approx J_0(0.3) = 0.9776 \tag{10.156}$$

$$J_1\left\{\frac{(1.85)^2}{4 + 2 \times 1.85^2}\right\} \approx J_1(0.3) = 0.1483 \tag{10.157}$$

Then

$$J_0\left\{\frac{(1.85)^2}{4 + 2 \times 1.85^2}\right\} - J_1\left\{\frac{(1.85)^2}{4 + 2 \times 1.85^2}\right\} = 0.9776 - 0.1483 = 0.8293 \tag{10.158}$$

From equation (10.121)

$$\gamma = \frac{E_{\mathrm{EB}}}{m_0c^2} = \frac{0.5 \times 10^9 \times 1.602 \times 10^{-19}}{9.109 \times 10^{-31} \times (2.9979 \times 10^8)^2} = \frac{0.5 \times 1.602}{9.109 \times (2.9979)^2} \quad (10.159)$$

$$\times\ 10^{9-19+31-16} = 0.00978425 \times 10^5 = 978.425$$

From equations (10.119), (10.158) and (10.159)

$$\rho = \frac{1}{2 \times 978.425}\left(\frac{20}{17.0455}\right)^{\frac{1}{3}}\left(\frac{1.85 \times 2.18 \times 10^{-2} \times 0.8293}{2 \times 3.14159 \times 1.414 \ \times 60 \times 10^{-6}}\right)^{\frac{2}{3}} \quad (10.160)$$

$$= \frac{1}{1956.85}(1.1733)^{\frac{1}{3}}(62.74)^{\frac{2}{3}} = 0.000511 \times 1.055 \times 15.79 = 0.0085$$

From equation (10.116)

$$L_{\mathrm{G}} = \frac{\lambda_{\mathrm{u}}}{4\pi\sqrt{3}\rho} = \frac{2.18 \times 10^{-2}}{4 \times 3.14159 \times 1.732 \times 0.0085} = 11.7837 \times 10^{-2}\ \mathrm{m} \quad (10.161)$$

$$= 0.11784\ \mathrm{m} = 0.11784\ \mathrm{m} \times 10^2\ \mathrm{cm} = 11.78\ \mathrm{cm}$$

From equation (10.54)

$$\lambda \approx \frac{\lambda_u}{2\gamma^2}\left(1 + \frac{K^2}{2}\right) = \frac{2.18 \times 10^{-2}}{2(978.425)^2}\left\{1 + \frac{(1.85)^2}{2}\right\} = 1.1386 \times 10^{-8} \times 2.711 \quad (10.162)$$

$$= 3.08674 \times 10^{-8}\ \mathrm{m} = 3.08674 \times 10^{-8} \times 10^9\ \mathrm{nm} \approx 30.9\ \mathrm{nm}$$

$$f = \frac{c}{\lambda} = \frac{2.9979 \times 10^8}{3.08674 \times 10^{-8}} = 0.971 \times 10^{16}\ \mathrm{Hz} = 9.71 \times 10^{15}\ \mathrm{Hz} \quad (10.163)$$

The FEL does not emit terahertz radiation. The emitted wavelength and frequency fall in the UV region of the electromagnetic spectrum (10^{-18}–4×10^{-7} m, 3×10^{16}–7.5×10^{14} Hz).

10.4.2 FEL oscillator

The basic FEL oscillator arrangement is the same as in SASE FEL. In addition, a Fabry–Perot resonator is used (figure 10.7). It consists of two opposing flat, highly reflecting mirrors. The radiation is coupled out through a hole in one of the mirrors. Sometimes a partially reflecting mirror in the cavity is employed for this purpose.

THz FEL oscillators generally have a low gain in single passage through the undulator. So, the laser is not saturated in a single pass. When the single-pass gain exceeds the round-trip loss, the electromagnetic field starts growing. Multiple round-trips across the cavity take place.

10.4.3 FEL amplifier

Operating with the SASE configuration, an input laser of suitable wavelength is used (figure 10.8). It is called the seed laser. There are neither any mirrors nor any resonant cavity.

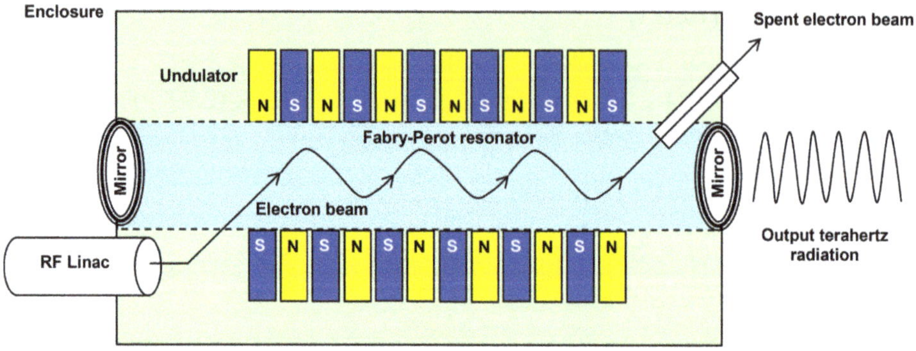

Figure 10.7. Free electron laser (FEL) oscillator.

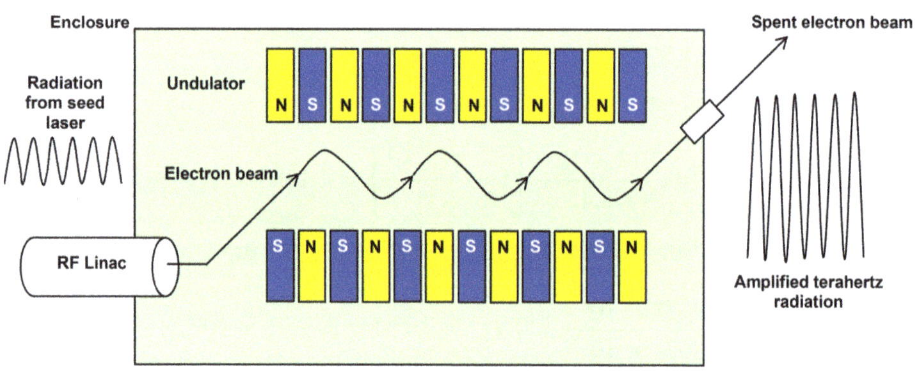

Figure 10.8. Free electron laser (FEL) amplifier.

10.5 Discussion and conclusions

The free electron laser is a large and expensive, specialized facility in which light is amplified in a periodic arrangement of magnets called undulator or wiggler by interaction with a high-energy relativistic electron bunch fed by a particle accelerator. It is capable of producing kilowatts of power at terahertz frequencies. The frequency of the emitted radiation is tunable by varying the energy of the electron beam or the strength of the magnetic field of the undulator (Biswas *et al* 2008). High-power FELs can be constructed providing frequencies from microwaves through terahertz to hard x-ray range. High brilliance, remarkable temporal and spatial resolution, wide spectral range and frequency/energy tunability, together with design flexibility bestow unrivalled merits on these intense electromagnetic beam systems. These exceptional properties provide chemical and structural analysis and morphological characterization capabilities to these systems fostering proliferation of applications in scientific disciplines such as chemistry, life and environmental sciences, microelectronics and other cross-disciplinary fields. The sophistication and extensiveness of free electron lasers can be ascertained from the fact that there are only a few such facilities worldwide.

References

Benson S 1999 High power free electron lasers *Proc. of the 1999 Particle Accelerator Conf. (New York, 27 March–2 April)* 212–6

Biswas B, Kumar V, Chouksey S and Krishnagopal S 2008 Design, construction and characterization of the compact ultrafast terahertz free-electron laser undulator *Pramana-J. Phys.* **71** (6) 1321–33

Dattoli G, Del Franco M, Labat M, Ottaviani P L and Pagnutti S 2012 Introduction to the physics of free electron laser and comparison with conventional laser sources *Free Electron Lasers* ed S Varró (Rijeka: InTech) pp 1–38

Dattoli G, Di Palma E, Pagnutti S and Sabia E 2018 Free electron coherent sources: from microwave to x-rays *Phys. Rep.* **739** 1–51

Huang Z and Kim K -Je 2007 Review of x-ray free-electron laser theory *Phys. Rev. ST Accel. Beams* **10** 034801

Margaritondo G and Ribic P R 2011 A simplified description of x-ray free-electron lasers *J. Synchrotron Radiat.* **18** 101–8

Nguyen D, Russell S and Moody N 2009 Theory and Practice of Free-Electron Lasers, US Particle Accelerator School, University of New Mexico—Albuquerque, LA-UR 09-02731 NM (Accessed on 8th March 2021) https://uspas.fnal.gov/materials/09UNM/Day%203.pdf

Pflueger J 2018 Undulator technology proceedings of the CAS–CERN accelerator school: free electron lasers and energy recovery linacs *CERN Yellow Reports: School Proceedings*Vol. 1/ 2018, CERN-2018-001-SP CERN *(Hamburg, 31 May–10 June 2016)*R Bailey (Geneva: CERN) pp 55–77

Walker R P 1996 Insertion devices: undulators and wigglers *CAS: CERN Accelerator School: Course on Synchrotron Radiation and Free-electron Lasers, Proc. (Grenoble, April 22–27)*S Turner (Geneva: CERN) pp 129–90

Xie M 1995 Design optimization for an x-ray free electron laser driven by SLAC linac *Proc. Particle Accelerator Conf.* vol 1 *(Dallas, TX, 1–5 May)* (New York: IEEE) 183–5

www.ingramcontent.com/pod-product-compliance
Ingram Content Group UK Ltd.
Pitfield, Milton Keynes, MK11 3LW, UK
UKHW051904100926
8556IPUK00005B/756

9 780750 331692